HIGH VOLTAGE ENGINEERING

Problems and Solutions

HIGH VOLTAGE ENGINEERING

Problems and Solutions

Rakosh Das Begamudre

Former Visiting Professor

Indian Institute of Technology

Kanpur, India

Publishing Globally

NEW AGE INTERNATIONAL (P) LIMITED, PUBLISHERS

New Delhi • Bangalore • Chennai • Cochin • Guwahati

Hyderabad • Kolkata • Lucknow • Mumbai

Visit us at **www.newagepublishers.com**

Published by New Age International (P) Ltd., Publishers
First Edition: 2010
Reprint: 2014

BRANCHES

- **Bangalore** 37/10, 8th Cross (Near Hanuman Temple), Azad Nagar, Chamarajpet, Bangalore-560 018 Tel.: (080) 26756823, Telefax: 26756820, **E-mail: bangalore@newagepublishers.com**
- **Chennai** 26, Damodaran Street, T. Nagar, Chennai-600 017 Tel.: (044) 24353401, Telefax: 24351463, **E-mail: chennai@newagepublishers.com**
- **Cochin** CC-39/1016, Carrier Station Road, Ernakulam South, Cochin-682 016 Tel.: (0484) 2377004, Telefax: 4051303, **E-mail: cochin@newagepublishers.com**
- **Guwahati** Hemsen Complex, Mohd. Shah Road, Paltan Bazar, Near Starline Hotel, Guwahati-781 008 Tel.: (0361) 2513881, Telefax: 2543669, **E-mail: guwahati@newagepublishers.com**
- **Hyderabad** 105, 1st Floor, Madhiray Kaveri Tower, 3-2-19, Azam Jahi Road, Near Kumar Theater Nimboliadda, Kachiguda, Hyderabad-500 027, Tel.: (040) 24652456, Telefax: 24652457 **E-mail: hyderabad@newagepublishers.com**
- **Kolkata** RDB Chambers (Formerly Lotus Cinema) 106A, 1st Floor, S.N. Banerjee Road, Kolkata-700 014 Tel.: (033) 22273773, Telefax: 22275247, **E-mail: kolkata@newagepublishers.com**
- **Lucknow** 16-A, Jopling Road, Lucknow-226 001 Tel.: (0522) 2209578, 4045297, Telefax: 2204098, **E-mail:lucknow@newagepublishers.com**
- **Mumbai** 142C, Victor House, Ground Floor, N.M. Joshi Marg, Lower Parel, Mumbai-400 013 Tel.: (022) 24927869, Telefax: 24915415, **E-mail: mumbai@newagepublishers.com**
- **New Delhi** 22, Golden House, Daryaganj, New Delhi-110 002 Tel.: (011) 23262368, 23262370, Telefax: 43551305, **E-mail: sales@newagepublishers.com**

ISBN: 978-81-224-2618-2

₹ 200.00

C-13-12-7269

Printed in India at Ajit Printers, Delhi.
Typeset at In-house, Delhi.

PUBLISHING GLOBALLY
NEW AGE INTERNATIONAL (P) LIMITED, PUBLISHERS
7/30 A, Daryaganj, New Delhi-110002
Visit us at **www.newagepublishers.com**

HIGH VOLTAGE ENGINEERING

Problems and Solutions

Rakosh Das Begamudre

Former Visiting Professor

Indian Institute of Technology

Kanpur, India

Publishing Globally

NEW AGE INTERNATIONAL (P) LIMITED, PUBLISHERS

New Delhi • Bangalore • Chennai • Cochin • Guwahati

Hyderabad • Kolkata • Lucknow • Mumbai

Visit us at www.newagepublishers.com

Published by New Age International (P) Ltd., Publishers
First Edition: 2010
Reprint: 2014

BRANCHES

- **Bangalore** 37/10, 8th Cross (Near Hanuman Temple), Azad Nagar, Chamarajpet, Bangalore-560 018 Tel.: (080) 26756823, Telefax: 26756820, **E-mail: bangalore@newagepublishers.com**
- **Chennai** 26, Damodaran Street, T. Nagar, Chennai-600 017 Tel.: (044) 24353401, Telefax: 24351463, **E-mail: chennai@newagepublishers.com**
- **Cochin** CC-39/1016, Carrier Station Road, Ernakulam South, Cochin-682 016 Tel.: (0484) 2377004, Telefax: 4051303, **E-mail: cochin@newagepublishers.com**
- **Guwahati** Hemsen Complex, Mohd. Shah Road, Paltan Bazar, Near Starline Hotel, Guwahati-781 008 Tel.: (0361) 2513881, Telefax: 2543669, **E-mail: guwahati@newagepublishers.com**
- **Hyderabad** 105, 1st Floor, Madhiray Kaveri Tower, 3-2-19, Azam Jahi Road, Near Kumar Theater Nimboliadda, Kachiguda, Hyderabad-500 027, Tel.: (040) 24652456, Telefax: 24652457 **E-mail: hyderabad@newagepublishers.com**
- **Kolkata** RDB Chambers (Formerly Lotus Cinema) 106A, 1st Floor, S.N. Banerjee Road, Kolkata-700 014 Tel.: (033) 22273773, Telefax: 22275247, **E-mail: kolkata@newagepublishers.com**
- **Lucknow** 16-A, Jopling Road, Lucknow-226 001 Tel.: (0522) 2209578, 4045297, Telefax: 2204098, **E-mail:lucknow@newagepublishers.com**
- **Mumbai** 142C, Victor House, Ground Floor, N.M. Joshi Marg, Lower Parel, Mumbai-400 013 Tel.: (022) 24927869, Telefax: 24915415, **E-mail: mumbai@newagepublishers.com**
- **New Delhi** 22, Golden House, Daryaganj, New Delhi-110 002 Tel.: (011) 23262368, 23262370, Telefax: 43551305, **E-mail: sales@newagepublishers.com**

ISBN: 978-81-224-2618-2

₹ 200.00

C-13-12-7269

Printed in India at Ajit Printers, Delhi.
Typeset at In-house, Delhi.

PUBLISHING GLOBALLY
NEW AGE INTERNATIONAL (P) LIMITED, PUBLISHERS
7/30 A, Daryaganj, New Delhi-110002
Visit us at **www.newagepublishers.com**

PREFACE

This book has been written in response to suggestions given by practising engineers in electrical power systems dealing with high voltage engineering in their every day work as well as taking decisions. After the publication of the author's book "Extra High Voltage A.C. Transmission Engineering" (Wiley Eastern and New Age International (P) Ltd., New Delhi) with the bachelor's and post-bachelor's degree student in mind, many practising engineers expressed the view to me that the mathematical content and rigour were too advanced and a compilation of questions and answers which they were getting would be more helpful in their professional work. This book is now presented with this aim in mind.

The questions were those which the author got during his long association with power systems, electrical machines and high voltage engineering (design, practice, teaching, research and laboratory measurements). The book is meant for academicians as well as for those preparing for professional examinations. Several excellent publications in the market cater for the needs of engineers and others who wish to write such examinations. Furthermore, it is my belief that engineers in their everyday professional activities do not have convenient and ready access to libraries which stack learned technical journals with research papers which are readily available to academicians in Universities and higher Institutes of Science and Technology. Therefore, no references are provided in this tract; an interested reader will find a long list of articles and books in a bibliography given by the author in his earlier work cited before. As such, it is quite possible that professors, research scholars and students in academic institutions may not find this book on questions and answers in high voltage engineering of a practical nature very helpful to them although they might obtain useful guidance on professional problems of a practical nature which engineers face in their professional activities everyday in high voltage engineering and transmission.

The book contains a total of 292 questions arranged in 17 chapters along with 167 line drawings to aid in understanding the material. The book is really a cross between a Handbook in High Voltage Engineering, an Encyclopaedia of information, and a scholarly textbook, but with the practising engineer primarily in mind. The work is entirely the author's and I am solely responsible for all shortcomings which engineers might find in the book. Therefore, I am responsible for all errors and omissions, if any and I would welcome suggestions for improvements from practising engineers for further editions, if any. It is my sincere hope that engineers will benefit by the contents of this book.

Vancouver, B.C., Canada **Rakosh Das Begamudre**

CONTENTS

LIST OF QUESTIONS IN CHAPTERS

CHAPTER 1: HIGH VOLTAGE AND ELECTRIC SHOCK

CHAPTER 2: CONDUCTORS AND INSULATING MATERIALS

Chapter 3: Electron Structure and Breakdown Mechanism

LIST OF QUESTIONS IN CHAPTERS

CHAPTER 1: HIGH VOLTAGE AND ELECTRIC SHOCK

CHAPTER 2: CONDUCTORS AND INSULATING MATERIALS

CHAPTER 3: ELECTRON STRUCTURE AND BREAKDOWN MECHANISM

CHAPTER 4: ELECTRICAL CONDUCTORS AND OVERHEAD LINES: ELECTRICAL AND MECHANICAL PROPERTIES

CHAPTER 5: ELECTRICAL CABLES

Chapter 6: A.C. and D.C. Transmission Systems: Overhead Lines, Cables and Insulators

Chapter 7: Three-phase Calculations in High Voltage Networks: Steady-state and Transient-state (Review)

CHAPTER 8: INDUCTANCE, CAPACITANCE AND VOLTAGE GRADIENTS OF OVERHEAD TRANSMISSION-LINE CONDUCTORS

Chapter 9: Lightning Phenomenon and Lightning Protection

CHAPTER 10: PARTIAL DISCHARGE (CORONA) AND ITS EFFECTS

CHAPTER 11: ELECTROSTATIC FIELD OF HIGH VOLTAGE TRANSMISSION LINES

Chapter 12: High Voltage Transformers and Circuit Breakers

Chapter 13: Voltage Control of AC and DC High Voltage Transmission Systems

CHAPTER 14: WAVE PHENOMENA IN HIGH VOLTAGE SYSTEMS

Chapter 15: Protection of High Voltage Systems

Chapter 16: Testing Sources in High Voltage Laboratories

Chapter 17: High Voltage Measurements and Techniques

CHAPTER 1

HIGH VOLTAGE AND ELECTRIC SHOCK

Q.1. What is the definition of High Voltage?

Ans. Any magnitude of voltage of any type which gives electric shock is considered as high voltage. There are other highly technical definitions which we will pick up as we proceed.

Q.2. What are the different types of voltages occurring in high voltage practice?

Ans. (*a*) Power-frequency a.c.;

(*b*) Direct voltage (also denoted as d.c.);

(*c*) Lightning, usually micro-second phenomenon;

(*d*) Switching voltages, milli-second phenomena; and

(*e*) Voltages caused by system faults, such as short-circuits.

Q.3. What are the salient time and other characteristics of the above voltages?

Ans. (*a*) Power-frequency adopted in the world is either 50 Hz or 60 Hz with periodic times (for 1 cycle) of 1/50 sec = 20 ms, or 1/60 sec = 16.67 ms.

See Fig. 1.1. Three important voltage values on this wave are:

(*i*) the peak, V_P;

(*ii*) Effective or root-mean-square (r.m.s. or rms) value, V; and

(*iii*) the average value, V_{av}.

$$[V_{av} = 2V_P/\pi = 0.637V_P \ , \ V = V_P/\sqrt{2} = 0.707 \ V_P].$$

(*b*) Direct voltage: Constant value for all time. Peak value only, see Fig. 1.2.

(*c*) Lightning voltage: It rises from zero to peak or crest value within 10 μs and falls to 50% of peak value in about 50–100 μs, and finally back again to zero value at greater than 200 μs, see Fig. 1.3. The timings given here show extreme variations. For testing of h.v. equipment, they are standardised on next page.

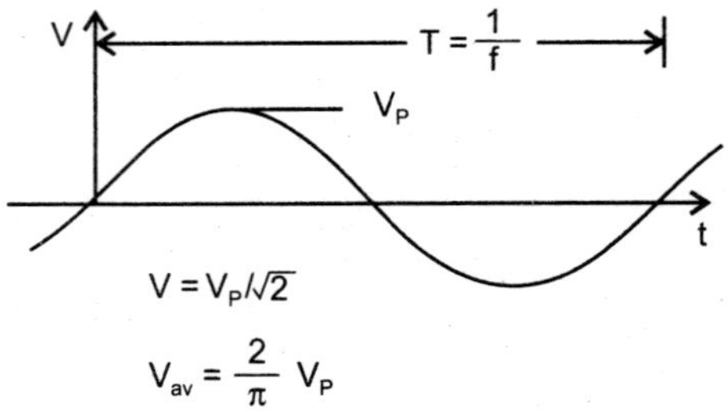

Fig. 1.1: Power-frequency sinusoidal voltage: Symmetrical in both half-cycles.

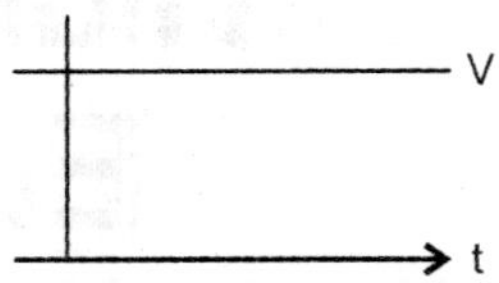

Fig. 1.2: Direct voltage or d.c. constant value with no ripple.

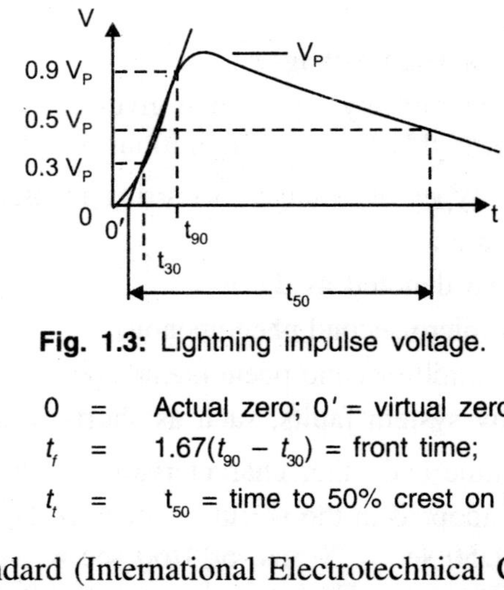

Fig. 1.3: Lightning impulse voltage.

0	=	Actual zero; 0′ = virtual zero;
t_f	=	$1.67(t_{90} - t_{30})$ = front time;
t_t	=	t_{50} = time to 50% crest on tail.

(*i*) I.E.C. Standard (International Electrotechnical Commission) also followed by India, U.K. and many countries: 1.2/50 μs, meaning that the front time is 1.2 μs and time to 50% value on the falling portion or tail is 50 μs from a point called the "virtual zero". These are further explained as follows:

We join the points on the nearly linearly-rising front of the wave at 30% and 90% of peak or crest value by a straight line. The point where this line intersects the horizontal axis through the actual zero of the wave is called the "virtual zero". The wavefront time is defined as $t_f = (t_{90} - t_{30})/0.6 = 1.67\ (t_{90} - t_{30})$. This is 1.2 μs as per Standard Specifications. The tail time is 50 μs from the virtual zero to 50% value on the falling portion of the voltage wave. See also Chapters 9 and 16 for a detailed description of lightning.

(*ii*) Other Standard Waves are 1.5/60 μs where t_f is defined as 1.25 ($t_{90} - t_{10}$), where t_{90} and t_{10} are timings at 90% and 10% of crest value. This is used chiefly in the U.S.A.

Lightning voltages are unidirectional. They can be both positive or negative with respect to ground.

(*d*) Switching-Surge voltages are caused when making switching operations in high voltage systems, such as the most important operation of energising a high voltage transmission line from a source by means of a circuit breaker. Switching surges are much longer in duration than lightning voltages, but unlike power-frequency voltages they are not periodic, see Fig. 1.3. They are usually unidirectional, either of positive polarity or negative polarity with respect to ground. The waveshapes and magnitudes show extreme variations, but for testing h.v. equipment other than transformers, the Standard waveshape according to I.E.C. Regulations has a front time of 250 μs = 0.25 ms, and time to 50% value on tail of 2500 μs = 2.5 ms. Details of other waveshapes are given under H.V. Testing in Chapter 16.

(*e*) Short circuits occur at normal voltage waveshapes and in a.c. systems they are of power-frequency. They are therefore 50 or 60 Hz phenomena.

Short circuits give rise to very heavy current flow in the shorted lines, but the voltage could also rise above normal peak value in healthy conductors not experiencing the fault. Any equipment used for protecting the system against dangers such as lightning arresters (also called Surge Absorbers) connected to the healthy phase or conductor and ground must be designed for these overvoltages, called the Dynamic Voltage Rise. See Chapter 9 for the values of Dynamic Voltage Rise and selection of lightning arresters for a given voltage level in a transmission system.

Q.4. What is the usual classification of voltages used in a.c. transmission?

Ans. There are five broad divisions in naming the voltage levels, based on their range of values. In a.c. power transmission when we refer to "voltage" of the system, it means the effective value or rms value between conductors. All a.c. high voltage transmission takes place by the 3-phase configuration where the line-to-line or conductor-to-conductor voltage is $\sqrt{3}$ times the voltage between a conductor and the neutral, which is usually called "ground". See Chapter 7 for discussion of Three-phase Systems. These five broad divisions of voltage are:

L.V. (Low Voltage) : Below 11 kV
M.V. (Medium Voltage) : 22 kV, 33 kV, 66 kV
H.V. (High Voltage) : 132 kV, 169 kV, 220 kV, 275 kV
E.H.V. (Extra High Voltage) : 345 kV, 400 kV, 500 kV, 750 kV
U.H.V. (Ultra High Voltage) : 1000 kV, 1150 kV, 1300 kV, 1500 kV

The above numbers or voltage values are known as Nominal Voltages and a range of values are allowed near the nominal values. For example, the 750 kV level can range from 735 kV to 765 kV.

In Sweden as well as in the CIGRE' usage, this level is also designated as the 800 kV level. See Chapter 6 also.

Q.5. What are the usual d.c. voltages used in the world for power transmission at high voltages?

Ans. Dr. K.R. Padiyar in his book "HVDC Power Transmission Systems" (New Age International (P) Ltd., 1990) has provided an extensive list of HVDC systems used all over the world to date. They range from a low of 150 kV to a high of 600 kV. (Voltages lower than 150 kV are usually utilised for "asynchronous ties" of 50 Hz and 60 Hz a.c. systems and do not have a cable or overhead line for transmission of bulk power). See Chapter 6 for details.

Usually these voltages are for bipolar transmission and are denoted as ± 150 kV or ± 500 kV or ± 600 kV, etc., and the voltage values are between one conductor and ground return, if any. In such cases they are called "bipolar" transmission lines. For example, in a ± 600 kV bipolar system, the voltage between two conductors or "poles" is 1200 kV, etc., while the voltage between any pole and ground is 600 kV, positive for one pole and negative for the other.

Q.6. What is meant by an electric shock?

Ans. Only a person who has experienced it can really tell what it feels like. But technically speaking, an electric shock is a disturbance to the normal tissue function in a body, whether human, animal or plant life. Its severity is measured by the current flow in the tissue caused by the voltage and body resistance.

Q.7. What are types of disturbances to the human body caused by electric shock?

Ans. Electric shock can be classified broadly as follows, depending upon its severity to the normal function of a life process.

(*a*) **Primary Shock Currents:** These cause direct physical harm when current flow exceeds about 6–10 mA. The danger is because of ventricular fibrillation, or vibration of the muscles controlling the ventricles or chambers of the heart. This causes disturbance to normal pumping function of the heart and arrests blood flow to the brain thus preventing flow of oxygen to the centres in the brain that support the normal functioning of life. Loss of life may be due to (*i*) arrest of blood circulation when current flows through the heart; (*ii*) permanent respiratory arrest when current flows in the brain; and (*iii*) asphyxia due to current flow across the chest thereby preventing muscle contraction controlling the lungs.

(*b*) **Secondary Shock Currents:** These current magnitudes do not cause direct physiological harm but may produce adverse reactions. Currents up to 1 mA to 6 mA are known as "let go" currents. At these levels, a human being has sufficient control of muscles and conciousness to let the conductors go as soon as a tingling sensation occurs. They may, however, cause a burn to the skin at the area of contact with the high voltage terminal or electrode.

Q.8. What are some of the current limits for human beings during an electric shock?

Ans. Steady-state flow of current up to 1 mA causes a tingling feeling in that part of the body coming into contact with a high voltage equipment, usually the fingers touching an exposed conductor or other live metallic objects. Currents from 1 to 6 mA are "let go" currents.

The "let go" current may increase to primary shock current if the contact is more firm such as the clasping or gripping a conductor by the fingers or palm of the hand. For men, this limit is 16 mA and for women 10 mA, at which levels there still exists a chance that no damage to the heart, brain or chest has occurred. In very rare cases, men can experience primary shock at 9 mA, women at 6 mA and children at 4.5 mA. This is called the probability that a "let go" current may become a primary shock current. Rare cases refer to 1 in 200 or 0.5% probability. Usual cases give a 50% probability level that a "let go" current may reach the primary shock current. For example, if at a level of 6 mA, the person has not let the live part go, then in 50% of the cases, the current may exceed 16 mA and become dangerous to life process. In the remaining 50% of the cases, there is still a chance that even at a current greater than 6 mA, the man will be able to let the conductor go. This depends entirely on the physiological constitution of the human body. Some people are more sensitive to electric shock than some other lucky ones.

Q.9. What are the values of resistance and capacitance of a human body which control the electric shock-level currents?

Ans. The normal human body has a resistance of about 2000 to 3000 ohms. For 10 mA to flow, this means that a voltage of 20 to 30 volts may be sufficient to give primary shock currents. [The reader might have used his/her tongue to test a battery of only 2–6 volts by feeling a tingling sensation of sourness!]. A voltage of 440 to 600 volts is certainly very dangerous and sometimes fatal. There is a rule that if a conductor of this voltage is touched with the knuckles on the back of the hand, there is a chance that the body will be thrown backwards and may escape death, but gripping the conductor is sure to lead to death or severe brain damage and cripple the person. A house voltage of 230 volts is fatal to children and very dangerous to adults at home, especially when the person is standing on a wet floor, as frequently happens in a bathroom with modern appliances. This is the reason that many countries, especially in North America, use 115 volts for normal home and shop distribution where a human being can come into contact with a live part. Also, all appliances must be properly grounded by using three-pin connectors. However, the current in lower-voltage systems is heavier than at higher voltages so that the cost of equipment is higher.

Higher the resistance of the body, lower will be the current level flowing in the body and higher will be the voltage at which the primary shock current is experienced. Normally, all Under Writers Companies who provide insurance against damage resulting from electric shock impose the condition that whenever a person (technician) is exposed to a voltage involving 600 volts or higher, he or she cannot work alone on the job. There must be a minimum of two people, so that if one of them experiences damage to his or her person, there must be another to telephone and obtain help immediately from hospitals.

Capacitance between a human body and a live equipment controls the charge acquired by the body. Naturally, this occurs when the person is not touching the live parts. When subsequently the body discharges, by coming into contact with a grounded object, a current flows in the body and a shock will result. This is also the case when a spark takes place between a human body and a charged body such as a large lorry or vehicle parked underneath or very close to a high voltage transmission line. The driver can experience a shock when he or she touches the truck or vehicle while opening the door and the vehicle has been parked under a high-tension overhead line on insulated tyres such as rubber ones.

A human body has a capacitance of 250 pF (pico-farad = pF = mmF = 10^{-12} farad) when standing on an insulated platform about 0.3 metre (1 foot). The charging current flowing at 50 Hz and 1000 volts will be $I = 2\pi \times 50 \times 250 \times 10^{-12} \times 1000$ amp = 80 μA = 0.08 mA. For 1 mA charging current, this requires 12,500 volts or 12.5 kV which is the lower limit for "let go" current. For 10 mA, the primary shock current, the voltage to ground will be 125 kV which corresponds to about 220 kV between conductors in a 3-phase transmission system ($125\sqrt{3}$). Minimum clearances of transmission lines to ground are based on these considerations as well as on insulation requirements.

A truck 8 m in length (40-ton truck),with body height of 3 m, width of 3 m, height of tyres 1.5 m, has a capacitance of about 100 pF when parked 6 metres from a h.v. transmission line which has conductors 13 m above ground at minimum clearance at mid-span usually, or 8.5 m above the roof of the truck, Fig. 1.4.

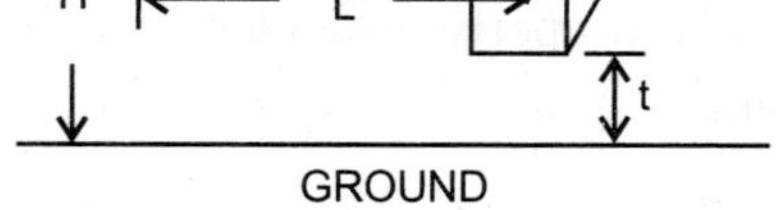

Fig. 1.4: Capacitance calculation of truck parked parallel to overhead line.

It therefore acquires a charge if its tyres are made of insulating material. The conductors of the line form one electrode of a capacitor while the metallic body of the vehicle forms the second electrode, so that the vehicle acquires an electric charge. When the driver of the truck (while opening the door of the cab, perhaps) touches the body of the vehicle, he discharges the capacitor through his own body and so experiences a shock.

Q.10. What are the broad methods of safety required against experiencing an electric shock?

Ans. The following list provides a very broad outline of the practices to be followed:

(*a*) A person handling high voltage equipment must develop a healthy respect for the h.v. equipment, and must habitually practice safety measures. There is no half measure in following safety precautions. No room should be allowed for taking risks. This is a daily habit and which is not developed by many individuals and societies, as can be seen in everyday life. In other words, technicians must be properly trained in safety precautions and not every person in uniform or otherwise should be turned loose as a technician on the high voltage work.

(*b*) The environment, particularly the air must be clean. Smoking in laboratories must be prohibited. Smoke contains carbon particles which are good conductors of electricity. If a smoker blows smoke towards the h.v. equipment, the air is filled with carbon particles and a conducting path is established between the h.v. parts and the smoker's mouth or other grounded objects in the area. A spark results which can be very intense and give a shock to the smoker's body.

(*c*) Since electric shock results from a flow of current in the human body due to the high voltage, any means adopted for increasing the resistance between the h.v. electrode and ground on which the person is located and through whom the current flows will remove the shock hazard.

This can be achieved in practice by: (*i*) wearing gloves made from an insulating material such as rubber; (*ii*) wearing insulated shoes such as rubber; (*iii*) standing on an insulated platform such as dry wood or a rubber mat; (*iv*) sit or stand in an insulated container such as a heavy-duty plastic gondola for overhead-line work.

Q.11. What is "static" and how can shock hazard be reduced?

Ans. The expression "static" refers to "static electricity" acquired by a person through friction. Usually, with synthetic materials increasingly used, a person walking with leather shoes on a carpet made of nylon or other synthetic fibre picks up electric charge due to friction of the soles on the surface of the rug. The person draws charge in that process and acquires a potential with respect to ground. A similar process occurs when a person wears synthetic materials such as nylon clothing. The cloth acquires a charge even when the person is merely walking in a dry atmosphere, such as a heated corridor in an office or hotel.

When the charged person touches a metallic grounded object such as a door handle to open a door, the charge is drained which is a current. A shock results. In order not to experience a shock to the fingers, it is advisable to have a key in the hand and touch the metallic knob with the key first. This discharges the body through an air gap by a spark which reduces the current, and no shock results.

Freshly-laundered clothing from a dryer also gives static due to the dry and hot air in which the clothes have been tumbling. There are anti-static sprays or papers available in the market which contain some metallic powders which keep discharging the drying clothes inside the dryer itself, or a metallic coat hanger can be run over the clothing before touching them. Naturally, wearing rubber gloves will prevent any of the unpleasant experiences. The crackling sound of sparks is familiar to any housewife or person doing the laundry!

Static electricity is responsible for many major accidents in industrial processes and material handling. There are instances where mixing of paints in plastic containers has developed such intense static due to the friction that the paint has caught fire, injuring the person mixing the paints. Grain elevators, where wheat is brought from the prairies of Canada by trains and stored for subsequent loading on ships have led to such intense static as to blow up the giant concrete structures. This occurs because very dry air at high speeds is utilised in huge pipes to suck the grain from the elevators and transfer it to the ship's holds. The friction combined with the dryness build up intense static and ignites the air causing the blow up of the huge concrete structure. The author has been witness to one such terrible explosive accident in Vancouver harbour where extensive damage occurred to the waterfront.

❑❑

CHAPTER 2

CONDUCTORS AND INSULATING MATERIALS

Q.1. What are the materials used for high voltage equipment and transmission lines?

Ans. Materials are broadly classified according to their electrical properties into three types:

(*a*) Conductors;

(*b*) Insulators; and

(*c*) Semiconductors.

Q.2. For what purposes are these used for in h.v. work?

Ans. (*a*) Conductors are used, as the name implies, for conducting currents in the desired paths of the system. They have very low resistance and do not give rise to high voltage-drops.

(*b*) Insulators are used for keeping the current flow in their regular pathways. They are required to separate two points or electrodes at different voltages or potentials, as otherwise there will be short circuits and currents will flow in undesired paths. They are absolutely necessary for safety to personnel.

(*c*) Semiconductors are used in h.v. devices and control circuits. High voltage d.c. transmission depends on the availability of high-current diodes (also called rectifiers and inverters) and thyristors (also called controlled rectifiers and inverters). Dr. K.R. Padiyar has given that the largest rating of diodes and thyristors utilised for h.v. d.c. transmission systems is at present in the range of 5 kV 3000 A per single unit. The highest transmission voltage has reached ± 600 kV (Itaipu Scheme in Brazil in South America). Therefore, many units have to be connected in series to withstand voltages above 5 kV, and in parallel for currents higher than 3000 A, if necessary.

Modern lightning arresters or surge absorbers are fabricated from semi-conducting materials and are called gapless ZnO arresters. Their conduction properties depend on change of energy levels brought about by the voltage applied during a lightning discharge or switching operation yielding high surges.

Q.3. What are the usual materials for conductors used in high voltage equipment and transmission lines?

Ans. Conductors are mostly fabricated from metals. Depending on the equipment, the following metals are found in h.v. work: Gold, silver, copper, aluminium, tin, lead, iron, nickel, chromium, tungsten and others.

Q.4. What are the salient characteristics of metals to consider for use in high voltage work?

Ans. These are:

(*a*) conductivity (or specific resistance) of material;

(*b*) temperature limit;

(*c*) fabrication or workability;

(*d*) chemical properties, chiefly for deterioration of primary properties (such as rusting of iron when exposed to air, etc.);

(*e*) availability; and most importantly,

(*f*) cost; and

(*g*) purity.

Some of the metals listed above are not directly used for conducting current but only as weather-proof coatings, chiefly nickel and chromium. Since high voltage equipments are prone to arcing and sparking, the metal should not be eaten away quickly. Obviously, gold and silver cannot be used in bulk but only for special purposes such as coatings on terminals of h.v. measuring equipment such as bridges, even though these two metals are the best available conductors. They reduce contact resistance in accurate instruments as also removing the problem of oxidation to which copper is subjected thereby causing a rectifier action.

Purity of metals is extremely vital when used in conjunction with vacuum as an insulator in circuit breakers and contactors. Similarly, the gas SF_6 (sulphur hexa fluoride) when used as an insulation, requires electrodes of the highest purity since any impurity in the metal will contaminate the gas and lower its insulating properties. This is also true of occluded gases which are released into the insulating medium under arcing or even under the high voltage stress.

Copper and aluminium are the most widely used conductors and are manufactured by electrolytic processes with purities in the neighbourhood of 99.9%.

Q.5. What are the salient properties of usual metals?

Ans. These are:

(*a*) Specific resistivity in ohm-metre;

(*b*) Temperature-resistance coefficient, ohm/°C;

(*c*) Specific gravity;

(*d*) Tensile strength when used in overhead transmission lines (for copper and aluminium only).

These properties are listed below in the Table.

Metal	Specific Res. (ohm-metre)	Temp.-Res. Coeff. (Ω/°C at 20°C)	Specific Gravity	Tensile Strength Young's Modulus (lbs/in²)
Symbol	r_S	α	δ	E
Gold	2.413×10^{-8}	0.0034	19.3	
Silver	1.625×10^{-8}	0.0038	10.5	9×10^6
Copper	1.72×10^{-8}	0.00393	8.89	17×10^6
Aluminium	2.8×10^{-8}	0.00403	2.7	10×10^6
Iron	11.5×10^{-8}	0.005	7.9	25×10^6
Nickel	10.75×10^{-8}		8.85	30×10^6
Lead	22×10^{-8}		11.4	

Q.6. How are insulating materials used in h.v. work classified?

Ans. Insulating materials (or insulations) are classified broadly into four categories: (1) Solids; (2) Liquids; (3) Gases; and (4) Vacuum.

Q.7. What are the most important types of solid insulation?

Ans. The list of solid insulations used is very long. But the most usual and important solid insulating materials are the following, along with an indication of usual purposes for which they are used:

Cotton Tape: Wrapping of conductors in transformers, generators, motors and such machines, with impregnation in liquid insulations such as oil, varnish and synthetic chemical fluids.

Paper: Mostly known as cellulose paper with trade name such as Kraft. Also in the form of pressboard. Used in cables and transformers impregnated with liquid insulation. All such impregnation of cotton tape and paper is carried out under vacuum.

Mica: For high-temperature work in generators, motors, cables adjacent to conductors, in tape or sheet form, impregnated in liquid insulating media.

Pressboard: In transformers impregnated with oil and other liquid insulations in order to separate low-voltage and high voltage windings. This is also called the major insulation, acts as supports also.

Resins: These are variously known as phenolic, urea, etc. They are used extensively for their high strength in transformers and machines for supporting terminals under oil or in air. They are also used as switchboards. Urea resins are free from carbon and can be used to withstand heavy arcing without loosing their properties or spewing carbon particles into the insulating media. Urea resin is used in arcing chambers of high voltage high-current circuit breakers.

Rubber: Natural rubber is one of the best solid insulations, although its use is restricted in h.v. work. It was used in wires and cables for its flexibility, but is now being replaced by synthetic rubbers, known as Butyl. The advent of plastics has replaced rubber by and large for all applications.

Porcelain: The best material for outdoor work in transmission lines and as supports, spacers, etc., for all h.v. installations, because of high strength, rigidity, electrical strength and capable of being moulded into any desired shape. The weakest spot is the cement which is required to join several pieces of smaller portions into a large insulator or to cement the pin and cap into a line insulator disc.

Glass: Also used for outdoor or indoor insulator for its high mechanical and electrical strengths. Plexiglass is a very popular insulator. Fibre glass is now extensively used. Optical fibres are used in many measuring situations to lead signals into measuring instruments as well as isolating two systems at high voltage such as in high voltage laboratories where certain instruments are kept at the high voltage end but the current has to be led into measuring instruments at near ground potential.

PVC and its Compounds: In this plastic age, it is but natural that Poly Vinyl Chloride forms a major portion of all insulations used in h.v. equipment, transmission systems for overhead and underground use and control circuitry. See Chapter 5 on Electrical Cables for list of compounds using PVC by itself or with other types of insulations such as kraft paper, etc. The use of Cross-linked Polyethylene (XLPE), a vulcanised form of PVC under high pressure and temperatures, and Polypropylene have become the most popular insulations for high voltage cables.

Wood: Dry or creosote-impregnated wood is used for towers where the insulating properties and high mechanical strength offer excellent material up to about 132 kV transmission and distribution work.

Q.8. What are the physical and electrical properties of important solid insulations suitable for high voltage work?

Ans. The three important properties to look for in solid insulation are:

(*i*) Dielectric strength which denotes kV/cm or kV/mm which the insulation can withstand and which a designer will use based on experience;

(*ii*) Dielectric constant; and

(*iii*) Loss angle called δ which denotes the heating loss in the insulation under the high voltage. The usual form in which this loss is denoted is by ($\tan\delta$).

Q.9. What are the definitions of these 3 above properties?

Ans. (*i*) Dielectric strength denotes the voltage which the insulation can withstand for a given thickness of the material. This is usually measured in volts/cm, kV/cm, MV/cm, or other units of voltage and thickness. This stress depends very much on the shape of electrode used which can vary over an enormous range. Therefore, Standard Specifications take this into account and have specified the shape of electrodes to be used when reporting the dielectric strength of any insulating material. We consider this aspect in more detail in Chapter 3.

(*ii*) Loss angle δ, or $\tan\delta$, is also called the "power factor". A perfect or ideal insulator should have no current flowing through it in phase with the voltage which produces heat and raises its temperature due to its I^2R loss. Any current which flows in the circuit connected to the insulation is a purely capacitive current in an ideal insulator which leads the applied voltage by an angle of 90°, see Fig. 2.1. However, insulators in practice are not the perfect. A small in-phase or loss component of current does flow so that the total current consists of a conduction component I_R and a pure capacitive component I_C, which we also call the charging current. The resultant current leads the applied voltage by an angle ϕ which is almost but not quite 90°. The angle $\delta = 90° - \phi$ is extremely small in practice. The ratio $I_R/I_C = \tan\delta$. The power factor of the dielectric is $I_R/I = \cos\phi$, the usual power factor of any electrical circuit. Now, $\cos\phi = \sin\delta \approx \tan\delta \approx \delta$, where normally $\delta < 1°$. The lower the angle δ is, the more loss-less the insulator is. See Chapter 17 for Bridge Measurements of loss angle ($\tan\phi$) and for other details.

(*iii*) Dielectric Constant, ε_r: This is a measure of the insulating strength of an insulation in comparison with air, normally speaking, whose dielectric constant is taken to be 1. In truth only vacuum, known as free space, has a dielectric constant of 1. The permittivity of vacuum is denoted by $\varepsilon_0 = 1000/36\pi\ \mu\mu\text{F/m} = 8.84194 \times 10^{-12}$ farad/metre. The permittivity of an insulating material is written as $\varepsilon = \varepsilon_0\ \varepsilon_r$. The capacitance of a parallel-plate capacitor with area of plates A and distance of separation d is

$$C = \varepsilon_0\ \varepsilon_r\ A/d \text{ farad,}$$

with A and d in metre units.

Thus, higher the dielectric constant ε_r, higher will be the capacitance.

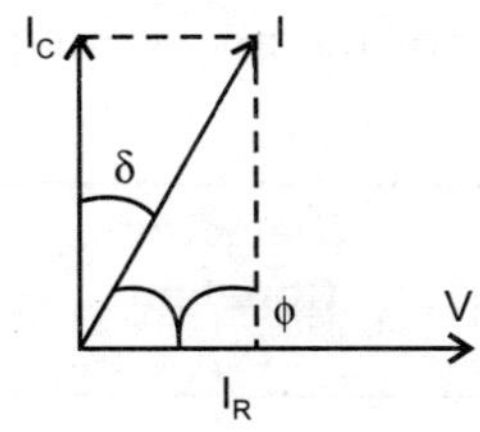

Fig. 2.1: Power factor cos ϕ, and loss angle δ of actual dielectric.

Q.10. What are some quantitative values of physical and electrical properties of important solid insulating materials?

Ans.

Material	Dielectric Strength, (kV/cm)	Dielectric Constant, (ε_r)	Loss Angle (δ or $\tan\delta$)
Ceramics		4.8–9.5	
Porcelain			
(High Strength)	150	6–8	
Glass	150	4	0.002
PVC	150	3.2–3.6	< 0.0005
XLPE	150	2.3	< 0.0005
Nylon		4–4.6	
Epoxy Resin		3–4.5	

Contd...

Phenolic Resin		4–15	
Silicones		3–4.5	
Mica	3–7	7	0.02
Asbestos	20		
Cellulose Paper	500	2.2	0.001–0.004

Q.11. What are the most usual liquid insulating materials used in high voltage equipment?

Ans. Insulating liquids are mostly natural or synthetic organic compounds. They can be classified as follows:

1. **Mineral Oil:** Used in transformers, cables and capacitors. These are hydrocarbons refined from crude petroleum. Many of these oils contain aromatic compounds (Benzene, Naphthalene). These absorb gases and oxidise which are harmful to the dielectric strength of the base oil.
2. **Askarel Liquids:** Mostly used in transformers and capacitors as compounds for impregnation. These are chlorinated aromatic compounds.
3. **Fluorocarbon Liquids:** These are non-flammable since hydrogen (H) is replaced by fluorine (F).
4. **Silicone Fluids:** Castor oil belongs to this group. It is used in energy-storage capacitors.

Q.12. What are the salient properties of the above liquid insulating materials?

Ans.

Liquid Insulating Materials	Dielectric Strength (kV/cm)	Dielectric Constant (ε_r)	Specific Resistance r_S (ohm-m)	Specific Gravity	tan δ
Transformer Mineral Oil	30	2.2	$(1-10) \times 10^{14}$	0.88	0.008
Transformer Askarel	Over 35	4.6	1×10^{13}	1.56	
Cable and Capacitor Mineral Oil	30		$(50–100) \times 10^{14}$	0.885	0.008
Silicone Fluids	>35	2.6			<0.008
Esters	Varies	4 – 7	Varies	Varies	

Q.13. What are the most usual gaseous insulating media used in high voltage equipment and lines?

Ans. 1. **Dry Air:** This is used at all values of pressure from near vacuum to extremely high pressures such as 200 lbs/sq.inch in circuit breakers for arc quenching. (1 atmospheric pressure = 14.7 lbs/sq.inch = 760 mm Hg = 101.38 kilopascals = 1.0138 bar; 1 Pa = 1 newton/sq. m. = 0.145×10^{-3} lb/sq.in.; 1 lb/sq.in. = 6895 N/sq.m. or pascals). Air is the insulation in high voltage transmission lines. Cables use solid insulation impregnated by liquids, and modern GILs or Gas Insulated Lines use higher dielectric strength gases such as SF_6. But compressed air is also used at some locations.

2. **Nitrogen:** Nitrogen can be manufactured in very pure form, and forms an extremely valuable insulating medium especially for very accurate high voltage capacitors used for bridge and other types of measuring situations. It forms the most widely used insulation for precision capacitors up to the highest voltage used in the world in bridge measurements, up to 500 kV. The pressure used in such cases is in excess of 200 lbs/sq.in.
3. **Sulphur Hexa Fluoride:** SF_6 is a toxic gas and is known as electronegative. Its molecules have very great affinity for absorbing free electrons, and thereby preventing conduction paths to form at the extremely high voltage stresses present in h.v. equipment such as cables, circuit breakers, transformers, etc. The gas is used in arcing situations like circuit breakers because of its excellent ability to absorb electrons and thereby aiding arc extinction or quenching. GILs using SF_6 as insulation between central conductor and grounded outer enclosure are used in long lengths in oil fields and bus ducts with pressures up to 5 atmospheres and for voltages up to 750 kV. The weakness is more in the solid spacers used for supporting the central conductor than in the insulating properties of the gas itself.

Q.14. What are the physical and electrical properties of important gaseous insulating media?

Ans. (*i*) **Air:** Normal air at sea level at 1 atmospheric pressure and 20°C is taken to be the standard, (NTP). At 0°C and 760 mm Hg, air has density of

$$\delta = 1.293 \text{ kg/m}^3 \text{ (273.15K, 101.38 kPa)}.$$

Using Boyle's law, $pV = \text{constant}$

and Charles's law, $V/T = \text{constant}$

we obtain $pV = nRT$

where $n = 22.414$ m³/mole at 0°C and 760 mm Hg pressure

and R = universal gas constant

$$= 8314.5 \text{ joule/ kilomole-K}$$

The density varies inversely as temperature so that

$$\delta = 1.293 \times 273.15/T$$

At $T = 243.15$ K (–30°C)

$$\delta = 112\% \text{ of } 1.293 \text{ kg/m}^3$$

and at $T = 303.15$ K $= +30$°C

$$\delta = 90\% \text{ of } 1.293 \text{ kg/m}^3, \text{ at the same pressure.}$$

δ varies with pressure according to the relation:

$$\delta = 0.4647\, p/T, \text{ with } p \text{ in mm Hg}$$

and

$$\delta = 3.4855\, p/T \text{ with } p \text{ in kPa.}$$

Temperature T is always in K or Å. The pressure of the atmosphere varies with altitude as follows: $p = 101.38 - 9.33\times 10^{-3} H$ mm Hg, with height H in metres.

Temperature of air varies with altitude H as follows:

$$T/T_0 = (p/p_0)^{0.286}$$

where $0.286 = (1.4 - 1)/1.4$

and $\sigma = 1.4$

the well-known specific heat of air under adiabatic (constant heat) condition.

Therefore, using the relation

$$p = 101.38 - 9.33 \times 10^{-3}\,\text{H}$$

the temperature of the atmosphere varies with altitude as follows with T_0 = temperature at $H = 0$, or at sea level:

$$T = [(101.38 - 9.33 \times 10^{-3}\,H)/101.38]^{0.286}\,T_0$$
$$= (1 - 92.1 \times 10^{-6}\,H)^{0.286}\,T_0$$

$$\therefore \quad \delta = 3.4855\,p/T$$
$$= 3.4855(1 - 92.1 \times 10^{-6}\,H)^{0.714} \times 101.38/T_0$$

The ratio of air density at a pressure p mm Hg and temperature t °C to that at 760 mm Hg and 20°C will be, since $\propto \alpha\, p/T$:

$$\delta = \text{air-density factor}$$
$$= (p/760)\cdot(273 + 20)/(273 + t)$$
$$= 0.3856\,p/T$$

The air-density factor δ given above has considerable influence on the electrical breakdown strength of air. This problem will be explained in more detail in Chapters 3 and 17 under "Breakdown Mechanisms" and "Testing of High Voltage Equipment".

(*ii*) **Nitrogen:** Nitrogen is an inert gas and does not enter into chemical reaction with other elements readily. It constitutes nearly 80% of the atmosphere by weight. It is not an electronegative gas. Nitrogen has the same electric strength as air at the same pressure and temperature. It liquifies at a temperature of –196°C = 77 K at 1 atmospheric pressure. Its dielectric constant is $\varepsilon_r = 1.000547$ at 20°C and 1 atm. (Air has $\varepsilon_r = 1.000536$). Its loss angle or tan $\delta = 0$.

The breakdown voltage of nitrogen varies directly with pressure between 1 atm and 15 atm. With a 12.7 mm gap (0.5 inch) and uniform electric field, the breakdown voltage values are 0.05 MV = 50 kV at 1 atm and 0.4 MV = 400 kV at 15 atm. This gives (400 – 50)/(15 – 1) = 25 kV/atm. At 10 atmospheric pressure the breakdown voltage is 275 kV for a 12.7 mm gap giving a breakdown voltage of 275/1.27 kV/cm = 216 kV/cm. The advantage in using nitrogen at high pressure is evident, see Fig. 2.2. This variation is nearly the same as for air of 225 kV/cm.

(*iii*) **Sulphur Hexa Fluoride, SF_6:** (See the excellent book by V.N. Maller and M.S. Naidu: "Advances in High Voltage Insulation and Arc Interruption in SF_6 and Vacuum", Pergamon Press). This gas is eminently suitable as insulation in Gas Insulated Cable or

Gas Insulated Line, GIL. Its arc-quenching properties have made this gas the most desirable medium in high voltage high-current circuit breakers.

SF_6 is 5 times heavier than air and has a specific gravity of about 6 kg/m³ or 6 gram/litre. (Air has 1.293 kg/m³ as discussed earlier). It liquifies at – 40°C at 3.6 atm and at –10°C at 10 atm. The usual pressure used in GIL is between 2 and 4 atm and therefore SF_6 is in gaseous phase at temperatures in excess of –35°C. It is colourless, odourless, mildly toxic, non-combustible, chemically extremely stable and inert up to 500°C. It is an electro-negative gas and captures free electrons readily in an electrical discharge through it, as in an arc in a circuit breaker. The dielectric strength is 2 – 3 times that of air at the same pressure and greater than that of oil above 2 atm, see Fig. 2.3. The dielectric constant of SF_6 is very nearly the same as that of air and nitrogen. $\varepsilon_r = 1.0021$ for SF_6, 1.00054 for air and nitrogen. $\tan \delta = 2 \times 10^{-7}$ at 1 bar and 4×10^{-7} at 20 bar, which therefore is practically zero.
(1 bar = 100 kPa = 100/101.38 atm = 0.987 atm = 750 mm Hg).

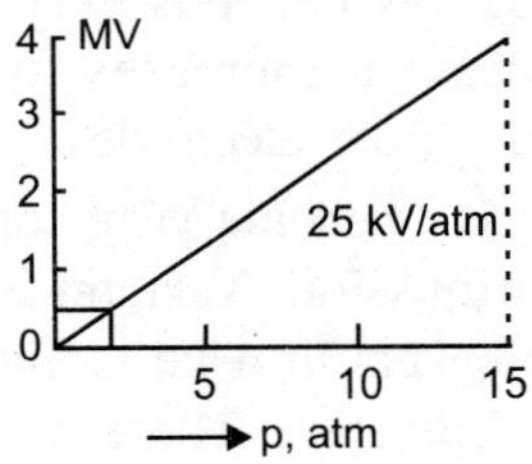

Fig. 2.2: Breakdown voltage of nitrogen with pressure.

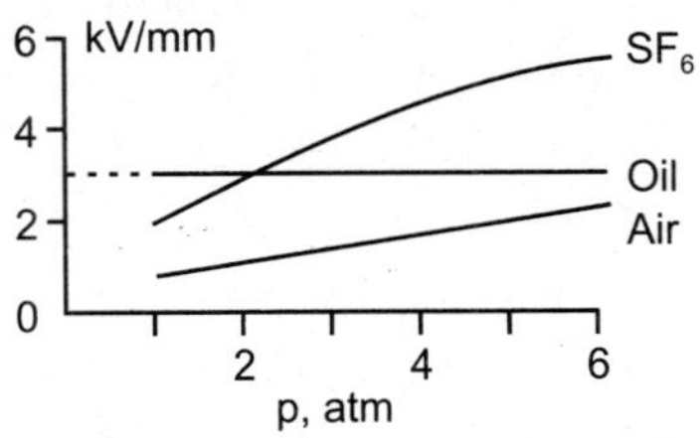

Fig. 2.3: Dielectric withstand strength of SF_6, oil and air variation with pressure.

Q.15. What are the broad properties of vacuum as an insulation?

Ans. Ideal or absolute vacuum has no electrons or ions and cannot conduct electricity. Therefore, its dielectric strength should be infinite. However, in practice, no vacuum is ideal and there will always be traces of gas molecules and other types of impurity which are the main sources of breakdown of the "vacuum". See Chapter 3 next for more details of Vacuum Breakdown. These impurities are adsorbed gases (a surface effect) in the metal electrodes and on insulating surfaces. Other causes which release impurities into a vacuum gap are known as field-effect emission of electrons from the metallic electrodes, micro-discharges and micro-particles ejected from the electrodes under the extremely high voltage stresses prevalent in vacuum gaps.

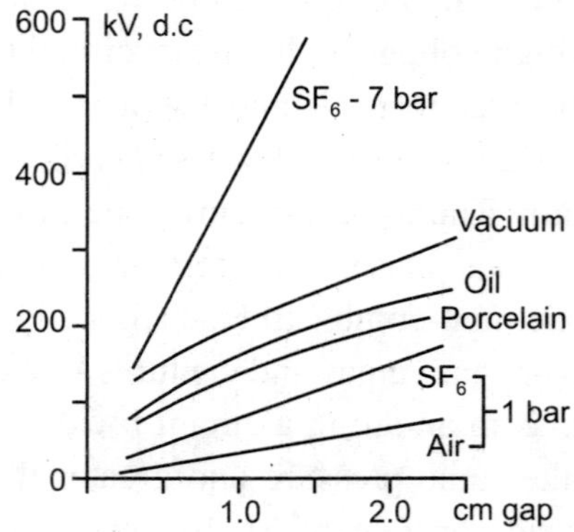

Fig. 2.4: Breakdown voltage (d.c.) of various solid, liquid, and gaseous dielectrics.

Small vacuum gaps with nickel electrodes have a breakdown strength of nearly 100 kV/mm = 1 MV/cm. But this depends very much on the material used for the electrodes. For steels (stainless) it increases to 120 kV/mm and lowers to nearly 40 kV/mm for aluminium and copper electrodes.

Vacuum circuit breakers have become very common from motor control contactors to power distribution and transmission. Vacuum interrupters are used with capacitors, transformers, and as neutral-breakers in what is known as 4-legged reactor-switching in very high voltage transmission practice. These breakers operate normally at 15 kV.

Figure 2.4 compares the d.c. breakdown strength of several types of insulating materials: solid, liquid, gas and vacuum.

❑❑

CHAPTER 3

Electron Structure and Breakdown Mechanism

Q.1. What is meant by Electron Structure?

Ans. The atom of any element consists of a central nucleus of radius equal to

$$R_n = R_0 \cdot A^{1/3}$$

where $R_0 = 1.2 - 1.35 \times 10^{-15}$ m

$= 1.2 - 1.35$ fermi or femtometer,

and A = atomic mass number

$= Z + N$

= number of protons + number of neutrons.

The number of protons Z inside the nucleus is called the atomic number. The protons carry positive electric charge of

$$e = 1.6022 \times 10^{-19} \text{ coulomb each.}$$

The mass of proton is 1.67265×10^{-27} kg

= 1.0072765 atomic mass unit

= 938.28 MeV

The neutron has no electric charge but its mass is

$$m_N = 1.67495 \times 10^{-27} \text{ kg}$$

= 1.008665 atomic mass unit

= 939.573 MeV

The radius of a proton or neutron is 1.2 – 1.35 fermi when we use $A = 1$ in the formula

$$R_n = R_0 \cdot A^{1/3}$$

The protons and neutrons are held very tightly inside the nucleus by strong nuclear forces against the electrostatic repulsion between the protons. This force is the strongest of all forces in nature. There are no electrons inside the nucleus in the stable state, but are given out during decay processes, such as radioactivity, called β-emission. Encircling the nucleus are Z electrons (the same number as protons inside the nucleus) each with a negative charge equal to that of a proton,

$$e = -1.6022 \times 10^{-19} \text{ C}$$

They are held in their stable orbits by the attractive force of the protons. The radius of an atom to the outermost electron is nealy 10^{-10} m, so that the atom is mostly empty space, much like the Solar System. [Radius of sun = 6.9626×10^5 km, radius of orbit of the outermost planet Pluto = 39.4 AU, 1 AU = 1 astronomical unit = radius of the orbit of the earth around the sun = 150×10^6 km. Therefore, ratio of (radius of Pluto's orbit/ Sun's radius)

$$= 39.4 \times 150/0.69626$$
$$= 8488 \approx 10^4.$$ Radius of atom/radius of nucleus $\approx 10^5$].

Every electron in its orbital positions is associated with an energy with discretc values, and these are named Energy Levels. The electron closest to the atomic nucleus has the lowest energy level as otherwise it can go further from the nucleus in spite of the attractive force. The electron occupying the farthest orbit has the highest energy level. By giving sufficient energy from an external source, such as electric or thermal or radiation, an electron can be moved from one orbit to another at a higher energy level. Similarly, when an electron falls from an orbit at a high energy to one of lower energy, a characteristic radiation is emitted according to the relation

$$E = hv,$$

where E = energy in joule

v = frequency in Hz

and h = the famous universal constant called Planck's constant or quantum of action

$$= 6.6262 \times 10^{-34} \text{ joule-second}$$

$$1 \text{ eV} = 1.6022 \times 10^{-19} \text{ joule}$$

$$1 \text{ MeV} = 10^6 \text{ eV}.$$

By being restricted by natural laws, the electrons around the nucleus arrange themselves in what are known as shells and sub-shells. These are governed by Pauli's Exclusion Principle, which states that no two electrons can be in the same state, or have the same property at the same instant of time in the atom.

This arrangement of electrons in shells and sub-shells is known as the Electron Structure of the atom.

Q.2. What is the detailed electron structure of the simplest of all hydrogen atoms?

Ans. In order to understand the ideas, we need the following units and universal constants or factors:

Charge of a proton or electron = 1.6022×10^{-19} coulomb (positive for the proton and negative for the electron); 1 electron-volt (eV) of energy = charge of 1.6022×10^{-19} C moved through a potential difference of 1 volt;

Electrical energy in joule = charge in coulomb × p.d. in volt;

So, 1 eV = 1.6022×10^{-19} joule,

and 1 joule = 6.24×10^{18} eV.

1 Angstrom = 1 Å = 10^{-8} cm = 10^{-10} m

Permittivity of free space $\varepsilon_0 = 8.84194 \times 10^{-12}$ F/m

Bohr's first radius, $r_b = 0.52858$ Å

$= 4\pi\varepsilon_0 (h/2\pi)/e^2 m_e$

where $h/2\pi = 6.6262 \times 10^{-34}/2\pi$, J-s

$= 1.0546 \times 10^{-34}$ J-s

$= 6.58066 \times 10^{-16}$ eV-s

Mass of electron, $m_e = 9.11 \times 10^{-31}$ kg

Einstein's mass-energy equation: $E = mc^2$, c = light velocity

Energy equivalent of rest mass of electron

$= 9.11 \times 10^{-31} \times 9 \times 10^{16}$ J

$= 511.7 \times 10^3$ eV

$= 0.5117$ MeV

Central force of attraction between positive charge of Ze and negative charge of electron, $F_e = Ze^2/4\pi\ \varepsilon_0\ r^2$

Centrifugal force of mass m moving on a circle of radius r with linear velocity v:

$F_c = mv^2/r$

Angular velocity of electron, $\omega = v/r$

Angular momentum of electron $= m\ v\ r = m\ \omega\ r^2$

Potential energy of electron, $E_p = -Ze^2/4\pi\varepsilon_0 r$

Kinetic energy of electron, $E_k = mv^2/2$;

n = integer = 1, 2, 3, ...

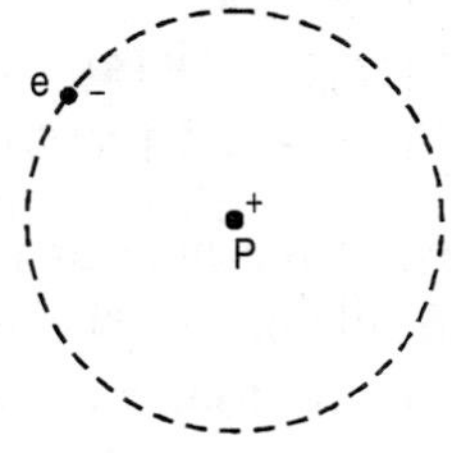

Fig. 3.1: Hydrogen atom.

The Hydrogen atom consists of one proton forming the nucleus and 1 electron orbiting around it, see Fig. 3.1. It is held in the steady-state orbit by the balance between the force of electrostatic attraction against the centrifugal force due to its velocity. Equality of the two forces gives the value of orbital radius, thus:

$$e^2/4\pi\varepsilon_0 r^2 = m_e v^2/r$$

or

$$r = e^2/4\pi\varepsilon_0 m_e v^2$$

Now, Niels Bohr hypothesised or postulated that when the electron is orbiting around the proton in the hydrogen atom, its radius cannot vary in a continuous fashion nor its energy, but the only orbits permitted are those for which the angular momentum $m_e vr = m_e \omega r^2$ is an integral multiple of Planck's constant $h/2\pi$.

Thus, $$m_e vr = nh/2\pi$$

Combining with $$r = e^2/4\pi\varepsilon_0 m_e v^2$$

we can eliminate v thereby giving the radius for any value of integer n to be:

$$r_n = [4\pi\varepsilon_0 \,(h/2\pi)^2/e^2 m_e]\, n^2$$

With known values of the constants ε_0, h, e and m_e, we obtain

$$r_n = 4\pi \times 8.842 \times 10^{-12} \times (1.0546 \times 10^{-34})^2 \,/\, [(1.6022 \times 10^{-19})^2 \times 9.11 \times 10^{-31}]\, n^2$$

$$= 0.528426 \times 10^{-10}\, n^2 \text{ metre}$$

i.e., $$r_n = 0.528426\, n^2 \text{ Å}$$

The first orbit for $n = 1$ has the radius, $r_1 = r_b = 0.528426$ Å, which is called the first Bohr radius.

Considering now the energies, the sum of potential energy and kinetic energy in any orbit with orbital number n is

$$E_n = E_{pn} + E_{kn}$$

The potential energy $$E_{pn} = -e^2/4\pi\varepsilon_0 r_n,$$

while the kinetic energy $$E_{kn} = m_e v^2/2 = \tfrac{1}{2}(e^2/4\pi\varepsilon_0 r_n).$$

Therefore, the total energy in any orbit with orbital number n is

$$E_n = -e^2/8\pi\varepsilon_0 r_n$$

$$= -[e^4 m_e/32\pi^2\varepsilon_0^2 (h/2\pi)^2]/n^2$$

Again, using the known values of e, m_e, ε_0 and h, the total energy is

$$E_n = -2.1862 \times 10^{-18}/n^2 \text{ joule}$$

$$= -13.645/n^2 \text{ eV}$$

For $n = 1$, the electron is closest to the nuclear proton, while for $n = \infty$, the electron is completely out of the influence of the proton because of its high energy level. The amount of energy 13.645 eV is known as the Ionization Potential of the hydrogen atom. When the total energy of the electron becomes this amount, the elctron is removed from the atom. The resulting atom, instead of being electrically neutral, is now in an ionised state with only a proton and therefore is now called a Positive Ion.

For atoms with more than 1 orbiting electron around the nucleus, this simple model does not apply and we resort to the Shell Model of Pauli and others. The above simple model of the hydrogen atom is known as Bohr's Atomic Model.

Q.3. What is meant by the Shell Model of a multi-electron atom?

Ans. An electron possesses several important properties, chief among them being:

(*a*) It has a spin around an axis of rotation, much like the earth spinning on its axis;

(*b*) Since the electron is an electrically charged particle, the spinning charge is equivalent to an electric current and therefore the electron behaves as a small bar magnet with north and south poles, much like the earth;

(*c*) When a magnetic field is applied, the spin axis can align itself parallel to the direction of the external magnetic field, or its axis can be anti-parallel. These two states are known as positive and negative and denote the two states of the electron. The spins are denoted with the numbers ($+\frac{1}{2}$, $-\frac{1}{2}$). The electron cannot be in any other intermediate state. The actual value of the spin amounts to $[(h/2\pi)/2]$;

(*d*) In the lowest energy state i.e., closest to the atomic nucleus, there can exist only 2 electrons with the two values of spin given above. No more electrons can be present according to the Pauli's Exclusion Principle, one of the most fundamental truths unearthed by a scientist to describe the workings of elementary particles;

(*e*) When the number of electrons in the lowest energy state is 2, they form a closed 1st shell. Thus, the first shell for $n = 1$ has 2 electrons and is closed. It could have only 1 electron in which case it is known as a partially-filled 1st shell. When the shell is closed, any further electrons must occupy states in the next shell which consists of sub-shells following certain rules, all governed by Pauli's Exclusion Principle.

(*f*) For $n = 2$, in the next shell,where n is termed the Principal Quantum Number, two possible arrangements exist: (*i*) In the first sub-shell,there can exist only 2 electrons; (*ii*) In the second or next sub-shell, there may be 6 electrons which depend on the two values of spin and three values of the "magnetic quantum number". These form the second shell.The second completed shell has a total of 8 electrons in its two completed sub-shells ($n = 2$).

(*g*) The rule thus far can be expressed as follows for electron numbers: For $n = 1$, number of electrons in completed shell = $2 = 2 \times 1$; For $n = 2$, in the first completed sub-shell, number of electrons = $2 = 2 \times 1$. In the second completed sub-shell, the number of electron states is $6 = 2 \times 3$. The total number of electrons in the second shell is thus $2 + 6 = 8 = 2 \times 2^2$. In general, it is not necessary that both sub-shells be complete i.e., completely filled. There can exist, and do, incomplete sub-shells.

(*h*) Similarly, in the 3rd shell, there will be 3 sub-shells with a completed sub-shell having electron numbers or energy states for the electrons to be: 2 × 1, 2 × 3, 2 × 5. In case all 3 sub-shells are filled, the total number of electron states in the 3rd shell amounts to $2 + 6 + 10 = 18 = 2 \times 3^2$. Again, it is not necessary that all sub-shells be completely filled, or in other words all energy states be occupied by electrons. There will always be incomplete sub-shells depending on the atomic number of the element, and consequently incomplete shells.

(*i*) For example, if the first two shells are completely filled, the total number of electrons will be 2 + 8 = 10. An atom with 10 orbiting electrons is the element neon with 10 protons and 10 neutrons inside the nucleus. It is denoted by the chemical symbol $^{20}Ne_{10}$. It is an inert gas and the electron structure is very stable.

(*j*) Similarly, if only the first shell is complete and the atom has only 2 electrons, the element is Helium $^4He_2$ which is another inert gas.

It is not necessary that a given shell should be completely filled with electrons for an element to be known as an inert one. For example, the inert gas argon has 18 orbiting electrons around a nucleus with 18 protons and 22 neutrons inside the nucleus. Its chemical symbol is $^{40}Ar_{18}$. The shell and sub-shell structure is as follows:

$n = 1$: First shell is complete with 2 electrons;

$n = 2$: Second shell is also complete with 8 electrons;

$n = 3$: First sub-shell is completely filled with 2 electrons; Second sub-shell is also completely filled with 6 electrons;

There are no more electrons to occupy any more energy levels. Since a sub-shell is completely filled even though the 3rd shell is not completely filled, the 18 electrons are tightly bound and it is very difficult to dislodge them from their orbits or energy states. [The reader should refer to books on Atomic Physics and the Periodic Table of the Elements for any further details about Electron Shell Structure of atoms].

Q.4. Describe the electron structures of good conductors such as aluminium and copper?

Ans. (*a*) The chemical symbol for aluminium is $^{27}Al_{13}$ so that the nucleus consists of 13 protons and (27 – 13) = 14 neutrons. There are 13 electrons arranged as follows in shells and sub-shells, as shown in Fig. 3.2.

Ist shell: $n = 1$: 2 electrons;

IInd shell: $n = 2$: Complete with 8 electrons in 2 sub-shells (2 + 6);

IIIrd shell: $n = 3$: First sub-shell complete with 2 electrons;

Second sub-shell contains only 1 electron.

Total number of electrons is (2 + 8) + (2 + 1) = 13.

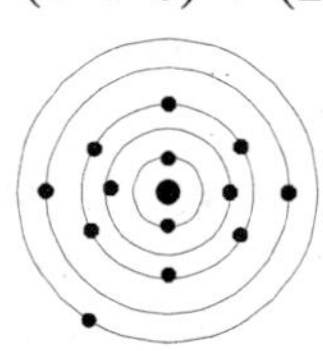

Fig. 3.2: Shell model of aluminium atom: $^{27}Al_{13}$.

The single electron in the incomplete 2nd sub-shell in the 3rd shell is very loosely bound, being far away from the nucleus as well as being subject to the force of repulsion of the 12 electrons in between the positive nucleus and itself. Very little energy is required to remove it from the orbit, even room temperature being sufficient. Therefore, a volume of aluminium containing millions of free electrons can be drained away by applying a very low voltage. This phenomenon is usually described as a sea of electrons in which the positive ion, forming the nucleus with 13 protons and 12 orbiting electrons, is embedded. This is the characteristic of all metals which are very good conductors. Sometimes, it is also referred to as Fermi gas model for conductors.

(*b*) Copper has the chemical symbol $^{63}Cu_{29}$ and the reader can figure out the number of protons and neutrons inside the nucleus. Obviously, there are 29 electrons orbiting the nucleus. The electron structure will be as follows: $2 \times 1^2 + 2 \times 2^2 + 2 \times 3^2 + 1 = 29$. There are, consequently, 3 completed shells with 28 electrons and 1 electron in the 1st incomplete sub-shell in the 4th shell, as shown in Fig. 3.3. It is this 1 electron which gives even better conduction characteristics to copper than aluminium by being farther from the nucleus and repelled by 28 electrons instead of 12 in $^{27}Al_{13}$.

The reader can practice the description of electron structure of any other element in this manner. However, there are some other rules that apply than the simple and straightforward ones given until now.

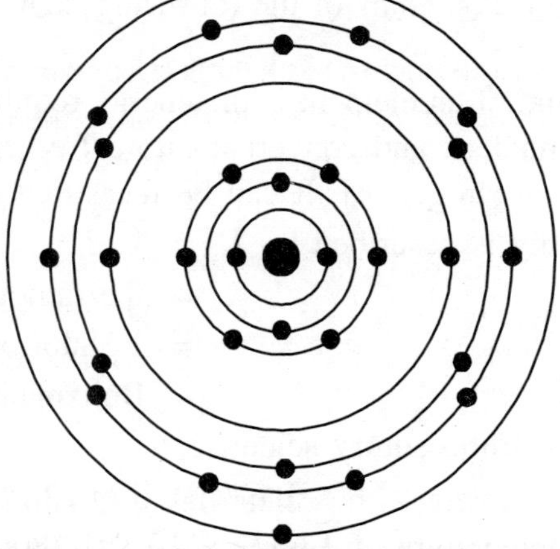

Fig. 3.3: Shell model of copper atom: $^{63}Cu_{29}$.

Q.5. What are the electron structures of the inert gases $^{84}Kr_{36}$, $^{132}Xe_{54}$ and $^{222}Rn_{86}$ (krypton, xenon and radon)?

Ans. These belong to the same group as $^{4}He_{2}$, $^{20}Ne_{10}$ and $^{40}Ar_{18}$ discussed in Question 3.

$$^{84}Kr_{36}: 2 \times 1^2 + 2 \times 2^2 + 2 \times 3^2 = 28.$$

The first two sub-shells in the 4th shell are completely filled with 2 + 6 = 8 electrons. There are no incomplete sub-shells; the gas is inert.

$^{132}Xe_{54}$: The first 3 shells are completely filled with 28 electrons as before. In the 4th shell the arrangement is that the first 3 sub-shells are completely filled with (2 + 6 + 10) = 18 electrons. There now remain 8 more electrons and the arrangement requires some explanation. If the simple rule given earlier is used, the 8 electrons could occupy 8 of the possible (2 × 7) = 14 states. But this is not a very stable arrangement since for an inert gas either a shell or sub-shell should be completely filled. Therefore, nature has ordered that these 8 remaining electrons go over to the next shell, the 5th in this case, and occupy the 8 states in the 1st and 2nd sub-shells (2 + 6). This is a case where the 5th shell has electrons even though the previous shell, the 4th, has not been completely filled.

$^{222}Rn_{86}$: The first 4 shells are now complete with (2 + 8 + 18 + 32) = 60 electrons. There now remain 26 electrons, 86 – 60. In the 5th shell, there are 5 sub-shells. The first 3 sub-shells are completely filled with (2 + 6 + 10) = 18 electrons. Once again, as in the case of Xe, we are also left with 8 electrons which go over to the next or 6th shell, filling the first two sub-shells completely. Thus the 6th shell has electrons even though the 5th shell is not completely filled. The element has all the properties of an inert gas and it is very difficult to dislodge any electron from the last incomplete shells. (By the word 'very difficult' implies that it requires an enormous amount of energy from an external source, such as temperature or voltage, to remove any of the electrons. We will come back to it under 'insulator'later on).

Q.6. What are the Ionization Potentials of the following elements: Li, Na, K, Cs, He, Ne, A, H, O and N?

Ans. The ionization potential of an element is the energy required to release the electron from the influence of the nucleus and convert it into a free electron. We found in Q. 2 that this is 13.645 eV for hydrogen, H. It can be measured in eV as well as *k* as follows:

The thermal energy is described by $E = k \cdot T$,

where k = the famous universal constant

= 1.38066×10^{-23} joule/K

= Boltzmann's constant,

after the very famous 19th century scientist.

Since $1 \text{ eV} = 1.6022 \times 10^{-19}$ J,

it corresponds to a temperature of $1.6022 \times 10^{-19}/1.38066 \times 10^{-23}$ = 1160K.

Thus, the ionization potential of the given elements in eV and K are as follows:

Element	Li	Na	K	Cs	He	Ne	Ar	O	N
E_i, (eV)	5.39	5.14	4.34	3.89	24.6	21.6	15.8	12	15.6
$T \times 10^{-4}$ (K)	0.6254	0.5964	0.5036	0.4535	2.854	2.506	1.833	1.392	1.81

[**Note:** The temperature numbers given in the above Table in row 3 must be multiplied by 10^4 = 10,000 to obtain the actual values. Thus, the ionisation temperature of element caesium Cs is 0.4535×10^4 = 4,535 K, corresponding to its ionisation potential of 3.89 eV; etc].

Note: that the alkali metals Li, Na, K, Cs have low ionisation potentials while the gases He, Ne, A, H, O and N have higher ionisation potentials. These latter gases make good electrical insulators. The temperature required to ionise them is in the range of

$$1.392 \times 10^4 \text{ to } 2.854 \times 10^4 \text{ K, or, } 13{,}920 \text{ to } 28{,}540 \text{ K.}$$

A material such as caesium Cs is used for converting a gas with high ionisation potential into one having a lower ionisation potential by including Cs in gaseous form into the parent gas. This is called 'seeding'. Such techniques belong to a field of energy generation called Magneto Hydro Dynamics and not to the insulation game which is of primary interest in h.v. engineering.

Q.7. What is meant by Electrical Breakdown?

Ans. When two electrodes are separated by an insulator and a potential difference is applied between them, it is normally expected that the insulator (solid, liquid, gas or vacuum) will support the voltage. This means that there are no free electrons in the insulating medium which cause a continuous stream of electron path or conducting path bridging the electrodes, see Fig. 3.4. If the voltage reaches or exceeds a value which pulls out electrons from their orbits in the atoms of the material and a continuous current flows through the insulation between the electrodes, then the insulation has broken down and becomes a conductor. This is called Electrical Breakdown.

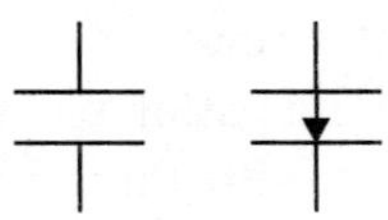

Fig. 3.4: Electrical breakdown between two electrodes.

We also observe that when once the insulation has suffered a breakdown, removal of the voltage might mitigate the damage. This is the case with "re-sealable" insulation such as gases and vacuum. But solids and liquids suffer more or less permanent molecular damage and cannot be re-used. The equipment has to be condemned. The mechanisms of breakdown are different for solid, liquid, gaseous and vacuum insulations but the end result is the same — the insulation is not good.

Q.8. What are the units for measurement of the breakdown strength of insulating materials?

Ans. This is measured as intensity of electric stress or voltage gradient described by potential per unit thickness or distance. The units can be usually volts/mm, or V/cm, or V/m, or kV/mm, or kV/cm, or kV/m, or MV/m. Since SI (International) unit of potential is the volt and distance is metre, the standard notation should be V/m. But this gives unmanageable numbers for insulation purposes. The most usual units used in practice are kV/mm or MV/m. Note that kV/mm and MV/m are the same. Higher this value, better is the insulation strength.

The breakdown strength is the minimum value of electric stress at which a conducting path is formed in the insulation between the electrodes. In nearly all cases of breakdown, the actual breakdown is preceded by pre-breakdown phenomena which are as important for evaluating the insulation strength as the main breakdown phenomenon. These are known as partial breakdown or corona-formation, etc.

Q.9. What is meant by the withstand strength of an insulation?

Ans. The withstand strength (also called the dielectric strength) is the maximum value of electric stress at which a breakdown process commences. Once this stress is reached, free electrons are released from the atoms of the insulating material with sufficient energy as to commence the further breakdown processes. The insulating medium looses its insulating properties beyond the breakdown strength.

Q.10. Are the withstand strength and breakdown strength constant for an insulating material?

Ans. No. They depend on the shape of the electrodes and the thickness of the insulation. This is the entire field of investigation of withstand and breakdown strengths of insulating materials.

As can be observed, experience in the field is absolutely necessary in order to be able to use these two values sensibly and intelligently, and providing sufficient margins or what are known in engineering as factors of safety. We can only indicate here some useful values.

Q.11. What are Standard Specifications for insulation testing?

Ans. This is a very wide field and the reader should consult them always. Experienced engineers in the profession, in manufacturing plants of insulation, equipment, inspectors for purchasers, and others concerned with this high voltage electrical engineering game always have these Standard Specifications in their bookshelves, and the more a person gets experience the more the engineer has them at the "finger tips", so to speak. A Standard Specification prescribes the shape of electrodes to be utilised and the insulation thicknesses to be used during testing in laboratories and testing departments. The engineer must be very conversant with these details before using the breakdown and withstand-strength values of the insulator used in equipment and high voltage transmission.

Q.12. What are the breakdown stress values of some important insulating materials?

Ans. The values given are for power-frequency a.c. voltage.

Solids:

Kraft paper : 30 kV/mm for thick paper
50 kV/mm for thin sheets and ribbon as in cables.

Mica : 5 kV/mm in oil.

PVC : 50 kV/mm.

XLPE : 75 kV/mm.

Liquids:

Transformer oils : 30 kV/cm. (Also for cables and capacitors).

Askarel : 35 kV/cm.

Silicone fluids : 35 kV/cm.

Gases:

Air	: 30 kV/cm (between 1-cm spheres).
N_2	: 30 kV/cm.
SF_6	: 20 to 50 kV/cm (at 1 atm to 5 atm).
Vacuum	: 1000 kV/cm with nickel electrodes.

Q.13. What are the shapes of electrodes in common use in high voltage equipment, and transmission lines?

Ans. See Fig. 3.5.

Q.14. What is meant by a Breakdown Mechanism?

Ans. The breakdown mechanism is the description of conditions occurring to the atoms or molecules of an insulating material that cause free electrons to be released to form a continuous conducting path between the two electrodes. These conditions vary from material to material. We can only discuss some basic processes in this book and leave the reader to consult the enormous amount of literature that exists for details on the physics and engineering aspects from other specialised sources.

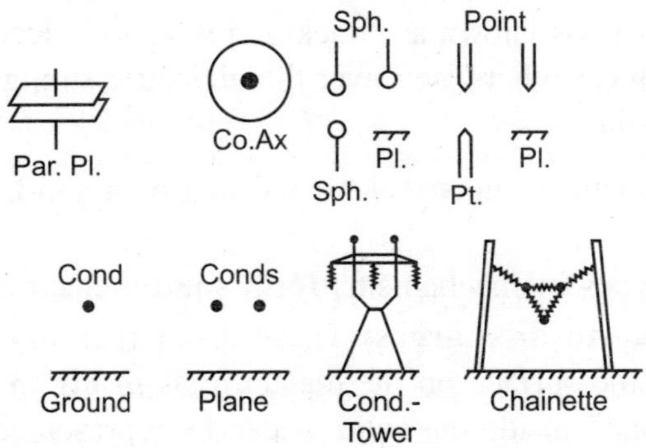

Fig. 3.5: Shapes of electrodes in high voltage work. Parallel plates; Co-axial cylinders; Sphere-Sphere; Sphere-Plane; Point-Point (Rod-Rod); Point-Plane (Rod-Plane); Conductor-Ground; Conductors above Plane; Skirt-type S/C Tower (Conductor-Tower); Chainette construction of e.h.v. and u.h.v. towers.

BREAKDOWN MECHANISM OF SOLID INSULATION

Q.15. What are the major factors causing electrical breakdown of solid insulation?

Ans. Any source of energy which can free electrons from the orbits in the atoms or molecules contributes to the breakdown. These sources or factors are: (*a*) Electrical stress due to high applied voltage; (*b*) Thermal breakdown due to high temperatures; (*c*) Chemical factors causing deterioration of material; and (*d*) a combination of all these factors.

Q.16. What are meant by (*a*) Partial ionisation or corona, and (*b*) Tracking, when referred to solid insulation?

Ans. (*a*) No solid insulation is free from gases or voids as they are called, which have a lower dielectric strength than the solid material. As the thickness of insulation increases, there will be more gaseous bubbles trapped in it. These are also known as occluded gases. Under an applied voltage, this gas ionises which gives out energetic free electrons that prove to be harmful to the chemical properties of the solid. This is partial discharge, also known as corona. It does not represent complete breakdown or failure of the insulation, but could indicate the commencement of breakdown. Tests required for cables and transformers using thick insulation for high voltages beyond a specified level prescribe detecting these partial discharges.

(*b*) **Tracking** is a surface phenomenon on a solid insulation. When the solid insulation is used in air or other gases or immersed in a liquid, there will exist small gaps on the surface of the solid which form potential areas for the commencement of discharges. For example, if a solid is in air, moisture in the air sticks to the surface as tiny particles of water. In the process of drying, they form small gaps in the conducting water which causes sparks. These sparks ionise the solid surface and makes it conducting. Ultimately, it will lead to flashover between the electrodes from the surface of the solid and not through the thickness of the solid. Every solid insulation has what is known as "tracking resistance" depending upon its vulnerability to the formation (or otherwise lower the dielectric strength) of foreign particles such as moisture or dust.

Q.17. What is the mechanism of electrical breakdown of a solid insulating material such as paper?

Ans. There are several types of mechanisms for a solid dielectric to breakdown electrically:

(*a*) **Breakdown due to discharges:** These discharges may occur in voids inside the dielectric or at the surface on the dielectric as in Fig. 3.6. Considering a void, or a gas bubble trapped inside the solid, it can be represented as a small capacitor C_v in series with a capacitor formed by the major solid insulation C_i. If the applied voltage is V between the electrodes across the solid insulation, the internal voltage appearing across the void will be such that $V_v C_v = V_i C_i$, with $V_v + V_i = V$. The voltages will then be inversely proportional to the capacitances since the product of voltage and capacitance is constant, which is of course the charge.

Therefore, $$V_v = (V - V_v)C_i/C_v.$$

This gives $$(1 + C_i/C_v)V_v = VC_i/C_v,$$

or, $$V_v = VC_i/(C_i + C_v).$$

When this voltage reaches a value required to commence a discharge, called the threshold voltage for corona in the void, free electrons are given out which affect the rest of the solid so that the void becomes larger. This causes the dielectric stress to increase, ultimately leading to breakdown.

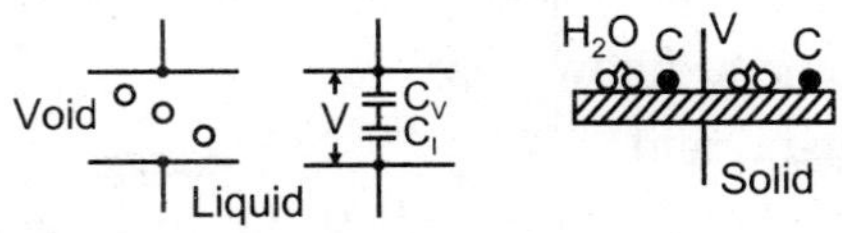

Fig. 3.6: Void inside and tracking on solid insulation.

(*b*) **Electro-mechanical breakdown:** When an electric stress is applied to a solid insulation, the solid is compressed which decreases the thickness of insulation and further increases the electric stress (volt/mm). This process is cumulative and leads to breakdown.

(*c*) **Thermal breakdown:** If a solid insulation is exposed to high temperatures, such as around current-carrying conductors in cables or in transformers, the dielectric properties are degraded and breakdown results. Temperature rise of all equipment encountered in practice is based on what the insulation can withstand, and solid insulations are classified according to the limiting temperatures.

(*d*) **Chemical breakdown:** In composite insulations such as paper or pressboard immersed in oils, chemical reactions can damage the inherent insulating properties of the insulation. This leads to breakdown.

(*e*) **Tracking or Surface breakdown:** Tracking is the formation of conducting paths on the surface of the insulating material leading to degeneration of its insulating properties. Surface breakdown usually occurs in the liquid or gaseous medium in which the major solid insulation is immersed. The presence of the solid with higher dielectric constant than the liquid or gas will distort the electric field sufficiently seriously so as to give rise to breakdown conditions in the medium with lower dielectric constant. It is, therefore, very essential to accurately plot the electrostatic field distribution in every case where insulations are used between metallic electrodes, including materials with different dielectric properties.

BREAKDOWN MECHANISM OF LIQUID INSULATION

Q.18. What are the mechanisms for breakdown of liquid dielectrics?

Ans. Liquids also suffer from the presence of voids because of gas bubbles present under pressure and temperature effects. They also contain impurities such as solids. The most serious of these is the formation of sludge in transformers in oil or synthetic liquids. These solid impurity particles distort the electrostatic field giving rise to dissolved gases to be evolved which form tiny bubbles. This gas-void ionises, thereby attacking the liquid. This is a cumulative process with more and more gas being evolved. Ultimately, the gas bubbles may bridge the electrodes and cause breakdown.

BREAKDOWN MECHANISM OF GASEOUS INSULATION

Q.19. What is the principal mechanism for breakdown of a gaseous insulation?

Ans. A gas has much less atoms in a given volume than either a solid or a liquid. The breakdown occurs due to a phenomenon called the Electron Avalanche. We will describe two fundamental phenomena known as:

(*i*) Townsend Mechanism, and

(*ii*) Meek and Loeb Mechanism also called the Streamer Mechanism.

Some of the breakdown mechanisms vary according as the electric field distribution is uniform or non-uniform, and convergent or divergent from the most highly-stressed electrode. Naturally, the former or uniform-field breakdown is simpler to understand than the latter.

The most spectacular insulation breakdown in nature which we witness is the lightning stroke. We will pay special attention to this phenomenon in Chapter 9.

Q.20. What is the Townsend Avalanche Theory of Breakdown of a gaseous medium under a uniform electric field?

Ans. [The theories of breakdown processes of gaseous insulation and data on insulation properties occupies an enormous amount of scientific literature. The reader is referred in particular to the monumental and authoritative work edited by J.D. Craggs and J.M. Meek: "Electrical Breakdown of Gases," John Wiley & Sons, Chichester, U.K., 1978].

Consider a uniform-field gap between an anode and a cathode with the electric stress $E = V/d$, volt/metre, as shown in Fig. 3.7(*a*). Under the applied electrical energy, an electron can be released from a molecule of the gas or from the cathode material which acquires kinetic energy due to the force exerted on it by the electric field. This electron collides with another molecule releasing an electron and in this process looses some if not all of its energy. Its velocity becomes practically zero in the collision process, but this assumption is not necessary here for understanding the mechanism. It is sufficient to know that one initial electron has given rise to two. These two free electrons acquire sufficient energy from the electric field to knock out two more electrons, and so on. The entire process is cumulative and after n collisions commencing from 1 electron there will result 2^n electrons. This is termed as Electron Avalanche.

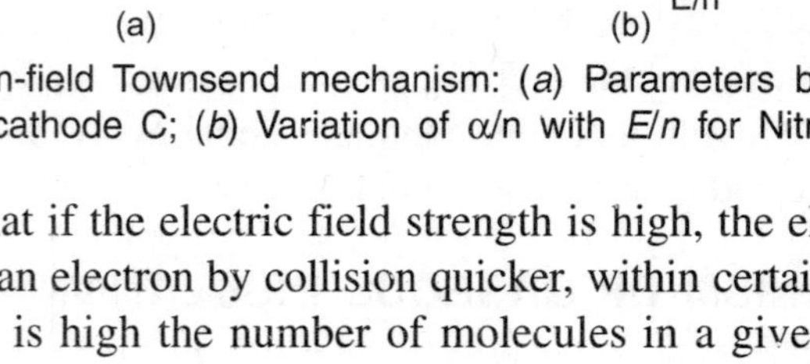

Fig. 3.7: Uniform-field Townsend mechanism: (*a*) Parameters between anode A and cathode C; (*b*) Variation of α/n with *E*/*n* for Nitrogen.

It is evident that if the electric field strength is high, the electrons acquire the energy required to release an electron by collision quicker, within certain limits. On the other hand, if the gas pressure is high the number of molecules in a given volume or the density is increased so that the free path of an electron is impeded or decreased before it encounters another molecule so that it has a chance of acquiring less energy and velocity. Thus, the probability that it will release another electron by collision with a molecule is reduced. Thus, the probability of creating new electrons by an incident electron from a gas molecule is directly proportional to E and inversely proportional to p, the gas pressure. This is expressed by Townsend's first ionisation coefficient, which is defined as follows:

α = mean number of collisions occurring in 1 cm of drift by 1 electron.

The value of α is determined experimentally for each gas, for different pressures, and for different applied fields in uniform gaps. For nitrogen, the experimentally-observed values of α are, with n = number density of gas molecules at a pressure p and temperature T:

E/n	100	200	300	400	500	600
α/n	10^{-19}	4×10^{-18}	1.5×10^{-17}	3×10^{-17}	4.5×10^{-17}	5×10^{-17}

This is shown in Fig. 3.7(b). The ion density n is related to the gas pressure and temperature as follows:

$$n = 2.69 \times 10^{19}\ (pT/760)\,.\,(273/T),\ \text{molecules/cm}^3$$

where pT = gas pressure in mm Hg,

and T = temperature of gas in K.

At 0°C, T = 273K,

and p_0 = 760 mm Hg,

the value of n = 2.69×10^{19} molecules/cm^3

The above values of (E/n) and (α/n) can be expressed by the equation

$$[\ \log_{10}(\alpha/n) = K_1/(E/n - K_2)],$$

where K_1 = 0.945 and

K_2 = 0.0597,

which are characteristic constants for nitrogen. The reader should derive this equation from the values given above, which is a mere mathematical exercise.

Now, n is directly proportional to p and inversely proportional to T. Thus, at a given temperature we have

$$(\alpha/n) \propto (\alpha/p) = K(\alpha/p),$$

and $$(E/n) \propto (E/p) = K(E/p),$$

where K is a constant and relates n to p. Thus, $\log_{10}(\alpha/n)$ can be converted to $\log_{10}(\alpha/p)$ and (E/n) into (E/p), since E and p can be measured directly in any experiment and not n.

Therefore, $\log_{10}(K \cdot \alpha/p)$ can be written as $K_1/(K \cdot E/p - K_2)$. Call this $(-\theta)$, say. This gives $\alpha/p = 10^{-\theta}/K$, where now θ is a function of (E/p). Thus, we obtain one of the most important relations that the Townsend α-coefficient can be written as $\alpha = p \cdot f(E/p)$, namely, that α is a function of pressure p multiplied by another function of (E/p), which is the ratio of electric field strength to the pressure.

For nitrogen,

$$K_1 = 0.945$$

and $$K_2 = 0.0597.$$

The value of $$K = 3.298 \times 10^{16}\, p_{20} \text{ at } t = 20°\text{C} = 293 \text{ K}.$$

$$[2.69 \times 10^{19} \times 273/(293 \times 760) = 3.298 \times 10^{16}].$$

When an avalanche of electrons is attracted by the positive electrode or anode, because of the negative charge at the avalanche head, it does not represent sparkover or breakdown of the gas, since breakdown requires complete bridging of the gap between the anode and the cathode by conducting particles. Thus, other mechanisms are required for breakdown. This occurs due to secondary mechanisms, which are as follows:

(*a*) An electron avalanche leaves behind positive ions which drift towards the cathode and acquire high enough energy which release more electrons from cathode surface due to thermionic emission;

(*b*) The excited gas molecules (having lost electrons) return to the ground or neutral state quickly by absorbing some of the free electrons and in this process emit radiation of sufficient energy as to cause ionisation of other molecules releasing more electrons;

(*c*) The positive ions near the cathode increase the electric field to sufficiently high value so as to emit electrons by collision with the cathode material.

The net result of these secondary processes left behind by the primary avalanche mechanism is to cause an increase in number of electrons. This process is described by Townsend's second ionisation coefficient β, which in turn is also a function of (E/p). From experimental results, the value of β is found to be:

$$\beta = 1.52 \times 10^{-6}\,(E/n)^{1.66}$$

$$= 5.782 \times 10^{-34}\,(E/p)^{1.66} \text{ at } 20°\text{C}.$$

The reader should prove this by using the relation between n and p.

The current intensifies due to both the a and b mechanisms in an exponential manner as follows in the gap distance d:

α – process only: $I = I_0 \exp(\alpha d)$;

β – process also: $I = I_0 \exp(\alpha d)/[1 - \beta\{\exp(\alpha d) - 1\}]$.

Now, Townsend established his criterion for sparkover of the gap by postulating that when conditions are such that $I \to \infty$, or, when the denominator is zero in the β-process also, sparkover occurs.

Thus, $1 - \beta[\exp(\alpha d) - 1] = 0$, or, $\beta[\exp(\alpha d) - 1] = 1$.

Since both α and β depend upon pressure p and $(E/p) = (V/pd)$ in a uniform field, the sparkover criterion depends upon p and d, the pressure and gap distance.

This breakdown mechanism and the breakdown criterion is also known as Townsend-Raether mechanism and criterion. We shall discuss more about this in Q. 23.

Q.21. What is meant by (*a*) Time lag of electron initiation, and (*b*) Time lag of breakdown of a spark gap?

Ans. (*a*) When a voltage is applied across a gap between two electrodes with an insulating gas in the gap, also called a spark gap or insulating gap, some electrons must be present in the medium to initiate the electron avalanche. If there are no initial electrons, it is clear that no avalanche can be produced and hence there will be no possibility of a breakdown of the gap. But the application of a voltage generates sufficiently high electric stress at the cathode to release electrons. There is a time lag between the application of the voltage and initiation of electron flow from the cathode surface, since these phenomena cannot occur instantaneously. This time interval is called the time lag of initiation of electrons. This is not a definite time for all cathode surfaces and gaps, but only an average which can be determined statistically.

(*b*) The time lag of breakdown refers to the time interval between application of a voltage across a spark gap and the complete breakdown of the gap. This is also called the formative time lag and usually refers to the time interval between generation of the initial electrons and the complete breakdown of the gap. Thus, it refers to the time required to build up the electron avalanche, its travel to the anode and the time required for the heavy positive ions to travel to the cathode, thereby giving rise to a current which bridges the spark gap.

Q.22. What is the practical application of the time-lag phenomenon in the design of spark gaps and insulation strengths?

Ans. As given in Chapter 1, there are several types of voltage in practice, namely, steady-state a.c., d.c., lightning and switching surges. In the case of the first two, namely steady-state a.c. and d.c., the voltage is held in the steady-state while lightning and switching surges are transients. They come very suddenly on the spark gap and vanish equally quickly, but can create havoc with the equipment. Spark gaps are used in high voltage equipment chiefly to protect the major insulation from suffering flashover inside the equipment under lightning impulse and switching surge since these two have extremely high magnitudes compared to the steady-state a.c. or d.c., nearly 2 to 3 times the normal peak value of a.c. or the d.c. This number is called the per-unit value of lightning or switching surge. Referring to Fig. 3.8, protective gaps, usually in air, are connected across transformer or circuit-breaker busings, when they are also known as spill gaps. It is absolutely necessary that such gaps in air breakdown or flashover prior to the major insulation.

Thus, the design of major insulation inside the equipment will depend on the time lag of breakdown of these spill gaps as well as on the magnitude of voltage at which they flashover. Normally, the gap is open and is able to withtrand the steady a.c. or d.c. voltage without a flashover. But on the occurrence of these high voltage short-duration surges due to lightning or switching, the protective gap breaks down thereby diverting the surge to ground. Since an air gap is re-sealable, it regains its insulating strength i.e., stays open after the surge passes away. Such spill gaps are also an essential part of a surge arrester or lightning arrester in those cases which use these protective devices containing gaps. Modern surge absorbers are becoming gapless, and are called Gapless Metal Oxide Arresters made of ZnO_2, Zinc Oxide.

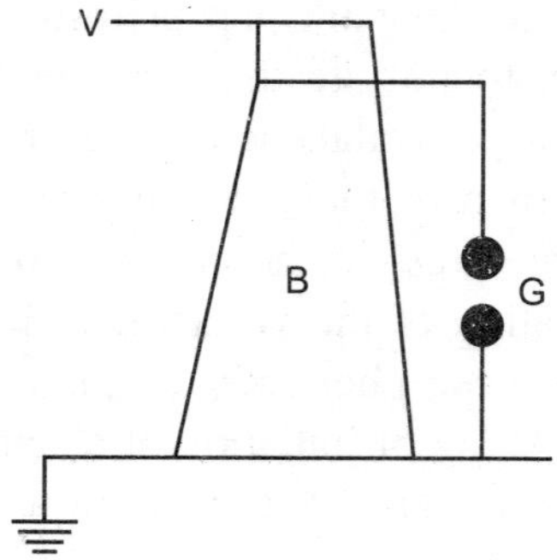

Fig. 3.8: Protective gap or spill gap across equipment bushing and internal insulation.

Q.23. Derive that (*a*) $I = I_0\, e^{\alpha d}$ for the Townsend-Raether α-mechanism of breakdown, and (*b*) $I = I_0\, e^{\alpha d}/[1 - \beta(e^{\alpha d} - 1)]$ for the α-β process.

Ans. (*a*) α = mean number of ionising collisions made by 1 electron in 1 cm of drift.

Let n_0 = number of electrons given out by the cathode due to any process per second, such as cosmic rays or temperature,

and n = number of electrons at a distance x from the cathode in the avalanche.

These n electrons at x in traversing a differential distance dx along the direction of the uniform field $E = V/d$ will generate new electrons.

This is: $$dn = \alpha \cdot n \cdot dx$$

so that $$dn/n = \alpha \cdot dx,$$

giving $$\log_e n = \alpha\, x + C_1,$$

or, $$n = C_2 \cdot e^{\alpha x},$$

where C_1, C_2 are constants. At $x = 0$, initially, there are n_0 electrons.

Thus, $C_2 = n_0$, and the number of electrons grows exponentially,

giving $$n = n_0 \cdot e^{\alpha x}.$$

At $x = d$, that is at the anode, the number of electrons striking it per second will be $n_d = n_0 \cdot e^{\alpha d}$. Each electron on the average produces $(e^{\alpha d} - 1)$ new electrons in traversing the gap-length from the cathode to the anode. This is the electron avalanche and is shown in Fig. 3.9. The heavy positive ions left behind drift slowly towards the cathode.

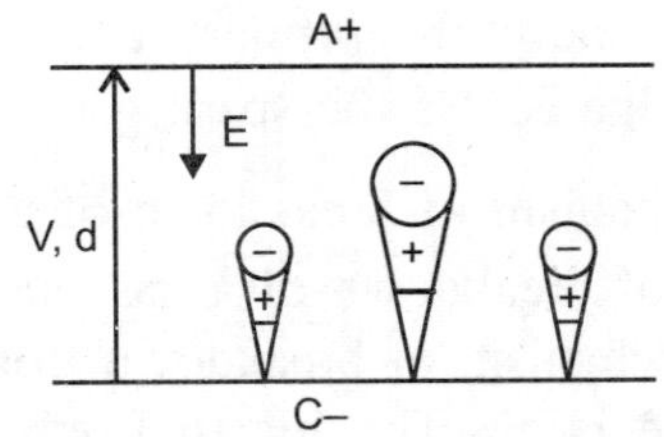

Fig. 3.9: Electron avalanches: Primary and secondary.

If the electrons at the cathode had been produced by some energy, they constitute a current I_0. Therefore, the current generated by the electron avalanche flowing towards the anode and the positive ions drifting towards the cathode constitute a total current of $I = I_0 \cdot e^{\alpha d}$. This current flows in the external circuit connected to the gap. If the charged particles constitute a continuous stream bridging the gap, then breakdown has occurred. But such is never the case. Secondary mechanisms enhance the generation of electrons prior to this condition.

(*b*) **α-β Process:** The Townsend-Raether α-process by itself does not represent a sufficiently powerful source of electrons to yield a continuous or sustained discharge current across the spark gap to constitute a flashover or breakdown of the gas. Secondary processes are essential. In effect, while the α-process depends on generation of an avalanche from 1 initial electron, the β-process generates many more electrons from the secondary or side effects of the main avalanche process, thereby enhancing the number of electrons initiating more and stronger electron avalanches.

The applied electric field generates n_{P0} electrons from the cathode per second which can and does give rise to n_{s0} secondary electrons. Thus, the total number of electrons initiating the electron avalanche will be

$$n_0 = n_{P0} + n_{s0}.$$

Due to collisions with gas molecules, each electron on the average produces $(e^{\alpha d} - 1)$ collisions in the gap length d. Therefore, the total number of collisions due to n_0 electrons is $n_0\ (e^{\alpha d} - 1)$ per second. Thus the number of secondary electrons generated is $n_{s0} = \beta n_0(e^{\alpha d} - 1)$.

Therefore, $n_0 = n_{P0} + \beta n_0(e^{\alpha d} - 1)$, giving $n_0[1 - \beta(e^{\alpha d} - 1)] = n_{P0}$, or, the total number of electrons ultimately giving rise to the electron avalanche by both the α- and the β-process is:

$$n_0 = n_{P0}/[1 - \beta(e^{\alpha d} - 1)].$$

These arrive at the anode intensified by $e^{\alpha d}$,

or,
$$n_d = n_0\ e^{\alpha d} = n_{P0} \cdot e^{\alpha d}/[1 - \beta(e^{\alpha d} - 1)].$$

The circuit current, as before, will be: $I = I_0 \cdot e^{\alpha d}/[1 - \beta\ (e^{\alpha d} - 1)]$, since n_{P0} electrons generate the current I_0. This is the total current generated by both the α- and β-processes, namely, the primary and secondary electrons generated by application of the voltage across the spark gap.

Q.24. What is the Streamer Mechanism of breakdown of a spark gap insulated by a gas?

Ans. The Streamer Mechanism of breakdown of a gaseous insulation gap or a spark gap is also called the "Kanal" mechanism of breakdown from the German word for Channel. This theory of breakdown is chiefly due to Loeb and Meek while the avalanche mechanism is due to Townsend and Raether. The streamer mechanism explains most of the properties of breakdown characteristics which the avalanche mechanism alone could not explain, chiefly the formative time lags for breakdown.

It was very soon determined experimentally after Townsend put forward the avalanche mechanism that spark gaps flashed over in less time than could be explained by the α-β mechanisms. Thus, some other mechanism is required to explain why the time lag should be less than only the transit time of avalanche-current across the gap length *d*. We still consider a uniform-field gap where the electric field strength is $E = V/d$, a constant everywhere.

According to this theory, in addition to the primary electron avalanche, and as a consequence of it, a streamer of intense charged particles is generated which bridges the gap as follows:

The head of the avalanche, Fig. 3.9, contains an immense number of electrons due to both the α- and the β-mechanisms and keeps growing as it moves towards the anode. Two phenomena occur now. The motion of the avalanche head causes intense photo-ionisation of the gas molecules in the gap instead of only from the cathode. Secondly, the approach of the intense negative cloud of electrons near the anode intensifies the electric field which in turn increases the velocity of electrons towards the anode. There results a distortion of the uniform field due to the presence of the negatively-charged cloud, being greater ahead of the electron avalanche, as well as behind it because of the positive ions formed which move towards the cathode.

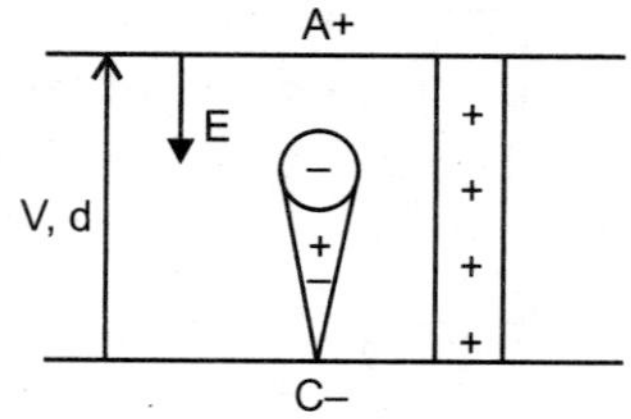

Fig. 3.10: Streamer mechanism of breakdown of Loeb and Meek.

Loeb and Meek hypothesised that the primary avalanche is swept into the anode leaving the heavy positive space-charge in the gap as shown in Fig. 3.10, having a conical shape. Photo-ionisation occurs. These photo-electrons generate further avalanches which once again leave behind positive charges. The electric field distorts further, and Loeb and Meek considered that when the electric field strength reaches the original value $E = V/d$, the externally applied field, the space charges are swept very rapidly to the electrodes and form a streamer which becomes a highy ionised gas between the electrodes. The gap thus breaks down.

The Streamer Theory is also known to be the principal mechanism initiating breakdown in non-uniform fields and in very large gas gaps. Such gaps are the more usual in highvoltage work than uniform-field spark gaps.

Q.25. What is Paschen's Law and Paschen Minimum?

Ans. We found in Question 20 that Townsend's first ionisation coefficient α is a function of pressure p and (E/p), where E = electric field strength, volt/m. Also, the second ionisation coefficient β is also a function of $E/p = V/pd$ for a uniform field spark gap. The Townsend-Raether breakdown criterion which resulted was:

$$I = I_0\, e^{\alpha d}/[1 - \beta(e^{\alpha d} - 1)] = \infty,$$

giving $$\beta(e^{\alpha d} - 1) = 1.$$

Let us write $$\alpha = p \cdot f_1(E/p) = p \cdot f_1(V/pd),$$

so that $$\alpha d = (pd) \cdot f_1(V/pd) \text{ and}$$

$$\beta = f_2(E/p) = f_2(V/pd).$$

Then the breakdown criterion becomes:

$$f_2(V/pd) \cdot [\exp\{pd \cdot f_1(V/pd)\} - 1] = 1.$$

Thus, the sparkover voltage $V = V_S$ is a function of the product of pressure and gap distance,(pd). This is Paschen's law for breakdown of a gap with uniform-field condition.

When the voltage V_S is plotted against (pd), the resulting curve looks like Fig. 3.11. The breakdown voltage reaches a minimum and increases on both sides as (pd) is decreased and increased. This minimum value of V_S and the corresponding value of (pd) are known as Paschen Minima.

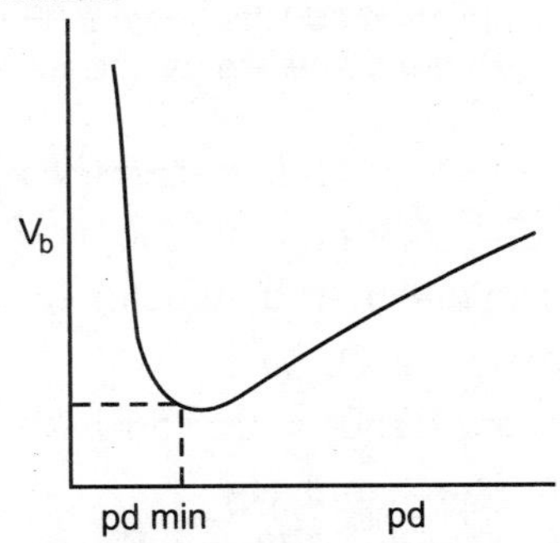

Fig. 3.11: Paschen's law for breakdown voltage V_b and (pd). Showing Paschen's minimum.

Q.26. What is the formula for sparkover voltage for an air gap in uniform field?

Ans. The accepted equation for design purposes is:

$$V_S = 24.22\,[(273 + 20)/(273 + t)] \cdot (pd/760)$$
$$+ 6.08\,[\{(273 + 20)/(273 + t)\} \cdot (pd/760)]^{1/2} \text{ kV}$$
$$= 9.34\,(pd)/T + 3.775\,[(pd/T)]^{1/2}$$

where t = temperature in °C,
T = $(273 + t)$ K,
p = pressure in mm Hg,
and d = gap length in cm.

At the standard conditions of $t = 20°C$ and $p = 760$ mm Hg, this equation yields, for $d = 1$ cm, $V_S = 24.22 + 6.08 = 30.3$ kV.

Q.27. What is meant by Corona in non-uniform field gaps? What are Peek's Formulae for corona-inception voltage-gradient for concentric cylindrical and parallel cylinder electrodes?

Ans. F.W. Peek, Jr, investigated very thoroughly after extremely painstaking experiments that in non-uniform electric fields, such as are encountered in practice for transmission lines and in laboratories, air breaks down locally on the most highly-stressed electrode. When a conductor is strung inside a large cylinder and concentric with it, the electric field strength is much higher on the conductor surface than elsewhere. See Chapter 5 on Cables. There results a partial ionisation of air around the conductor which is termed "Corona Discharge". This is usually bluish in colour because of the Oxygen molecules being ionised.

The following two very famous formulae are due to Peek for the electric field strength required for the commencement of corona, which are also known as Corona Threshold gradient or Corona Inception gradient, see Fig. 3.12.

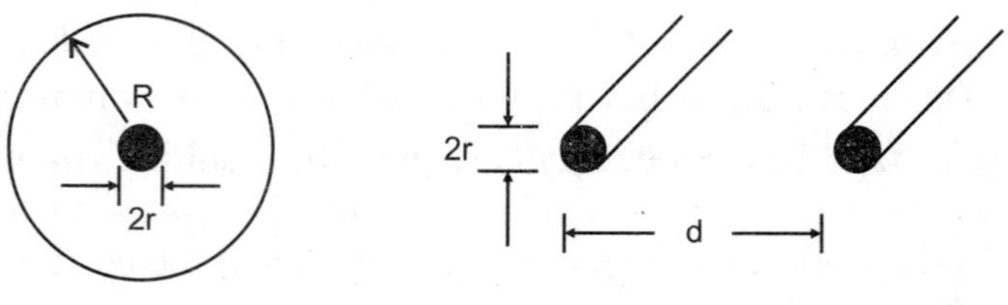

Fig. 3.12: Corona-inception gradients. Peek's Formulae for Co-axial Cylinders and Parallel Cylinders.

Concentric Cylinders: $E_0 = 31\delta[(1 + 0.308/\sqrt{(r\delta)}]$ kV/cm peak;
Parallel Cylinders: $E_0 = 30\delta[(1 + 0.301/\sqrt{r\delta})]$ kV/cm peak.

Here, r = radius of inner cylinder or conductor in cm,
δ = air density factor, see Q. 14
$= [293/(273 + t)] \times (p/760) = 0.3856\ p/T$,
p = atmospheric pressure, mm Hg
and t = temperature in °C. ($T = 273 + t$ K).

For example, at an elevation of 1000 metres above sea level, such as the city of Bangalore in India where major high voltage facilities are located, and at a temperature of 0°C, and $r = 1$ cm radius of conductor, the corona-threshold or inception gradient will be calculated as follows:

$$p = (760 - 70) \times 10^{-3}H$$
$$= 690 \times 10^{-3} \times 1000 = 690 \text{ mm Hg},$$
$$T = 273$$

Therefore,

$$\delta = (293/273) \cdot (690/760)$$
$$= 0.9744,$$

so that $\sqrt{(r\delta)} = 0.987$

For concentric cylinders:

$$E_0 = 31 \times 0.9744 \ (1 + 0.308/0.987)$$
$$= 39.63 \text{ kV/cm peak value.}$$

As a comparison, at sea level and 20°C,

$$\delta = 1$$

and $E_0 = 40.55$ kV/cm peak value.

For parallel cylinders:

$$E_0 = 30 \times 0.9744 \ (1 + 0.301/0.987)$$
$$= 38.15 \text{ kV/cm peak value.}$$

At sea level and 20°C,

$$E_0 = 30 \times (1 + 0.301)$$
$$= 39.03 \text{ kV/cm peak.}$$

Q.28. What are the mechanisms of breakdown of the composite gas sulphur hexa fluoride (SF_6)?

Ans. The dielectric strength of SF_6 is nearly 2.5 times that of air at the same temperature and pressure. This high dielectric strength is the result of the high affinity of its molecules with free electrons, preventing a conduction path from forming and current flowing easily through the gas. The breakdown behaviour of SF_6 under various shapes of applied voltage is essential for practical design applications of this insulation in circuit breakers, GIL, etc.

In an non-electron-attaching gas like nitrogen, the electrons in an avalanche can proceed all the way to the anode from the cathode. But in SF_6, which is an electro-negative gas, this will not be normally possible easily. The process starts as the Townsend-Raether α-β mechanism described in Question 20 earlier, but the avalanche is quenched before it reaches the anode so that we might expect the current to be carried only by the heavy positive ions left behind. The current is therefore reduced, because it depends on the product of the charge and the velocity of the charged particles. Since the positive ions are much heavier than the electrons, the velocity acquired by them in the electric field is also much lower. As can be visualised, a minimum value of field strength is necessary (in a uniform field) before the Townsend current can be initiated. In air, this is about 30 kV/cm, but in SF_6 it is about 90 kV/cm. Also, SF_6 is normally used at high pressures. Therefore, the Streamer Theory comes closer to the mechanism of breakdown of SF_6.

Since the molecule of SF_6 attracts electrons strongly, some of the electrons generated by the avalanche process or ionisation by collision are lost for further generation of electrons. We therefore have to introduce an attachment or annihilation coefficient μ in addition to Townsend's coefficient. We recall that:

α = number of electrons produced in the path taken by the single electron in 1 cm along the direction of the applied electric field,

β = number of secondary electrons produced at the cathode per electron in the primary process.

Other electrons are also produced through secondary processes such as photo-ionisation in the gas caused by photons from excited atoms. We define:

μ = electron attachment coefficient which is the number of attachments per electron per cm (unit length) of path in the direction of field.

The Townsend-Raether criterion for breakdown was found to be, for a non-electron-attaching gas, in Questions 20 and 23, $\beta(e^{\alpha d} - 1) = 1$. But with attachment introduced the net number of electrons generated is lower. In 1 cm of ionising collision, it is $(\alpha - \mu)$ so that the breakdown criterion becomes instead $\beta[\alpha/(\alpha - \mu)] \cdot [\exp\{(\alpha - \mu)d\} - 1] = 1$. When $\mu = 0$, it reduces to the previous equation.

It is clear that if the attachment coefficient μ equals the Townsend first ionisation coefficient α, all electrons generated by collision of an electron are lost and no avalanche build-up takes place. For this to occur, it is experimentally established that the electric field strength must be about 90 kV/cm at 1 atmospheric pressure.

Q.29. What are some of the more important factors affecting the breakdown voltage of a gap insulated with SF_6?

Ans. The following are important for design of high voltage equipment using SF_6 as the insulating medium:

1. Electrode geometry and distribution of electric field stress;
2. Types of voltage applied. It is found that the electric strength of SF_6 is lowest for lightning impulse voltages;
3. Mixture of other gases with SF_6. The dielectric strength of air and nitrogen can be increased by the addition of SF_6, but not to the same extent as pure SF_6. In such cases, cost is less than using only pure SF_6;
4. Dust and conducting particles in SF_6 reduce the dielectric strength as in every other gas;
5. Electrode material and conditioning. Conditioning refers to manufacture and cleaning processes prior to assembly of the high voltage equipment. Gases occluded in the electrode surfaces are primarily responsible for this which are released in the gas under the very intense applied electric field. Where the electrodes experience arcing, such as in circuit breakers, a further cause is electrode pitting which releases metal

or other conducting particles into the gas. Electrode surfaces become rough with the result that there will be local field intensification and the dielectric strength of the gap will not be the designed value but lower;

6. Coating of electrode surfaces with a dielectric material has been found to increase the dielectric strength of the gap;
7. Any insulating spacers used to provide mechanical strength to the equipment will affect the insulation strength, because of surface flashovers on these. This is a problem in gas-insulated lines (GIL) when long lengths of conductors have to be supported at intervals by spacers from the grounded outer cylinder or enclosure;
8. Moisture has a major role to play in any high voltage electrical equipment. Condensation of moisture on insulating surfaces seriously undermines the dielectric strength of the gas;
9. Gas pressure is a major factor. Some idea of this can be seen below where the dielectric strength in kV/mm is shown at different pressures in atmospheres:

Pressure	1	2	3	4	5	6	Atmospheres
SF_6	2	2.9	3.68	4.5	5	5.4	kV/mm (nonlinear)
Air	1	1.3	1.6	1.9	2.2	2.5	kV/mm (linear)
Oil	3	3	3	3	3	3	kV/mm (constant)

Normally, SF_6-insulated equipment uses pressures above 1 atmosphere in order to prevent leak of air into the equipment.

BREAKDOWN MECHANISM OF VACUUM INSULATION

Q.30. Where is vacuum used as insulation in modern high voltage equipment?

Ans. Vacuum is widely used as an insulation in: (*a*) Electronic valves of high power; (*b*) Very high frequency components such as capacitors; (*c*) Vacuum contactors for motors and other equipment; (*d*) Circuit interrupters for high powers; (*e*) Electrostatic generators in MV range and also lower voltages if cost permits. By far the most important of these is in vacuum circuit breakers.

Q.31. Since vacuum by definition has no atoms of any element, how does breakdown of a vacuum gap take place?

Ans. Breakdown of a gap insulated by vacuum can only occur if there are atoms and molecules of any elements present. Therefore, we have to seek for reasons why such gas and other particles have a chance to contaminate the vacuum gap. There are various reasons for this, each of which contributes to the breakdown and which have been thoroughly investigated very extensively all over the world.

(*a*) Firstly, there can exist no such thing as an ideal vacuum. The best vacuum obtainable is in the order of 10^{-9}–10^{-10} torr or mm Hg.

(*b*) Since vacuum can support very high voltages, the electric field stresses are also high on the electrodes. The usual value used is in the neighbourhood of 100 kV/mm. This is one of the main reasons for the presence of particulate material in the gap which initiates and causes the breakdown. These high stresses give rise to the following effects: (*i*) Field emission; (*ii*) Thermionic emission; (*iii*) Field-assisted thermionic emission; (*iv*) Micro-discharges; (*v*) Transfer of micro particles.

(*c*) Electrode conditioning and purity contribute other particles. Seals must be very carefully manufactured which also are major factors.

(*d*) Spacers of solid insulating materials inside the gap are also factors causing surface breakdown or giving rise to discharges at high electric fields.

Q.32. What are the effects of the above factors in contributing to the breakdown mechanism of a vacuum spark gap?

Ans. (*a*) *Field Emission* of electrons occurs at electric stresses in excess of 20 kV/mm from electrode material. No electrode surface can be made free from tiny or micro projections. These become hot spots. Oxide films on electrode surfaces also contain trapped particles or adsorbed particles which might have a low work function. (The work function is the energy in eV required to release an electron from the atom of the material. We found that for hydrogen, this is 13.645 eV. We also gave certain values of ionisation potential of alkali metals and gases in Question 6).

(*b*) *Micro Discharges* are bursts of charged particles released from adsorbed particles on electrode surfaces. These discharges do not give rise to complete breakdown of the gap by themselves but the photoemission of electrons from these ions under the high electric field causes the breakdown.

(*c*) *Micro Particles* are also torn out of the electrode surfaces very violently and contribute to the breakdown. Since the anode is the site of violent bombardment of electrons and negative ions, the anode surface also contributes particles which are ejected into the gap. This is known as anode-initiated breakdown in conjunction with cathode-initiated breakdown which is present in every type of gap.

(*d*) Steel Electrodes are found to be superior to all other metals in controlling the dielectric strength of a vacuum spark gap. Some values of breakdown-voltage stresses in the gap are as follows:

Steel electrodes	:	120 kV/mm;
Nickel electrodes	:	100 kV/mm;
Aluminium electrodes	:	40 kV/mm;
Copper electrodes	:	35 kV/mm.

(*e*) Insulated coatings on electrodes improve the breakdown voltage of a vacuum-insulated gap by reducing the cathode-emitted current by a factor of 10^{-4} times and the breakdown voltage is nearly 175% of that with electrodes without such coatings. This is due to supression of field emission at the high electric stress on the cathode. These insulating materials are usually high-temperature synthetics such as teflon, epoxy and glass.

(*f*) Pressure and temperature also have a role to play in affecting the breakdown strength of a vacuum gap. Electrodes cooled to very low temperatures with liquid nitrogen are frequently used when the cost of equipment permits the investment in the refrigerating equipment.

Q.33. How is the phenomenon of arc interruption in circuit breakers related to insulation breakdown phenomenon?

Ans. When a circuit has to be opened and the current interrupted, the electrodes which were originally in contact with each other must be mechanically separated. While this is being carried out, an arc is struck which is an intense stream of charged particles at very high temperature. These temperatures are in the range of 5000–10000°C. [Under the relation $E = kT$, the temperature is a form of energy, where k = Boltzmann's constant equal to 1.38066×10^{-23} joule/K. Thus, the temperature of any body is an indication of the energy of the particles composing the body]. The arc must be cooled down by some means to reduce the temperature, that is, the charged particles must be made to have low energy before finally they cease to remain self-sustained in the breaker arcing chamber. When finally the arc is quenched, which occurs in vacuum only by lengthening it and exposing a very large area, the full voltage appears across the contacts or the separated electrodes. The medium in which the arcing has taken place must therefore withstand this full voltage, which in extra high voltage work can rise to a million volts. The insulation must withstand this voltage without breaking down and causing an arc to strike again, thereby re-establishing the circuit current. This voltage is also called the "restriking voltage". For successful arc interruption, the rate of increase of dielectric strength of the medium (whether it be air, oil, SF_6, or vacuum, etc.) must be higher than the rate of increase of restriking voltage. Thus,the breakdown strength of insulation is directly responsible for successful interruption of the arc in a circuit breaker.

Circuit-breaker operation is discussed in Chapter 12 under High Voltage Equipment.

❑❑

CHAPTER 4

Electrical Conductors and Overhead Lines: Electrical and Mechanical Properties

Q.1. What are the materials used for overhead line conductors?

Ans. Copper and aluminium.

Q.2. What are the usual cross-sections of copper conductors?

Ans. Obviously, round conductors are the simplest. They are either solid for small sizes or made of a number of small conductors twisted or stranded to make up a large size, see Fig. 4.1. Solid conductors are usually limited to a diameter of 12.7 mm (0.5 inch). Beyond this, stranded conductors have to be invariably used in order to be flexible. Copper conductors for overhead-line use are known as hard-drawn copper. A copper rod is drawn in cold state through a steel die at a tremendous force which converts the larger rod into a required smaller size. The copper is not heated since it becomes annealed when cooled down and loses its strength.

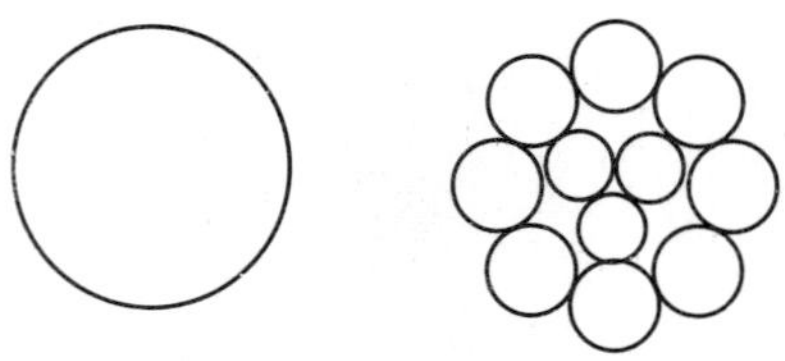

Fig. 4.1: Usual cross-sections of copper conductors.

Many small copper wires are twisted in a machine to manufacture a stranded cable. The distance between two similar positions of any one strand after twisting is called the "lay". Because of this, a strand has about 5% more length than the length of an unstranded cable.

Round solid copper conductors are designated by numbers. An example is provided in the following Table as per Indian Standard Specifications (I.S.S.):

Size: #	7/0	6/0	5/0	4/0	3/0	2/0	1/0	1	
Dia. mm:	12.7	11.79	10.97	10.16	9.45	8.84	8.23	7.62	
inch:	0.5	0.464	0.432	0.4	0.372	0.348	0.324	0.3	
Size: #	2	3	4	5	6	7	8	9	10
mm:	7.01	6.4	5.9	5.39	4.9	4.47	4.06	3.66	3.25
inch:	0.276	0.252	0.232	0.212	0.192	0.176	0.16	0.144	0.128

Taking size #1 as the reference, we note that as the number increases from 1/0 (0) to 7/0 (0,000,000), the diameter and area increase. On the other hand, on the opposite side, as the size number designation increases from 1 to 10, the diameter and area decrease. Some other numbers and diameters can be seen:

7/0–0.5"; 4/0–0.4"; 1–0.3".

Stranded conductors usually have 1, 3, 7, 19, 37 strands for convenience of manufacture. With these numbers, the strands can be embedded properly, an upper strand occupying the depression between strands in the lower layer. Use of copper conductors is restricted to distribution lines and for transmission lines up to 220 kV. By far the more popular conductors are made from aluminium, which no doubt has lower electrical conductivity and lower mechanical strength than copper in tension. However, it is much lighter than copper and requires smaller tower structures. Specific gravities are: copper-8.92, aluminium-2.7.

Q.3. What are the types of aluminium conductors used for high voltage transmission lines?

Ans. Aluminium conductors are invariably stranded ones to make up a large cross-section. Since it is mechanically weaker than copper, several innovations have come into the industry in order to strengthen a stranded aluminium conductor. These schemes have given rise to the following types:

(*a*) ACSR: Aluminium Conductor Steel Reinforced;

(*b*) ACAR: Aluminium Conductor Alloy Reinforced;

(*c*) AAAC: All Aluminium Alloy Conductor.

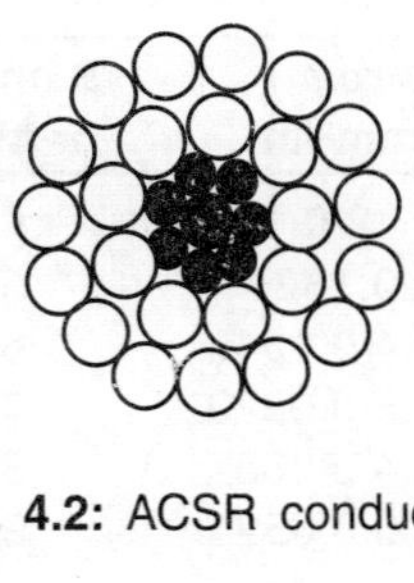

Fig. 4.2: ACSR conductor.

A brief description is given of these three.

(*a*) **ACSR:** A typical cross-section is shown in Fig. 4.2. It consists of 7 inner strands of steel wire, called the steel rope, and 24 strands of aluminium conductors in 2 layers.

The inner steel rope strengthens the overhead cable but because of its excellent magnetic properties, does not carry any of the line current. All the current flows only in the outer aluminium strands because it is non-magnetic. ACSR conductors are designated by names such as birds and animals. A typical list taken from I.S.S. is given in the Table below, which consists of the standard sizes of manufacture by the cable industry.

Name	Outer dia, mm/inch	Aluminium area, sq.mm.	Name	Outer dia, mm/inch	Alum. area,mm^2
Cayote	15.86/0.6244	128.5	Goat	25.97/1.0224	316.5
Tiger	16.52/0.65	128.1	Sheep	27.93/1.0996	366.1
Wolf	18.13/0.7138	154.3	Deer	29.89/1.1768	419.3
Lynx	19.52/0.769	179	Elk	31.5 /1.24	465.7
Panther	21/0.8268	207	Moose	31.77/1.2568	515.7
Lion	22.26/0.8764	232.5			

(*b*) **ACAR:** In the ACSR conductor, the steel strands are provided solely for mechanical strength and do not carry any of the line current. Thus, some metal is lost as far as current-carrying capacity is concerned. New high strength aluminium is now used to replace the steel as well as carry its portion of the line current. This is more effective than ACSR. ACAR = Al. conductor Al. reinforced.

(*c*) **AAAC:** This conductor has all strands made from an aluminium alloy, which consists of Al, Mg, and Si. Its resistance is lower than either ACSR or ACAR by 10–15% and is an improvement over the other two. AAAC = All Al. Alloy Cond.

Q.4. What are some properties of copper and aluminium conductors used in North American practice?

Ans. The standard used in North America is called AWG, or, *B* and *S*. These mean respectively American Wire Gauge, and Brown and Sharpe Gauge. (*B* and *S* should not be confused with BS which stands for British Standard, on which many Indian Standards are also based).

(*a*) Some representative AWG solid copper sizes are as follows:

(Note: Aluminium wires also have the same AWG numbers, sizes and cross-sectional areas).

B and S or AWG	Diameter mm/inch	Area mm^2/in^2	B and S or AWG	Diameter mm/inch	Area mm^2/in^2
4/0 (0000)	11.68/0.46	107.2/0.1662	4	5.19/0.204	21.15/0.0328
3/0 (000)	10.4/0.41	85/0.132	5	4.62/0.182	16.8 /0.026
2/0 (00)	9.27/0.365	67.4/0.1045	6	4.11/0.162	13.3/0.0206
1/0 (0)	8.25/0.325	53.5/0.0829	7	3.665/0.144	10.55/0.0163
1	7.35/0.29	42.4/0.0657	8	3.264/0.1285	8.37/0.013
2	6.54/0.258	33.6/0.0521	9	2.91/0.114	6.63/0.0103
3	5.83/0.23	26.7/0.0413	10	2.59/0.102	5.26/0.0081

(*b*) **ACSR:** These are by far the most commonly used for overhead transmission lines. Some representative conductors and their properties are as follows:

Code Word or Name	Al. Strand No. x inch	Steel Details No. x in.	Outside Dia. of Complete Cond. in.	Dia. of Steel Core only inch	Rated Strength Pounds
1. Turkey	6 × 0.066	1 × 0.066	0.198	0.066	1,120 (½ T)
2. Robin	6 × 0.118	1 × 0.118	0.355	0.118	3,300
3. Penguin	6 × 0.118	1 × 0.118	0.563	0.188	7,500
4. Chickadee	18 × 0.149	1 × 0.149	0.743	0.149	9,500
5. Lark	30 × 0.115	7 × 0.115	0.806	0.345	19,000
6. Hen	30 × 0.1261	7 × 0.1261	0.883	0.3783	22,000
7. Flamingo	24 × 0.1667	7 × 0.1111	1.000	0.333	22,000
8. Drake	26 × 0.175	7 × 0.136	1.108	0.408	29,000
9. Tern	45 × 0.1329	7 × 0.09	1.063	0.266	21,000
10. Canary	54 × 0.1291	7 × 0.1291	1.162	0.387	30,000

There are nearly 50 different sizes of ACSR manufactured in North America. These sizes are supplied by any manufacturer and are also listed in Handbooks and Reference Books. The reader should consult the following two standard works, among many other excellent books:

1. Standard Handbook for Electrical Engineers: Editor-in-Chief: D.G. Fink;
2. Westinghouse Transmission and Distribution Reference Book. In general, the reinforcement consists of either 1 or 7 strands of steel on which the aluminium strands are twisted to yield the current-carrying cross-section. Tables of ACSR conductors also provide the resistance values and a current-carrying capacity at 20°C = 68°F. But these will vary with the temperature of the conductors and must be calculated for each case, since the ambient temperature cannot always be 20°C, especially in India where the summer temperatures can reach 50°C or more in the sun.

(*c*) AWG wire sizes follow a simple law between their AWG numbers and sizes (diameter and area). The ratio of any diameter to the diameter of a wire 6 numbers higher = 2.005 » 2. This also means that for 3 numbers higher, the diameter is $\sqrt{2}$ times and the area is doubled. This also doubles the weight of conductor while the resistance is halved.

An increase of 10 gauge numbers decreases the area and weight by a factor of 10, while the resistance is increased to 10 times.

Example: No. 10 wire has a diameter of 2.59 mm and area of 5.26 mm². The diameter of No. 4 wire is 5.19 mm = 2 × 2.59 mm, and area of 21.15 mm² = 4 × 5.26 mm².

No. 4/0 wire has diameter and area of 11.68 mm and 107.2 mm², respectively. No. 7 wire (10 gauge numbers down in size) has diameter of 3.655 mm and area of 10.55 mm² = 10% of 107.2 mm², nearly.

Q.5. What are the specific resistivities, ρ, of copper and aluminium, and α, their temperature-resistance coefficients?

Ans. **ρ: copper:** 1.7241 μΩ-cm = 1.7241×10^{-8} Ω-m

aluminium: 2.8 μΩ-cm = 2.8×10^{-8} Ω-m

α: copper: 0.00393 ohm/°C; **aluminium:** 0.00403 ohm/°C.

These values are usually denoted as being at 20°C.

Q.6. What are the resistances of the following representative conductors per km? (*a*) IS 4/0 copper; (*b*) Drake ACSR conductor (At 20°C).

Ans. Resistance $R = \rho \cdot l/A = l/\sigma A$, $\sigma = 1/\rho$ = specific conductivity in mho/m or siemens/m.

(*a*) Area

$$A = \pi d^2/4$$
$$= \pi \times 10.16^2/4 \text{ mm}^2$$
$$= 81.08 \text{ mm}^2$$
$$= 81.08 \times 10^{-6} \text{ m}^2$$
$$l = 1 \text{ km} = 1000 \text{ m}$$
$$\rho = 1.7241 \times 10^{-8} \text{ ohm-m}$$

So, $R = 1.7241 \times 10^{-8} \times 1000/81.08 \times 10^{-6} = 0.21265$ ohm at 20°C.

(*b*) Drake ACSR conductor has 26 strands of aluminium, each strand having a diameter of 0.175 inch.

Area of each strand is thus $= \pi \times 0.175^2/4 \text{ in}^2$

$$= 0.024 \text{ in}^2$$
$$= 0.024 \times 2.54^2 \text{ cm}^2$$
$$= 0.155 \text{ cm}^2$$
$$= 0.155 \times 10^{-4} \text{ m}^2$$

There are 7 steel strands in the centre for reinforcement only. These do not carry any current, so that the resistance of Drake ACSR is due only to the 26 aluminium strands. Because of stranding or twisting, the average length of each strand is taken to be 5% longer than the overall length of the cable. This is only an average since the outer layers are longer than the inner layers.

Therefore, length for resistance calculation will be $l = 1.05 \times 1000 = 1050$ m.

Thus, the resistance for 1 km of conductor is:

$$R = 2.8 \times 10^{-8} \times 1050/(26 \times 0.155 \times 10^{-4}) = 0.073 \text{ ohm at } 20°\text{C}.$$

We also note that the average resistance of each Al strand is

$$r_S = 2.8 \times 10^{-8} \times 1050/0.155 \times 10^{-4} = 1.897 \text{ ohms}.$$

Since the 26 strands are electrically in parallel, each strand carrying 1/26-th of the total current, the resistance of the overall conductor is also calculated as

$$R = r_S/26 = 1.897/26 = 0.073 \text{ ohm}.$$

Q.7. What are the resistances of No. 4/0 copper conductor and the Drake ACSR conductor at 65°C?

Ans. Taking temperature t_1 as the reference and α_{t1} = resistance-temperature coefficient at t_1, the resistance at any other temperature t_2 is calculated from the relation:

$$R_{t2} = R_{t1} [\ 1 + \alpha_{t1} (t_2 - t_1)],$$

where R_{t2}, R_{t1} = resistances at the two temperatures, and $t_2 - t_1$ = temperature difference, rise or fall as the case may be.

(*a*) For 4/0 copper conductor, taking

$$t_1 = 20°C$$
$$R_{20} = 0.2126 \text{ ohm/km}$$
and
$$\alpha_{20} = 0.00393 \text{ ohm/°C}$$
$$t_2 = 65°C$$

we obtain:

$$\begin{aligned} R_{65} &= R_{20} [\ 1 + 0.00393\ (65 - 20)] \\ &= 0.21265 \times 1.17685 \\ &= 0.25 \text{ ohm/km} \end{aligned}$$

This is an increase of 17.685% when the temperature increases by 45°C.

(*b*) For the Drake ACSR,

$$R_{20} = 0.073 \text{ ohm/km at } 20°C$$
and
$$\begin{aligned} \alpha_{20} &= 0.00403 \text{ ohm/°C at } 20°C \\ R_{65} &= 0.073 [\ 1 + 0.00403\ (65 - 20)] \\ &= 0.073 \times 1.18135 \\ &= 0.08624 \text{ ohm/km} \\ &= \text{increase of } 18.135\% \end{aligned}$$

We can observe that the temperature-resistance coefficients, (also written as resistance-temperature coefficients) of copper and aluminium at 20°C are nearly 0.004 = 0.4% per degree centigrade change in temperature.

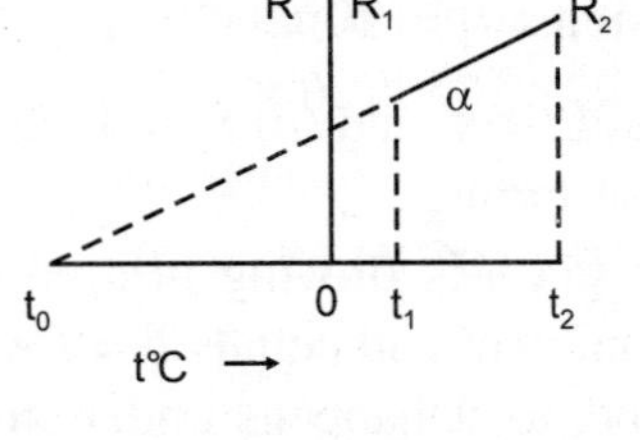

Fig. 4.3: Linear variation of resistance of metal between temperatures t_1 and t_2. α = slope = ohm/°C.

As shown in Fig. 4.3, if the variation of resistance between temperatures t_2 and t_1 °C is linear with the slope 0.004, then the resistance will become zero at

$$-(1/0.004) + 20 = -230°\text{C},$$

from the relation $R_{t2} = R_{t1}\,(t_0 + t_2)/(t_0 + t_1),$

where t_0 = the temperature in °C between the zero of resistance and 0°C.

If $R_{t2} = 0,$

then, $t_0 + t_2 = 0$

giving $t_2 = -\,t_0$

Since $R_{t2} = 0$

$= R_{t1}\,[1 + \alpha(t_2 - t_1)],$

this relation yields $[1 - \alpha(t_0 + t_1)] = 0,$

which gives $\alpha = 1/(t_0 + t_1),$

or, $t_0 = 1/\alpha - t_1$

$= 1/0.004 - 20 = 230°\text{C}$

Therefore, we obtain the ratios of resistances at two different temperatures in an alternative form:

$$R_{t2}/R_{t1} = (t_0 + t_2)/(t_0 + t_1)$$

$$= (230 + t_2)/(230 + t_1).$$

For copper, the actual value of $\alpha = 0.00393$

This gives $t_0 = 1/0.00393 - 20$

$= 234.45°\text{C}$

For aluminium, $\alpha = 0.00403$

giving $t_0 = 228.14°\text{C}$

Therefore, we can use the following equations for more correctness, for resistance value R_{t2} at temperature t_2:

(*a*) For copper: $R_{t2} = R_{t1}\cdot(234.45 + t_2)/(234.45 + t_1)$

(*b*) For aluminium: $R_{t2} = R_{t1}\cdot(228.14 + t_2)/(228.14 + t_1)$

where in both expressions, $t_1 = 20°\text{C}$.

Q.8. What are meant by: (*a*) Skin Effect; and (*b*) Skin Effect Ratio? Use only single solid round conductor to illustrate these terms.

Ans. (*a*) When a direct current (d.c.) is flowing in a conductor of area A, for the current I, the current density is uniform and equals $\delta = I/A$, amperes/unit area. However, if the current varies with time, as it happens under steady-state power-or high frequency, or under lightning impulses and switching surges, etc., the current does not flow uniformly in the entire cross-section of the metal of the conductor. The higher the frequency, the more the current crowds on the outer portion leaving the inner portion free of current. This crowding of current in the filaments of material near the surface of conductor is known as "Skin Effect". This is illustrated in Fig. 4.4.

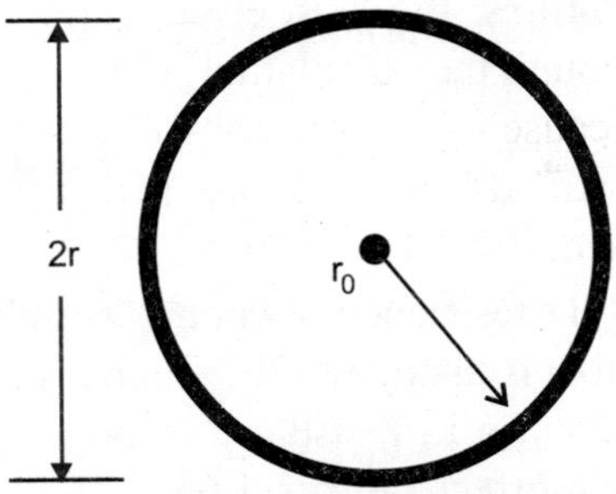

Fig. 4.4: Skin effect in round conductors (Ideal).

The technical reason for this phenomenon to occur is that any tubular conductor carrying current in the axial direction along its length creates a magnetic field which is circular and coaxial with the tube. The inner filaments link more flux than the outer filaments. The increase in flux linkages causes a higher inductance (inductance = flux linkage/ampere), which offers a higher impedance to current flow as the frequency increases. The current therefore has a tendency to flow in the outer filaments of the wire or conductor. In ACSR conductors, the steel reinforcing rope is not only a good magnetic substance but also occupies the inner portion of the overall cable. Therefore, the circuit current flows only in the outer aluminium strands which are not only non-magnetic but also occupy the outer portion of the entire cable.

(*b*) Because the current flows in a thin tube which is shown as having a thickness $(r - r_0)$, where r = outer radius of conductor and r_0 = the radius of a tube in which practically no current is assumed to be flowing, under the assumption that the current density is uniform in this tube of thickness $(r - r_0)$, the effective resistance of the overall conductor assumes the value $R_{ac} = (1/\sigma) \cdot l/\{P(r^2 - r_0^2)\}$.

If the current were to flow uniformly over the entire cross-section of radius r, the d.c. resistance is $R_{dc} = (1/\sigma)l/\pi r^2$, for the same length of conductor l and same outer radius r. Consequently, the ratio of a.c. resistance to the d.c. resistance is $R_{ac}/R_{dc} = r^2/(r^2 - r_0^2) > 1$.

Therefore, the a.c. resistance is higher than the d.c. resistance. This ratio is called the "Skin-Effect Ratio". In general, the skin-effect ratio does not follow this simple relation. The mathematical complexity is considered too difficult to describe here. (It involves what are known as Bessel Functions). Tables of conductor details give this ratio not only as the outer radius of a conductor varies but also as a function of frequency of current.

The parameter that controls the skin-effect ratio is denoted by

$X = 2\pi r \sqrt{[2 f \mu \sigma]}$ for a round conductor,

where r = outer radius of conductor,

f = frequency of applied voltage and current flowing,

μ = magnetic permeability of the conductor material,

and σ = specific conductance of material of conductor,

= reciprocal of specific resistivity of conductor material.

For **stranded conductors**, the skin-effect ratio is found only from experimental results at the factory, although the calculation is of a more highly specialised nature than for a round conductor because of what is known as "Proximity Effect," where the closeness or proximity of the strands creates more difficulties for handling the distribution of the resulting magnetic field.

In general, the 50/60-Hz resistance of copper and aluminium conductors for overhead lines (using solid conductor) is about 10–20% higher than their d.c. resistance values. For ACSR conductors, the increase in resistance at power frequency is 5% higher than the d.c. value, that is, the skin-effect ratio is 1.05.

For conductors in the slots of electrical machines or in transformers, the skin-effect ratio is about 1.4 giving an increase of 40% under power-frequency a.c. over the d.c. resistance.

Q.9. What are these units of measurement: mil, circular mil, and their relation to mm, inch, mm^2 and inch^2?

Ans. Because wire diameters are very small numbers, their diameters are measured and denoted in mils. 1 mil = 1/1000-th of an inch, so that 1 inch = 1000 mils.

The area of a round conductor, when its diameter is given in mils, is: Area in circular mils = square of the diameter in mils. Circular mils = cir-mils = c-m = CM = (mils)^2.

[The reader is cautioned not to confuse c-m with centimetres, cm. Also, some books and handbooks use CM for circular mils. But in Roman numeral system, $M = 1000$, $C = 100$, $L = 50$, $X = 10$, and so on. Under such a system, MCM = 1900. But when this lettering is used for describing conductor areas, MCM stands for 1000 Circular Mils, and not for Million Circular Mils.]

Let us consider a round conductor of 1 inch in diameter = 1000 mils.

Its diameter = 25.4 mm

(and its area is) $\pi d^2/4 = \pi/4$ sq.in.

$= \pi \times 25.4^2/4$ sq.mm

$= 506.7$ mm^2

$= 5.067$ cm^2

But its area in circular mils $= 1000^2$ c-m $= 10^6$ c-m

Therefore, 10^6 c-m $= \pi/4$ sq.in.

(or, 1 sq.in. $= 4 \times 10^6/\pi$ c-m) $= 506.7$ mm^2 (or, 1 mm^2 = 1973.55 c-m).

This also gives 1 cm^2 = 100 mm^2 $= 0.197355 \times 10^6$ c-m.

Example: The diameter of each strand of the Drake ACSR conductor is 0.1749 inch with 26 strands in the overall cable (neglecting the inner steel core). What is the cross-sectional area of the aluminium in:(*a*) circular mils, (*b*) square inches, and (*c*) square centimetres?

Diameter of each Al strand $= 174.9^2$ c-m = 30,590 c-m.

Area of 26 strands = 795,340 c-m. In Tables of ACSR conductors, the aluminium area of the Drake conductor is given as 795,000 c-m = 795 MCM.

$$\begin{aligned}
\text{Area of Al in square inches} &= 795{,}340 \times \pi \times 10^{-6}/4 \\
&= 0.62466 \text{ sq.in.} \\
\text{Also,} \quad 26 \times \pi \times 0.1749^2/4 &= 0.62466 \text{ sq.in.} \\
&= 0.62466 \times 2.54^2 \text{ sq.cm} \\
&= 4.03 \text{ cm}^2 \\
&= 403 \text{ mm}^2 \\
\text{Also,} \quad 795{,}340/1973.55 &= 403 \text{ mm}^2
\end{aligned}$$

Note that the largest size of ACSR manufactured is the Canary which has an aluminium area of 900,000 c-m = 900 MCM = 0.7068 sq.in. in 54 strands each with diameter of 0.1291 inch = 129.1 mils. The steel stranding is 7 × 0.1291". The total area of conductor is 0.7984 in² and the overall diameter of the conductor is 1.162 inch = 1162 mils. The Lark, Drake and Canary are very extensively used in North America. But refer to Westinghouse Transmission and Distribution Reference Book for a list of many transmission lines and the conductors used.

In India, the most popular ACSR conductor is the MOOSE used for the 400-kV transmission lines in the country.

Q.10. What are bundled conductors and their usual configurations?

Ans. For power transmission at 345 kV and above, namely, for Extra High Voltage and Ultra High Voltage transmission, the line current is extremely high. Also, severe highvoltage effects exist at these very high voltages. A single conductor cannot carry the line current under normal circumstances. Therefore, two or more conductors per phase in a.c. and poles of d.c. transmission are found necessary. Some examples of current-carrying capacity (or Ampacity as it is known) for single ACSR conductors are as follows:

Lark : 397,500 c-m, 575 amps, 0.806" overall diameter;
Dove : 556,500 c-m, 710 A, 0.927";
Flamingo : 666,600 c-m, 790 A, 1.00";
Drake : 795,000 c-m, 890 A, 1.108";
Canary : 900,000 c-m, 955 A, 1.162".

These current-carrying capacities are based on the final temperature of the conductor being equal to 80°C when a cross-wind of 2 ft/sec = 1.36 mph = 2.2 kmph is blowing. Above this temperature the strength of the conductor is reduced. But one must consult the relevant National Standards for the final temperature.

The normal loading of e. h.v. lines per circuit is as follows, with line loading decreasing in proportion to its length in inverse proportion, which is determined by stability considerations.

Line Length, km	400 kV	750 kV	1000 kV	1200 kV
400	670	2860	6000	8625 MW/circuit
600	450	1900	4000	5750
800	335	1430	3000	4310
1000	270	1140	2400	3450
1200	225	950	2000	2875

The corresponding phase currents can be calculated from the equation: $P_{3ph} = \sqrt{3}\, V \cdot I \cdot \cos\varphi$. For a unity power factor, $\cos\varphi = 1$, and the current is $I = P_{3ph}/\sqrt{3} \cdot V$, V being the system line-to-line voltage magnitude. For example, consider a 750-kV line of 400 km length from Upper Ganga and Yamuna areas to Delhi. The current is $2860 \times 10^3/\sqrt{3} \times 750 = 2200$ amps/circuit. A MOOSE ACSR conductor can carry about 900 amperes. We observe that each phase will have to be made up of more than 2 conductors to carry 2200 amperes. Other high voltage considerations eventually require 3 conductors to be used per phase. The present 400-kV lines in India use a two-conductor MOOSE bundle per phase. This conductor has an aluminium area of 515.7 mm² having an outer diameter of 31.7 mm, which is the equivalent of the Drake conductor of North American manufacture.

The use of more than one single conductor per phase or pole is called a bundle conductor. Figure 4.5 shows several arrangements of bundled conductors comprising 2 to 6 sub-conductors arranged on a circle with equal spacing between adjacent sub-conductors. There are some exceptional cases where the conductor spacings may be unequal, but are rare.

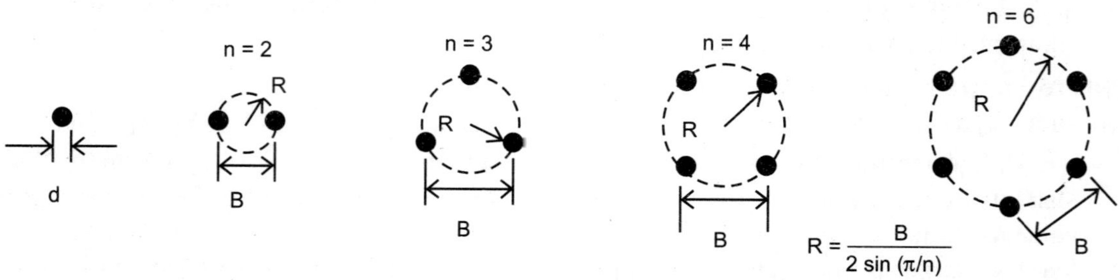

Fig. 4.5: Single- and bundled conductors. Cond. radius, $r = d/2$; bundle spacing = B; bundle rad. = R; No. of sub-conductors = $n \cdot R = B/[2 \sin(\pi/n)]$.

For a d.c. transmission line carrying 3000 amps per pole, it is evident that a single conductor of even the largest size manufactured is inadequate. 4 sub-conductors of at least the size of Drake or Moose may be necessary.

The sub-conductors of the bundle are all at the same potential and are electrically parallel throughout their long length Due to severe mechanical considerations such as oscillations and vibration during wind under normal working conditions, the sub-conductors must be mechanically separated. Also, since the charges are of the same polarity, there is a force of repulsion among them, whereas since the currents are in the same direction there is electro-magnetic attraction between the sub-conductors. The separation of sub-conductors is achieved by giving suitable metallic spacers made of phosphor-bronze at intervals of about 100 metres. The largest number of sub-conductors used for a bundle in the world has gone up to 12, and the reader can imagine what engineering difficulties arise in designing the spacers that keep the sub-conductors apart.

MECHANICAL CONSIDERATIONS

Q.11. What are the types of oscillation and vibration that take place during wind for transmission-line conductors?

Ans. There are three main types of wind-induced oscillations and vibrations: (1) Aeolian Vibration; (2) Galloping; and (3) Wake-Induced Oscillation.

Q.12. What are the sources and characteristics of aeolian vibrations?

Ans. The word "Aeolian" is the Greek term for the God of Wind, similar to the Sanskrit word Vayu Dev or Marut. Of course, there is nothing divine or imaginary about the vibration caused by wind on an actual transmission-line conductor, as every designer knows.

When a long slender conductor is under tension and a steady wind blows across it, because the wind is obstructed by the conductor even if it be a single conductor of only a few centimetres in diameter, the flow is disturbed sufficiently for vortices to be formed on the leeward side of the conductor, away from the direction of the wind. These vortices are released from the conductor surface periodically at finite intervals, and when they do they give rise to vibration of the conductor. Such a phenomenon is very common on an aircraft wing when wind at high speed for keeping the plane aloft is absolutely vital. These are known as Karman Vortices, named after the famous aeronautical engineer and authority on vibrations, Theodor van Karman.

The resulting vibrational forces cause fatigue of conductor strands and supporting structures, such as towers and spacers as well as the insulators. This vibration is collectively known as Aeolian Vibration. In order to prevent excessive vibration, weights known as dampers are given at towers suspended from the conductor. There are several such devices on the market known as armour rods, stockbridge dampers, or just lead weights attached to the conductor. Some spacers to keep sub-conductors apart in a bundle also help to damp aeolian vibration in which case they are called spacer-dampers.

Normally, aeolian vibrations occur below a wind speed of 25 km/hr and principally in terrains which are flat. Thus, trees and hills do not present as severe a situation as plains. The amplitude of vibration is normally less than 2.5 cm and the frequency of vibration is 20 Hz.

Q.13. What is galloping?

Ans. A very long slender conductor, as can be imagined, acts like a powerful whip when shaken by wind. When the conductor is coated with ice in hilly areas of the north India during winter, or in cold countries, the ice becomes heavy enough to fall off the conductor. This sudden release of weight causes the conductor to act as a whip and it oscillates violently. The conductor during this motion exhibits several loops and appears to "gallop".

Also, when wind blows across an iced conductor, it presents an uneven geometry which gives rise to a drag force and lift force, in exactly the same manner as an aircraft wing. This is usually a torsional force and there is galloping action of the conductor.

Galloping is usually a problem at wind speeds between 15 and 50 km/hr. The oscillation is very large and two adjacent phases or poles can touch each other at even 10 metres apart so as to cause a severe short-circuit on the transmission system.

Q.14. What is a wake-induced oscillation?

Ans. This type of oscillation is limited to bundled conductors only and does not occur with single conductor as is the case with the previous two types of vibrations, the Aeolian and galloping. The term WAKE denotes the effect of the leeward side of an obstruction. Consider a bundle with two sub-conductors (which is sometimes termed two-bundle or twin-bundle, etc.). When wind is blowing across the bundle, one conductor on the windward side creates a wake on the leeward side in which the second conductor may be located if it is too close. Thus, the first conductor shields the second aerodynamically and an oscillation is induced in both conductors. If the leeward conductor is spaced beyond the wake generated by the windward conductor, then no oscillation takes place since this conductor also experiences the same aerodynamic conditions. It is for this reason that the spacing between sub-conductors should not be less than 15*d*, where *d* is the diameter of each sub-conductor.

Bundle-spacing or spacing between adjacent sub-conductors in the bundle is standardised throughout the world at 40 cm, 45 cm, and 50 cm, (16", 18" and 20"). The most usual spacing is 45 cm. This allows spacers to be designed and manufactured easily in bulk if their length is standardised. The spacers prevent sub-conductors from clashing during wake-induced oscillations, although this does not represent a short-circuit. Wake-induced oscillations occur at wind speeds in the range 25–65 km/hr and the frequency of oscillation is below 3 Hz. Spacers can fail due to fatigue at the points of attachment to the conductor, as also due to fatigue to upper strands in the conductor caused by constant friction or rubbing. Usually spacers are provided with hard rubber or plastic liners on the inside in which the conductor is able to move.

❑❑

CHAPTER 5

ELECTRICAL CABLES

Q.1. Where are cables used in high voltage work?

Ans. Cables were the first type of electrical conductors used with a central current-carrying part wrapped with insulation and separated from ground so as not to come into contact with human beings. As can be imagined, distribution of electrical power from generating stations to load centres was first started in large cities with heavy population and traffic such as London, Paris, New York, Mumbai, Kolkata, etc. Thus, underground cables had to be used which were properly insulated from ground. Cable transmission of power started nearly 130 years ago in 1870s and have improved continually ever since, the voltage increasing to 1200 kV between two conductors (or, $1200/\sqrt{3} = 693$ kV between conductor and outer grounded sheath).

Cables are used wherever and whenever power has to be transmitted without impediment to traffic, etc. This is particularly true in large cities, large industrial establishments such as iron and steel mills, petroleum refineries, to name only a few. Interconnection of equipment in electrical generating and receiving stations requires cables up to the highest voltages in use.

Q.2. What are the shapes of conductor used in cables?

Ans. Round solid conductors are never used in cables because cables must be flexible, and round solid conductors are rigid and easily break. Thus, even the smallest size of cable (which is also termed as insulated conductor) for low-voltage house-wiring is made from stranded copper or aluminium wires. Since cables are laid on proper supports always, they do not need reinforcement at the centre like overhead conductors. But because of the heat generated and the compactness of the geometry, cooling ducts may be provided in the central portion inside the current-carrying conductor itself to carry away a major portion of the heat generated by the conductor. Thus, flexibility, large conductor cross-section and cooling determine the shape of the cross-section.

Some representative shapes are shown in Fig. 5.1. They can be classified as follows: (*a*) Single core; (*b*) Three core; (*c*) Segmented core; (*d*) Inner-cooled core; and (*e*) Gas Insulated Line.

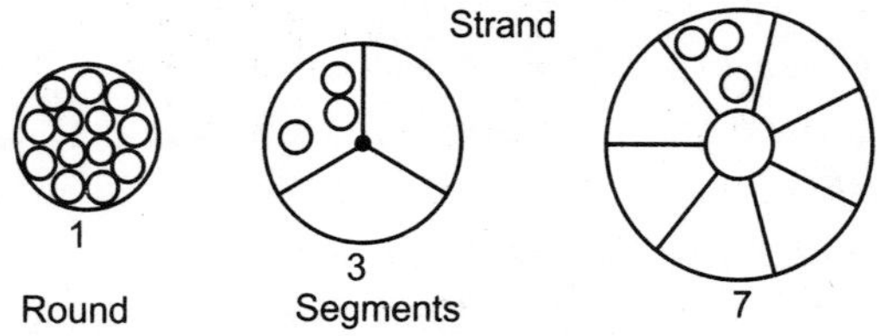

Fig. 5.1: Cross-sections of representative cables.

Q.3. What are representative cross-sections of cables manufactured in the world?

Ans. Figure 5.2 shows complete cross-sections of 5 types of cables with central cooling duct, copper conductors, insulation, sheath and armour.

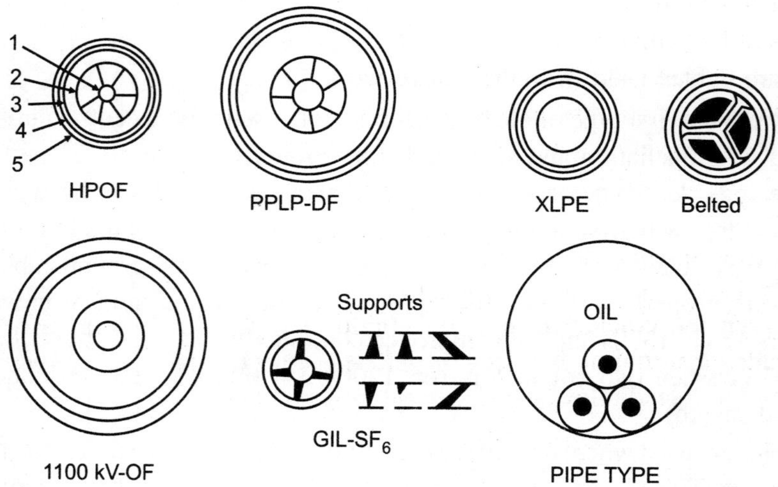

Fig. 5.2: Components and construction of representative cables.

Q.4. What are the components of representative cables?

Ans. A high voltage cable can consist of some or all of the following components whose functions are explained. Refer to Fig. 5.2, please.

1. **Central Oil Duct:** When the current becomes very high and the copper temperature rises above 90°C, which is the safe value for the insulation, a central oil duct inside the hollow conductor is provided. These are normally 12.7 mm (½ inch) or 25.4 mm (1") in diameter. The oil is kept circulating either by pumps or by gravity, and carries away a large portion if not all of the heat. These ducts are usually made from spiral steel conduits over which the conductor is wrapped in a stranding machine.
2. **Conductor:** These are always made of stranded copper wire. They can be round or segmented with 3 or 7 segments which bed together to yield a rounded outer

surface on which the insulation is tightly wound. Each segment is in turn made of stranded wire, properly shaped to size.

3. **Carbon Black Tape or Metallised Paper or Semi-Conducting Layer:** The rough outer surface presented by a stranded conductor is not as beneficial as a smooth round surface from the point of view of electric stress control. The carbon-black tape or metallised paper tape or the semi-conducting layer given over the conductor surface is wrapped prior to or at the same time as the major insulation, or extruded along with the insulation in plastic insulated cables. These provide the smooth outer surface on the stranded conductor.
4. **Major Insulation:** The types of major insulation used are described in the next Question 5. These are always in tape form if made of cotton tape, Kraft paper tape and some plastics (PPLP) if different thicknesses of tape have to be used. For example, in Cable No. 2 of Fig. 5.2, the PPLP insulation (Polypropylene Laminated Paper) is used in thicknesses of 120 μm to 200 μm in different layers. XLPE (Cross-Linked Polyethylene) is usually extruded under heat and pressure in a triple-extrusion process along with the semi-conducting layer on the central conductor and the external shield over the major insulation. This will eliminate voids in the insulation, and the resulting very smooth surface on the conductor will be responsible for eliminating what are known as "water trees" which are the main sources of damage to the insulation.
5. **External Shield:** Over the major insulation for equalizing of potential along the length of the cable, this conducting shield is necessary. The applied voltage on the cable is between the conductor and this shield. This could be semi-conducting tape, or extruded material such as lead, metallised paper in tape form, copper tape, corrugated aluminium, lead alloy, stainless steel tape, etc.
6. **Sheath:** This is used as protection against wear during cable laying, for weather-proofing, strength, corrosion proofing, etc. The usual materials are: High-Density or Heavy Duty (HD) PVC jackets extruded over the external shield, Neoprene-fibre glass tape protected by Polyethelene jacket, lead, etc.
7. **Outer Cooling Jacket or Pipe:** Certain extra high voltage or ultra high voltage (EHV and UHV) cables require external cooling as well as internal cooling with oil. The external cooling may be, however, with water or oil or both. North American cables include a special scheme for three-phase cables where three large single-core cables are laid inside an oil pipe and called Pipe Type Cable.

Q.5. What are the different types of cables used for high voltage work?

Ans. High voltage cables are classified according to the type of major insulation used in them. There are four major types which are as follows:

1. High-Pressure Oil-Filled (HPOF) Paper Insulated Cables;
2. Cables with Extruded Insulation such as Cross-Linked Polyethylene (XLPE), Ethylene Propylene Rubber (EPR);

3. Composite Laminated Insulation having plastic films laminated with Kraft cellulose paper in 40–60% ratio. They have names such as Polypropylene Laminated Paper (PPLP);
4. High-Pressuer SF_6 Gas-Insulated Cable or Line (GIL).

Q.6. What are some typical insulation thicknesses for cables? (Voltages are line-to-line values. Conductor-Sheath voltage = Line-to-line voltage/ $\sqrt{3}$).

Ans.

(a)	HPOF: (ε_r = 3.5)	138 kV 13 mm	275 kV 20 mm	400 – 500 kV 25 mm	1100 kV 35 mm
(b)	XLPE: (ε_r = 2.3)	69 kV 16 mm	275 kV 27 mm	500 kV 32 mm	
(c)	PPLP: (ε_r = 2.75)			500 kV 25 mm	765 kV 30 mm
(d)	SF_6: (Gas) (ε_r = 1)		275 kV 125 –160 mm	500 kV 200 mm	750 kV 275 mm

Q.7. Describe typical cable construction for 500 kV level using different types of insulation?

Ans. We will consider a HPOF cable, OF cable with composite insulation and an XLPE insulated cable.

(*a*) **HPOF Cable:** See Fig. 5.2.

Part No.

1. Oil duct: 12 mm inside the conductor;
2. Conductor: 7-segment Milliken type stranded Copper 5000 MCM cross-section (2500 mm²);
3. Major Insulation: Oil-impregnated Kraft cellulose paper 33.8 mm in thickness;
4. Sheath: Corrugated aluminium;
5. Jacket: High Density Polyethylene (HDPE).

(*b*) **OF Cable with Composite Insulation:** See Fig. 5.2.

Part No.

1. Oil Channel: Inside conductor 25 mm diameter and 1.5 mm thickness. Total diameter 28 mm;
2. Conductor: 7-segment Milliken type. Outer diameter 68 mm;
3. Conductor Shield: Carbon paper 2 sheets and metallised paper 1 sheet. Total thickness 0.375 mm.
4. Major Insulation: PPLP tape of several thicknesses as follows:

120 μm thick	:	5.5 mm
170 μm	:	8 mm
220 μm	:	7 mm
200 μm	:	4 mm

Total thickness of major insulations

$$= 5.5 + 8 + 7 + 4 = 24.5 \text{ mm}$$

5. Insulation Shield: Carbon and metallised papers: 3.75 mm.
6. Copper wire woven with cloth tape: 0.5 mm thickness.
7. Aluminium Sheath: 3 mm thickness.
8. Anti-Corrosion PVC Jacket: 6 mm thickness.

Overall diameter of single-core cable: 153 mm = 6 inches.

(*c*) **XLPE Cable:** See Fig. 5.2.

Part No.

1. Conductor 2000 mm² stranded Copper with extruded semi-conducting shield;
2. Major Insulation: XLPE 35 mm thick. (Some cables impregnate XLPE with SF_6 gas. Otherwise, XLPE need not be impregnated, but extruded all by itself and remain in the cable unimpregnated);
3. External Shield;
4. PVC Jacket.

The semi-conducting shield over conductor, together with the major XLPE insulation and the external shield, are extruded at the same time from a suitably-shaped nipple die by a triple-extrusion process and not separately extruded as in cables (*a*) and (*b*) which have the insulation in tape form wrapped in layers.

Q.8. How is the electric stress in the insulation of a concentric cable calculated? What is the capacitance of this cable per unit length?

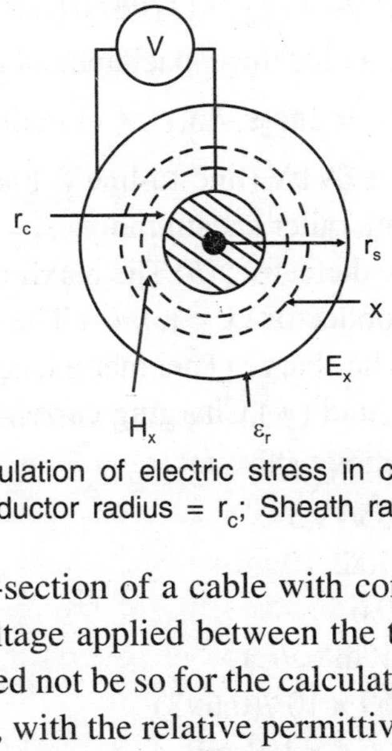

Fig. 5.3: Calculation of electric stress in cable dielectric. Conductor radius = r_c; Sheath radius = r_s.

Ans. Figure 5.3 shows the cross-section of a cable with conductor radius r_c and outer metallic sheath of radius r_s. The voltage applied between the two is *V*, with the sheath usually at ground potential but this need not be so for the calculations. The dielectric or the insulation has the thickness $t_I = r_s - r_c$, with the relative permittivity ε_r. If the cable has a capacitance of *C* farad/metre, the charge is $Q = CV$ coulomb/m.

Let E_x = dielectric stress in volt/metre, (V/m), at radius *x* from the centre of the conductor.

Gauss's law in electrostatics says that the surface integral of the dielectric flux density $D = \varepsilon E$, over a closed surface equals the total charge enclosed inside the volume bounded by the surface. We consider a circle of radius x on which E_x is the same everywhere due to symmetry. The dielectric flux density is $D = \varepsilon E_x = \varepsilon_0 \varepsilon_r E_x$, where ε_0 = permittivity of free space = 8.842×10^{-12} F/m. The surface area is $(2\pi x)$ for 1 metre length of cable, and the charge is Q.

$\therefore$ $(2\pi x) \cdot (\varepsilon_0 \varepsilon_r E_x) = Q$, or, $E_x = (Q/2\pi\varepsilon_0 \varepsilon_r)/x$ V/m.

Now, considering a differential thickness dx of the insulation at x, the p.d. across it is $dV_x = E_x \cdot dx = (Q/2\pi\varepsilon_0 \varepsilon_r) \cdot (1/x) \cdot dx$. Since the potential difference between the conductor surface and outer sheath is V, we must have

$$V = \int_{r_c}^{r_s} dV_x = (Q/2\pi\varepsilon_0\varepsilon_r) \int_{r_c}^{r_s} \frac{1}{x} \cdot dx = (Q)/2\pi\varepsilon_0\varepsilon_r) \cdot \ln x \Big|_{r_c}^{r_s}$$

$$= (Q/2\pi\varepsilon_0\varepsilon_r) \cdot \ln(r_s/r_c)$$

When the applied voltage is specified and $(r_s, r_c, \varepsilon_0, \varepsilon_r)$ are known for a cable, the charge Q per unit length of the cable will be:

$$Q = 2\pi\varepsilon_0\varepsilon_r \cdot V/\ln(r_s/r_c)\ ,$$

or,
$$Q/(2\pi\varepsilon_0\varepsilon_r) = V/\ln(r_s/r_c)\ .$$

Using this value of $(Q/2\pi\varepsilon_0\varepsilon_r)$, we obtain the dielectric stress at any radius x to be:

$$E_x = (Q/2\pi\varepsilon_0\varepsilon_r)/x = \{V/\ln(r_s/r_c)\}/x \text{ volt/metre.}$$

We also obtain the equation for the capacitance of the cable per metre length to be:

$$C = Q/V = 2\pi\varepsilon_0\varepsilon_r/\ln(r_s/r_c) \text{ farad/metre.}$$

As an example of a 3-phase 69 kV (line-to-line voltage) XLPE single-core cable, having conductor radius r_c = 19.3 mm, outer sheath radius r_s = 35 mm, calculate: (*a*) The electric stress at any radius x inside the dielectric, (*b*) The maximum value of electric stress occuring on the surface of the inner conductor ($x = r_c$), (*c*) The electric stress at the sheath on the outside of the insulation, (*d*) The charge Q per metre length of the cable, (*e*) The capacitance of the cable per metre length, and (*f*) Charging current per km of cable at 50 Hz.

The voltage to use is $V = 69/\sqrt{3}$ kV,
rms value.

$$r_s/r_c = 35/19.3 = 1.82$$

So
$$\ln 1.82 = 0.6$$

(*a*)
$$Q/(2\pi\varepsilon_0\varepsilon_r) = V/\ln(r_s/r_c) = 69 \times 10^3/(0.6\sqrt{3}) = 66.4 \times 10^3 \text{ volt}$$

$\therefore$
$$E_x = 66.4 \times 10^3/x \text{ volt/metre}$$

(*b*) The maximum value of electric stress in dielectric occurs when x is least i.e., on the conductor surface for $x = r_c = 19.3$ mm.

$$\therefore \quad E_{max} = 66.4 \times 10^3/19.3 \times 10^{-3}$$
$$= 3.45 \times 10^6 \text{ V/m}$$
$$= 3.45 \text{ kV/mm}$$

(*c*) At $x = r_c = 35$ mm, on the outside of the insulation, the value of x is maximum so that the electric stress in the insulation is minimum.

$$\therefore \quad E_{min} = 66.4 \times 10^3/35 \times 10^{-3}$$
$$= 1.89 \times 10^6 \text{ V/m} = 1.89 \text{ kV/mm}$$

Note also that the ratio of $E_{min}/E_{max} = r_c/r_s = 1/1.82 = 0.5495$

We can also observe that there is a large variation in the electric stress in the insulation, being highest on the conductor surface and least at the inner surface of the sheath (55%). In certain cable designs, insulating materials with higher dielectric strength are used on the conductor, as in the 500-kV PPLP cable of Q. 7(*b*) where the thinner PPLP tape is used on the inside.

(*d*) The charge Q per metre length of the cable is:

$$Q = 2\pi\varepsilon_0\varepsilon_r\, V/\ln(r_s/r_c)$$
$$= 2\pi \times 8.842\times 10^{-12} \times 2.3 \times 69 \times 10^3/(0.6\sqrt{3})$$
$$= 8.493 \times 10^{-6} \text{ C/m}$$
$$= 8.493 \ \mu\text{C/m}$$
$$= 8.493 \times 10^{-3} \text{ C/km}$$

(*e*) Capacitance per metre length of cable is

$$C = Q/V = 2\pi\varepsilon_0\varepsilon_r/\ln(r_s/r_c)$$
$$= 8.493 \times 10^{-6} \times \sqrt{3}/69 \times 10^3$$
$$= 212 \times 10^{-12} \text{ farad/metre}$$
$$= 212 \ \mu\mu\text{F/m}$$

As a check, $C = 2\pi \times 8.842 \times 10^{-12} \times 2.3/0.6 = 212 \times 10^{-12}$ F/m

(*f*) Charging current per km at $69/\sqrt{3}$ kV and 50 Hz:

$$I_c = 2\pi f CV = 2\pi f Q$$
$$= 2\pi \times 50 \times 8.493 \times 10^{-3} \text{ amperes}$$
$$= 2.668 \text{ amperes}$$

Q.9. What are the expressions (equations) for maximum electric stress and minimum electric stress in the dielectric? What is the average value of stress?

Ans. From the results of Q. 8 before, there are:

$$E_{max} = [V/\ln(r_s/r_c)(1/x)\big|_{x = \min = r_c}$$

$$= [V/\ln(r_s/r_c)]/r_c$$
$$E_{\min} = [V/\ln(r_s/r_c)]/r_s$$
$$E_{av} = V/t_I = V/(r_s - r_c).$$

In Q. 8 for the 69 kV XLPE cable, the average stress in the dielectric is

$$69/(35 - 19.3)\sqrt{3} = 2.5374 \text{ kV/mm} = 25.374 \text{ kV/cm}.$$

Q.10. What is the ratio of r_s/r_c = sheath radius/conductor radius for the maximum stress E_{max} to have the least possible value?

Ans. From Question 8, the electric stress on the conductor surface is $E_c = [V/\ln(r_s/r_c)]/r_c$. Also, insulation thickness $t_I = r_s - r_c$. The conductor radius r_c is determined always from the point of view of current-carrying capacity and has nothing to do with voltage, so that it has a fixed value for a given cable for the power transmitted and the voltage level used. Thus we observe that the electric stress on the inner surface of the insulation will change if the insulation thickness t_I is changed. In other words E_c is a function of sheath radius $r_s = t_I + r_c$. There is a certain value or ratio r_s/r_c for which E_c is optimum, that is least. Designers of cable insulation usually wish to minimise the electric stress on the conductor surface, if possible.

In order to find the ratio (r_s/r_c) for the minimum value of E_c to occur, let us keep the sheath radius r_c fixed and find the value of (r_s/r_c), i.e., the value of r_c for the minimum value to occur, since the calculation is simpler. Then we can write:

$$E_c = [V/\ln(r_s/r_c)]/r_c = (V/r_s) \cdot (r_s/r_c) \cdot [1/\ln(r_s/r_c)].$$

Let us denote $r_s/r_c = y$, a number. Then, $E_c = (V/r_s) \cdot y/\ln(y)$.

Differentiating E_c with respect to y, there results:

$$dE_c/dy = (V/r_s) \cdot [\ln y - 1]/(\ln y)^2.$$

For E_c to be minimum, $dE_c/dy = 0$, which will result when $(\ln y - 1) = 0$. This condition yields $y = r_s/r_c = e^1 = 2.71828$ = base of Naperean or natural logarithm.

Thus, for optimum design of insulation, the sheath radius to the inside must be 'e' times the conductor radius.

In practice, this ratio is difficult to achieve, since cables are manufactured for several current ranges at the same voltage level. Thus, a cable for larger current will have a larger r_s but the voltage may be the same for both large and smaller currents which may require nearly the same insulation thickness.

A Table is provided here describing details of many types of cables for several voltage levels manufactured in the world. The reader should calculate all the details given in the Table to obtain familiarity with the calculations and to get an idea of cables designed in practice. [This Table is taken from the author's book: "Extra High Voltage A.C. Transmission Engineering", Third Ed. (New Age International (P) Ltd.) New Delhi, 2006, p. 496.]

Table: Dimensions and Electrical Characteristics of Cables

Electrical Stresses**

Voltage Class	Insulation Type	Cond. Radius r_c, mm	Sheath Radius r_s, mm	$\ln(r_s/r_c)$	t_I mm	E_{max}	E_{av}	E_{min}
						⟵	kV/mm	⟶
69 kV	XLPE	19.3	35.2	0.6	15.9	3.43	2.5	1.88
138 kV	O-F	25.25	38.5	0.42	16	7.86	5.2	5.15
150 kV	EPRM*	13.2	27.5	1.045	24.3		6.3	
275 kV	HPOF	20	40	0.693	19.5	12	8.5	5.36
	XLPE	20	47	0.854	27	9.7	6.14	4.13
400 kV	HPOF	23	49.4	0.765	26.4	13.8	9.2	6.43
500 kV	HPOF	34	58.5	0.543	24.5	17.2	13	10
	XLPE	21.1	53.1	0.923	32	16.3	9.9	6.5
765 kV	KP-OF	17.9	54.7	1.12	36.8	22	12	8
	PPP-OF	17.9	48.4	0.995	30.5	24.8	14.5	9.2
1100 kV	HPOF	29	64	0.8	35	29	19	13

Optimum: 1.0

* EPRM — Ethylene Propylene Rubber Moulded

** Electrical stresses are calculated at the following maximum working voltages (line-to-line):

Nominal Voltage:	69 kV	138 kV	275 kV	400 kV	500 kV	765 kV	1100 kV
Maximum Voltage:	69	145	287	420	550	765	1100

The voltages to be used for calculation of stresses are:

V = Maximum line-to-line voltage given above/√3 .

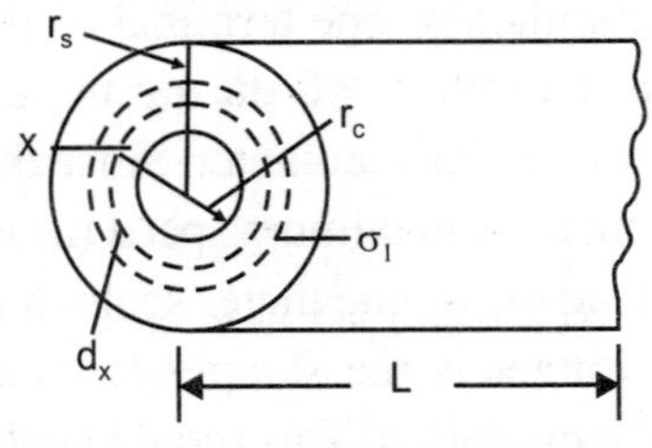

Fig. 5.4: Calculation of insulation resistance of concentric cable.

Q.11. How is the insulation resistance of a cable calculated and experimentally measured?

Ans. Figure 5.4 shows a length L of a single-core cable. There is very little voltage drop along the length of the conductor since the conductor is usually of large cross-section and has very low resistance. The current flowing in the insulation is entirely from the inner conductor to the outer sheath in the radial direction. The length of path for this current to flow is $t_I = r_s - r_c$, the insulation thickness.

Since the area for the current flow is increasing from $(2\pi r_c L)$ at the conductor to $(2\pi r_s L)$ at the sheath, we have to use the following procedure for determining the total insulation resistance of the cable for a length L:

Let $\beta_I = 1/\sigma_I$ = specific resistance or resistivity of the material in ohm-m. (This is usually in excess of 10^8 while for good conductors it is only about 10^{-8}).

At any radius x from the centre of the cable, the area presented for current flow is $2\pi xL$. Consider a thickness dx along the radial direction which is the direction of current flow in the insulation. The resistance of this cylinder is:

$$dR_I = l/A\sigma_I = dx/2\pi \times L\sigma_I$$

∴ The total insulation resistance is, for $x = r_c$ to r_s:

$$R_I = (1/2\pi\ L\sigma_I) \int_{r_c}^{r_s} dx/x = (1/2\pi L\sigma_I)\cdot \ln(r_s/r_c)$$

We found that the capacitance of the same cable for length L was

$$C = 2\pi L\ \varepsilon_0\varepsilon_r/\ln(r_s/r_c)\ .$$

Therefore, the insulation resistance decreases as the length L of the cable is increased, while the cable capacitance increases with L. This decrease in resistance with length occurs because the current flowing in the insulation increases in proportion to the length L. Since the voltage V remains constant, the resistance value will decrease as the length increases. This has great significance when testing a cable for its insulation resistance.

In order to determine insulation resistance experimentally, an instrument known as MEGGER (Mega Ohm Meter) is used. Usually, it has a hand-cranked magnetic dynamo for 600 volts, but may be 1000 V in some cases. Its two terminals are connected across the core and the sheath of the cable. Normally, for a long cable, the two ends are connected together on the drum by a thin wire and then to one terminal of the MEGGER in order that a high voltage is not experienced at the other end during the experiment. The indicating meter inside the MEGGER is a sensitive micro-ammeter. Since the voltage output of the instrument is fixed when the dynamo is run at its maximum speed (as indicated by a slipping clutch), the insulation resistance can be read off on the meter scale directly as the μA-flow is calibrated in megohms. Naturally, the longer is the sample of cable chosen, the lower will be the indicated resistance value of the insulation. This should not be taken to mean that the insulation is poor. An intelligent operator or testing technician will convert the meter reading properly if the property that the insulation resistance is lowered as the length of cable is increased is properly understood.

Q.12. What is meant by the time-constant of a cable and what is its value?

Ans. The time constant refers to the behaviour of a transient operation of the voltage, as for example, when a cable is switched on, or a lightning impulse hits the cable, and the voltage distribution on the insulation is to be determined. It takes a little while (usually some μs) for the cable to reach the steady state after the impact of the transient voltage.

For any given electric circuit consisting of a resistance and a capacitance, the product $R{\cdot}C$ is called the time constant. It has units of seconds and is the time required for the transient to rise to $(1 - e^{-1}) = 0.632 = 63.2\%$ of the final value, which is the peak of the impressed voltage.

For a concentric cylindrical-conductor-sheath cable, we had

$$R_I = (1/2\pi L\sigma_I)\cdot \ln(r_s/r_c),$$

and

$$C = 2\pi L\varepsilon_0\varepsilon_r/\ln(r_s/r_c).$$

Their product is $\tau = R_I\cdot C = \varepsilon_0\varepsilon_r/\sigma_I$, which is a constant for a given cable-insulation. This quantity is independent of cable dimensions (r_s and r_c) and length L, but only depends on the permittivity $\varepsilon_0\varepsilon_r$ and specific conductivity σ_I (reciprocal of the specific resistivity), of the material used for insulation.

Q.13. What are meant by Surge Impedance and Velocity of Propagation of a surge in the cable and what are their values?

Ans. When a surge impinges on a cable between the conductor and the grounded outer sheath, such as lightning, the ratio of voltage to the resulting current in the insulation is called the surge impedance or characteristic impedance. The lightning voltage propagates inside the cable between conductor and grounded sheath, and the resulting impedance has nothing to do with the insulation resistance of the cable. The surge impedance is purely the relation between the voltage and current.

Its value is

$$Z_c = 1/vC,$$

where v = velocity of propagation of the travelling wave.

This velocity is given by

$$v = 1/\sqrt{(\varepsilon_0\varepsilon_r\mu_0)}$$

where μ_0 = permeability of free-space and of non-magnetic materials
$= 4\pi \times 10^{-7}$ henry/metre.

Also, ε_0 can be written as

$$(1000/36\pi)\times 10^{-12} = 8.842 \times 10^{-12} \text{ F/m}.$$

[When a radio wave propagates in free space, $\varepsilon_r = 1$,

its velocity is:

$$\begin{aligned} c &= 1/\sqrt{(\varepsilon_0\mu_0)} \\ &= 1/\sqrt{[(1000/36\pi) \times 10^{-12} \times 4\pi \times 10^{-7}]} \\ &= 3 \times 10^8 \text{m/s} = \text{well-known velocity of light in vacuum.} \end{aligned}$$

The characteristic impedance of free space presented to the electro-magnetic wave is:

$$\begin{aligned} Z_0 &= \sqrt{(\mu_0/\varepsilon_0)} \\ &= \sqrt{[4\pi \times 10^{-7}(1000/36\pi) \times 10^{-12}]} \\ &= 120\pi \text{ ohms} \\ &= 377 \text{ ohms} \end{aligned}$$

These are well-known properties of free space for wave propagation].

For a cable with an insulation having a dielectric constant ε_r, the velocity of propagation is

$$v = [1/\sqrt{(\mu_0\varepsilon_0)}]\cdot(1/\sqrt{\varepsilon_r}) = c/\sqrt{\varepsilon_r}.$$

The velocity varies inversely as the square root of the dielectric constant. For XLPE, with $\varepsilon_r = 2.25$, the velocity of surge becomes 2/3 the velocity of light.

The surge impedance, on the other hand, is:

$$\begin{aligned} Z_c &= 1/vC \\ &= \sqrt{(\mu_0 \varepsilon_0 \varepsilon_r)} \cdot (1/2\pi\varepsilon_0\varepsilon_r) \cdot \ln(r_s/r_c) \\ &= (1/2\pi) \cdot \sqrt{(\mu_0/\varepsilon_0)}\ (1/\sqrt{\varepsilon_r}) \cdot \ln(r_s/r_c) \\ &= (1/2\pi) \times 120\pi \cdot (1/\sqrt{\varepsilon_r}) \cdot \ln\ (r_s/r_c) \\ &= (60/\sqrt{\varepsilon_r}) \cdot \ln\ (r_s/r_c). \end{aligned}$$

The surge impedance is independent of cable length but depends only on the dielectric constant and the ratio (r_s/r_c) = sheath radius/conductor radius.

For $\varepsilon_r = 2.25$ and $(r_s/r_c) = e = 2.7183$, the surge impedance is $60/\sqrt{2.25} = 40$ ohms. Cables usually are designated by their surge impedance, especially in high voltage measurement situations where the output of a voltage divider is to be matched to measuring instruments mainly oscilloscopes. The most usual coaxial cables come with impedances of 50 or 75 ohms.

The reader is urged to calculate the velocity v and surge impedance Z_c for the cables given in Table in Question 10 for practice, using the ε_r values given in Question 6 for some of the insulation materials.

Q.14. What are meant by: (*a*) Charging Current; (*b*) Loss Angle Current; (*c*) Power Factor; (*d*) Loss Angle δ; and (*e*) $\tan \delta$ of a cable? What are their significances?

Ans. (*a*) The charging current is the capacitance portion of the total current flowing in the insulation. This is

$$I = 2\pi f C V,$$

where f = frequency of applied voltage,

C = capacitance of cable for any given length,

and V = rms value of voltage between central conductor and outer conducting sheath.

This current being a pure capacitance current, leads the applied voltage by exactly 90°. The current flows even when no load is connected to the cable. Under load conditions, it has to be added vectorially to the load current. As the cable length increases, the charging current could become a considerable portion of load current. The source of power to the load has to supply both currents and thus become overloaded, or, in other words, the load current has to be limited to the safe value for the cable. The conductor temperature also rises due to this charging current even if the cable is not supplying any useful load.

This is the only reason why high voltage power transmission by cables is restricted to short distances. The longest cables used are for transmitting d.c. under the sea, since under d.c. the charging component of current is absent and only the loss component is present. This will be dealt with in the next Chapter 6.

Associated with the charging current is the charging reactive power denoted as Q, vars. (The symbol Q has also been used for charge). For a 3-phase cable, this is: $Q = 3VI$, vars 3-phase. If V is in kV and I in kA, then Q is in MVAR. The following Table gives typical values of C, I, Q for several types of cable. The reader should calculate all these values on her/his own to become familiar with the calculation as well as having an idea of the magnitudes of these quantities.

(*b*) In addition to the charging current, the finite value of insulation resistance draws an in-phase component of current I_R, as shown in the phasor diagram of Fig. 5.5. The total charging

Table: C, I, Q of Typical Cable Designs (See also Table in Q. 5.10)

Category I: HPOF:		$C = 3.6/18 \ln(r_s/r_c)$;		$\varepsilon_r = 3.6$	
138	kV	(145 kV Max.):	C = 0.46 μF/km,	I = 12.12 A/km,	Q = 3.05 MVAR/km
275	kV	(287 kV)	0.28	14.5	6.95
500	kV	(550 kV)	0.36	35.7	34
1100	kV	(1150 kV)	0.246	51.3	102
Category II: XLPE:		$C = 2.3/18 \ln(r_s/r_c)$;		$\varepsilon_r = 2.3$.	
275	kV	(287 kV)	C = 0.15 μF/km,	I = 7.89 A/km,	Q = 3.87 MVAR/km
500	kV	(550 kV)	0.1384	13.8	13.17
Category III: PPLP:		$C = 2.7/18 \ln(r_s/r_c)$;		$\varepsilon_r = 2.7$.	
765	kV	(765 kV)	C = 0.151 μF/km,	I = 20.925 A/km,	Q = 28.14 MVAR/km

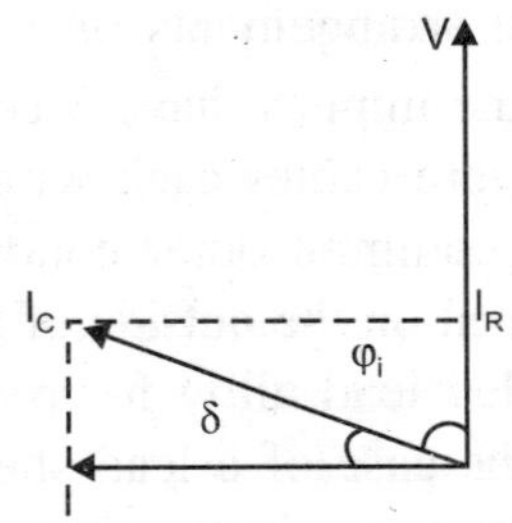

Fig. 5.5: Charging current I, Loss current I_R, Total current I_c of lossy dielectric. Loss angle δ, power factor angle φ_i.

current I_c = phasor sum of I and I_R, which are at right angles to each other. The angle φ between V and I_C is almost 90° in practical cases. For an ideal insulation without loss component I_R of the current, only the pure capacitive current $I = 2\pi fCV$ is supplied by the source. We denote $\delta = 90° - \varphi$, which is called the Loss Angle. The loss component of current is I_R.

(*c*) The Power Factor of the cable as a load on the voltage source $\cos \varphi_I = I_R/I_c$, and the power loss which heats the insulation and raises its temperature is $P_I = V\, I_c \cos \varphi_I = V{\cdot}I_R$, watt.

(*d*) Now, $\cos \varphi = \sin \delta$ and if δ is less than 1°, then $\sin \delta \approx \tan \delta \approx \delta$. The magnitude of $\tan \delta$ or δ denotes how lossy the insulation is. We also have $\tan \delta = I_R/I$. In this form, it is easy to measure I_R and the capacitive component I, as will be discussed when considering the Schering Bridge in Chapter 17.

We also note that:

(*a*) $I = 2\pi fCV = (2\pi)^2\, V f\, \varepsilon_0\varepsilon_r/\ln(r_s/r_c)$, and

(*b*) $I_R = V/R_I = 2\pi\, \sigma_I\, V/\ln(r_s/r_c)$

$\therefore \tan \delta = I_R/I = \sigma_I/2\pi\, f\, \varepsilon_0\varepsilon_r$

Thus, $\tan \delta$ is inversely proportional to ε_r in contrast to velocity v and surge impedance Z_c, which are inversely proportional to $\sqrt{\varepsilon_r}$.

Average values of ε_r and $\tan \delta$ for several insulations are as follows:

Kraft Paper in Oil	: $\varepsilon_r = 3.5$,	$\tan \delta$ = 0.002 (0.2%);
XLPE	: 2.3	0.0002 – 0.0005;
Composite PPLP	: 2.75	0.0007 (0.07%);
SF_6	: 1	0

For XLPE and PPLP, the lower value of ε_r yields lower $\tan \delta$ than for Kraft Paper in Oil. This reflects in rise of temperature of the cable dielectric for the same value of $1/\sigma_I$. In general, because of the compactness of extruded insulation, σ_I or the conductivity is low i.e., the specific resistance is higher for the plastics than for paper, which also contributes to lowering of $\tan \delta$.

Q.15. What are some representative arrangements of 3-phase cables in practice?

Ans. 3-phase a.c. cables, as the name implies, have 3 central conductors, one for each phase. These can be either 3 single-core cables each with its own outer conducting sheath, or the 3 conductors can have a common outer conducting sheath. By sheath we usually mean a metallised paper wrapped on the outside of the insulation and protected by a PVC or Neoprene jacket or extruded lead alloy for prevention of corrosion, wear, and for mechanical strength, etc. In the case of a lead sheath it will also act as current carrier as well as providing necessary protection against mechanical and chemical damage.

(*a*) When 3 single-core cables are used for the 3 phases, they can be laid in the following manner:

(*i*) Laid directly in soil side by side or in troughs. Natural cooling carries the heat away;

(*ii*) They can also be force-cooled by installing water pipes adjacent to cable, or installing the cable inside a larger water pipe, or with internal cooling with oil and cable placed inside a water pipe.

A Pipe-type cable is installed inside a large pipe which carries cooling oil. Either gravity or oil pumps keep oil circulating for better cooling.

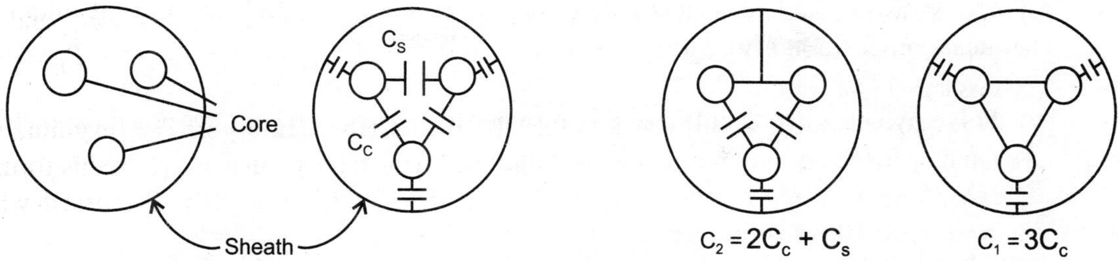

Fig. 5.6: Capacitance network of 3-core cable. Two test conditions.

(*b*) For lower voltages up to 66 kV, 3-core cables are used in addition to 3 single-core cables. A cross-section is shown in Fig. 5.6 along with the capacitance network. The three cores are separately insulated, placed in position and all the three cores are wrapped by further insulation. A common weather protection sheath and a conducting sheath are provided on the outside.

Cooling methods and cable-laying methods are similar to 3 single-core cables, although the lower voltage does not require liquid cooling in most cases.

It is shown in Fig. 5.6 that the capacitance network consists of C_c = core-core capacitance and C_s = core-sheath capacitance. It is quite difficult to calculate these unlike for a single-core cable. Therefore, experimental methods using the Schering Bridge (Chapter 17) is the more usual method.

Two tests are made:

1. The 3 cores are connected electrically together and the capacitance is measured between the common and the sheath.

   Let this be C_1

   Then, $C_1 = 3C_c$

2. Two of the 3 cores are connected together and to the sheath. The third is left free. Capacitance is measured between the common and the free core. Let this be C_s.

   Then, $C_2 = 2C_c + C_s$

   From the above two measurements, we can calculate C_c and C_s individually as follows:

$$C_s = C_1/3;\ C_2 = 2\ C_c + C_1/3$$

giving $$C_c = (C_2 - C_1/3)/2$$

Q.16. How is XLPE manufactured and what are its superior properties over other insulations for high voltage cables?

Ans. XLPE stands for Cross-Linked Poly Ethylene. The search for an insulating material with low ε_r, low tan δ, high dielectric strength without oil or other impregnation, easy installation, better fire-resistant property and attack against chemicals has resulted in discovery of this excellent insulating material. It is of U.S. origin, although PVC itself was invented by I.C.I. in Great Britain after World War II. Cross-linking means simply vulcanising PVC under high temperature and pressure. However, the process is more than this: it alters the chemical composition also.

Polyethylene is a molecule chain represented as 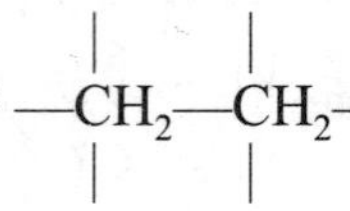 . During cross-linking, many bonds between C and H_2 are replaced by C thus: 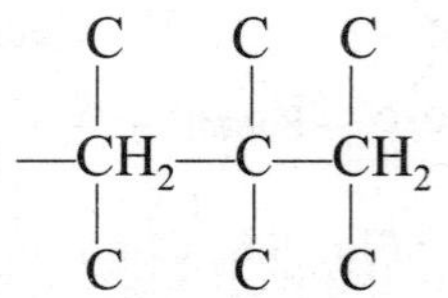 .

Several methods are used for this process. These are: (*a*) Chemical linking; (*b*) Nuclear irradiation; and (*c*) Vulcanising. One such process is shown in Fig. 5.7 where the conductor is first covered with extruded polyethylene in a machine and then pressed through a special cross-linking machine. Pressure is applied by Nitrogen gas at high pressure, and heat is applied by external electric heaters. The process is continuous.

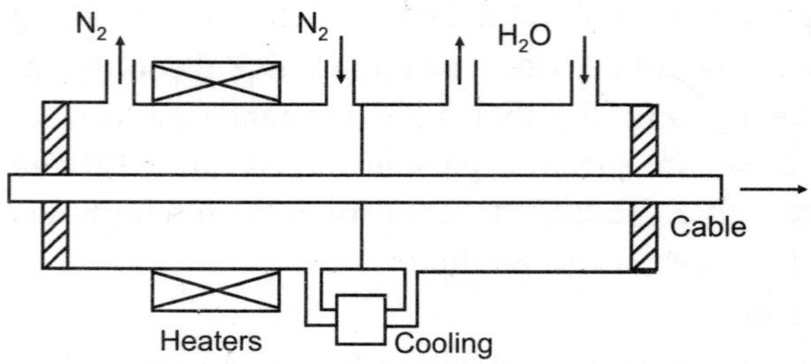

Fig. 5.7: Schematic of manufacture of XLPE extruded insulation.

Q.17. A parallel-plate capacitor has plate area *A*. Its dielectric is made from two different materials having dielectric constants ε_1 and ε_2, and thicknesses t_1 and t_2. The applied voltage is *V*. Calculate the following:

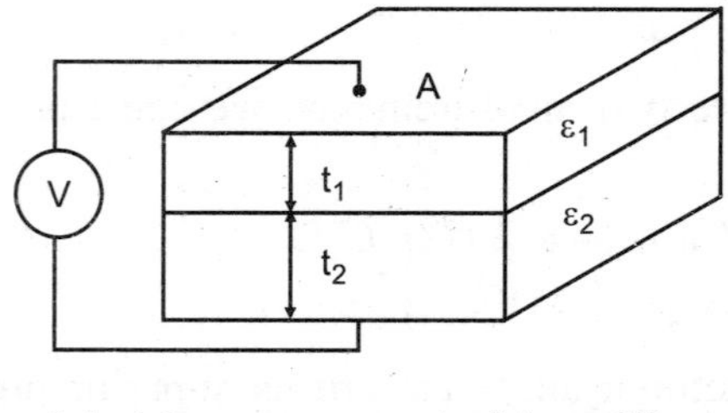

Fig. 5.8: Parallel-plate arrangement of two different dielectrics.

(*a*) The capacitance of each dielectric and the total capacitance of the arrangement;

(*b*) The voltage across each dielectric; and

(*c*) The dielectric stresses in V/m in the two dielectrics for the applied voltage *V*. (Note: All quantities are in SI units).

Ans. Let the partial capacitance be C_1 and C_2, and the voltages across them V_1 and V_2. Then, since the two dielectrics are in series, the charges across them are equal, and $V_1 + V_2 = V$.

$\therefore \quad Q = C_1V_1 = C_2V_2 = CV$

or, $\quad V_1 = CV/C_1$

and $\quad V_2 = CV/C_2$

Therefore, $V_1 + V_2 = V = C[(1/C_1) + (1/C_2)]\,V$

or, $\quad 1/C = 1/C_1 + 1/C_2$

which is the well-known relation for the equivalent capacitance of two capacitors in series.

(*a*) Now, $\quad C_1 = \varepsilon_0\varepsilon_1\, A/t_1$

and $\quad C_2 = \varepsilon_0\varepsilon_2\, A/t_2$

$\therefore \quad 1/C = (1/\varepsilon_0 A)[(t_1/\varepsilon_1) + (t_2/\varepsilon_2)]$,

giving $\quad C = \varepsilon_0\varepsilon_1\varepsilon_2\, A/(t_1\, \varepsilon_2 + t_2\, \varepsilon_1)$

(*b*) The total charge is $Q = CV$.

Then,

$$V_1 = Q/C_1 = V\,[t_1\,\varepsilon_2/(t_1\,\varepsilon_2 + t_2\,\varepsilon_1)]$$

and
$$V_2 = V - V_1 = V\,[t_2\varepsilon_1/(t_1\,\varepsilon_2 + t_2\,\varepsilon_1)]$$

Example: Let $t_1 = 2$ mm

$t_2 = 1$ mm

$\varepsilon_1 = 1$ (air)

$\varepsilon_2 = 2$ (plastic)

Then, $V_1 = V_{air} = V\,[2 \times 2/(2 \times 2 + 1 \times 1)]$

$= (4/5)\,V$ across 2 mm of air;

$V_2 = V_{plastic} = (1/5)\,V$ across 1 mm of plastic.

(*c*) The dielectric stresses in the two dielectrics are:

$$E_1 = V_1/t_1, \text{ and } E_2 = V_2/t_2$$

For the numerical example,

$$E_{air} = (2/5)V, \text{ and } E_{pl} = V/5 = E_{air}/2.$$

Let us assume that the electric stress in air is to be limited to E_{air} = 30 kV/cm = 3 kV/mm. Then the applied voltage can be 7.5 kV. This gives the stress in the plastic of 1.5 kV/mm = 15 kV/cm. Note that the insulation in the parallel-plate capacitor is not properly designed, since plastic can be stressed to nearly 10 kV/mm.

This example shows how to calculate stresses in composite insulations and where the designs can be improved to obtain optimum results.

Example: If $t_1 = t_2$, then $V_1/V_2 = \varepsilon_2/\varepsilon_1$, showing that the voltages are inversely proportional to the dielectric constants. The same is true for the electric stresses. The general expression for the ratio of voltages and stresses are:

$$V_1/V_2 = (t_1/t_2)\cdot(\varepsilon_2/\varepsilon_1); \text{ and } E_1/E_2 = \varepsilon_2/\varepsilon_1$$

The insulation with the lower dielectric constant bears higher stress in it than the one with the higher dielectric constant.

Q.18. Repeat Question 17 for a concentric-cylindrical arrangement of the dielectrics, with conductor radius r_c, thicknesses of insulations t_1 and t_2, and dielectric constants ε_1 and ε_2. See Fig. 5.9.

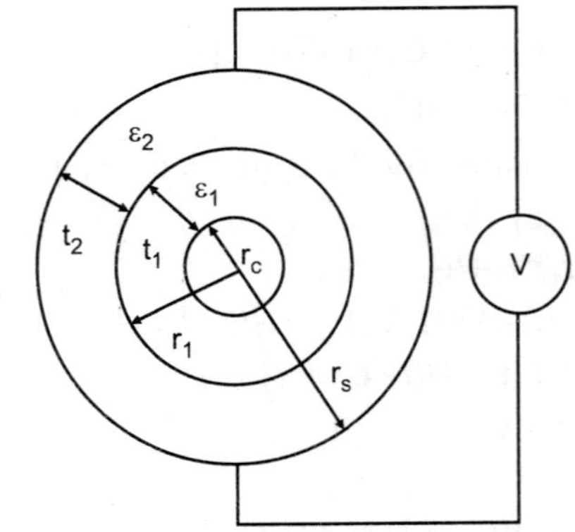

Fig. 5.9: Concentric-cylindrical arrangement of two different dielectrics.

Ans. Referring to Question 8, the capacitance of a concentric cylindrical arrangement with inner conductor radius r_c and outer cylinder radius r will be:

$$C = 2\pi\varepsilon_0\varepsilon_r/\ln(r/r_c)$$

Therefore, if $r_1 = r_c + t_1$ and $r_s = r_1 + t_2$,

we have $$C_1 = 2\pi\varepsilon_0\varepsilon_1/\ln(r_1/r_c)$$

and $$C_2 = 2\pi\varepsilon_0\varepsilon_2/\ln(r_s/r_1)$$

Once again, since C_1 and C_2 are in series across the voltage $V = V_1 + V_2$, the total capacitance of the composite-insulation cable arrangement is given by

$$1/C = 1/C_1 + 1/C_2 = [1/(2\pi\varepsilon_0\varepsilon_1\varepsilon_2)] \cdot [\varepsilon_2 \cdot \ln(r_1/r_c) + \varepsilon_1 \cdot \ln(r_s/r_1)].$$

Since the charges on both dielectrics are equal, we have:

$$V_1C_1 = V_2C_2 = VC$$

$\therefore$ $$V_1 = V\,C/C_1 = V \cdot \varepsilon_2 \cdot \ln(r_1/r_c)/[\varepsilon_2 \cdot \ln(r_1/r_c) + \varepsilon_1 \cdot \ln(r_c/r_1)]$$

Similarly, $$V_2 = V \cdot \varepsilon_1 \cdot \ln(r_c/r_1)/[\varepsilon_2 \cdot \ln(r_1/r_c) + \varepsilon_1 \cdot \ln(r_s/r_1)]$$

The ratio of the two voltages is:

$$V_1/V_2 = (\varepsilon_2/\varepsilon_1) \cdot \{\ln(r_1/r_c)\}/\{\ln(r_s/r_1)\}$$

where $r_1 = r_c + t_1$ and $r_s = r_1 + t_2$.

❑❑

CHAPTER 6

A.C. AND D.C. TRANSMISSION SYSTEMS: OVERHEAD LINES, CABLES AND INSULATORS

Q.1. What are the components of a high voltage a.c. transmission system?

Ans. A power transmission system consists of: (*a*) Generating Station; (*b*) Step-up transformers; (*c*) Overhead Line or Underground Cable; (*d*) Step-down transformers at receiving stations; (*e*) Voltage Control Equipment: Synchronous Condensers, or Switched Capacitors and Controlled Reactors; and (*f*) Loads.

There are several other secondary but vital components for operation and protection of the system. These are so numerous and varied that it is not possible to enumerate them and describe them here. Generally speaking, they fall into these broad categories: (*a*) Governors and voltage regulators for maintaining constant frequency and improving or maintaining stability without losing synchronism of the a.c. system during load changes and faults; (*b*) Relays for protection of the system from damage during system faults such as mainly short circuits; (*c*) Lightning arresters or surge absorbers; (*d*) Communication and protection channels using the high voltage line, called Carrier Communication and Protection; (*e*) Tap-changing equipment in transformers; (*f*) Potential and Current Transformers for measurement and relaying.

Q.2. What are the components of a high voltage d.c. transmission system?

Ans. The term "D.C. Transmission System" is usually called a "D.C. Transmission Link". Here, the generation is still a.c. as well as the loads. The generated a.c. is rectified into d.c., transmitted by overhead lines or underground cables, inverted back to a.c. and fed to the loads. Thus, the line or cable only acts as a link between a.c. generation and a.c.

loads. Nevertheless, the entire system consisting of generation, transmission and utilisation is known as a D.C. Transmission System. [Please see specialised books on HVDC Systems, such as (1) K.R. Padiyar: HVDC Power Transmission Systems; (2) J. Arrilaga: HVDC Transmission; (3) E. Uhlmann: Power Transmission by D.C.].

A d.c. transmission system consists of nearly all the components of a corresponding a.c. system of comparable voltage, current and power levels, but in addition has the following vital characteristic or unique equipments, not present or required for a.c. transmission: (*a*) Filters to smooth rectified a.c.; (*b*) Reactors to supply reactive power; (*c*) High voltage high-current controlled rectifiers which form the heart of the system; (*d*) Inverters to convert d.c. into a.c. at the load, and (*e*) Ground electrodes.

Q.3. What is a D.C. Back-to-Back Link?

Ans. There are many situations where a country may have a.c. systems operating at two different frequencies, such as in Japan (Sakuma Station), USSR-Finland Tie (Vyborg Station), Austria, USA (Big Eddy and others). These are usually 50/60 Hz ties. The a.c. power at both ends is converted into d.c. and the systems tied back-to-back inside the same station-house, and power can be interchanged between the two systems at will. There is no transmission link using overhead lines or underground cables as such. These systems are technically known as "Asynchronous Ties" since by a Synchronous Tie is meant two systems interconnected at the same frequency. Synchronous Ties using a.c. are also used and are known with the more popular and generic name of Grid System. It is one of the aims of power systems in a vast country (or even internationally) to be connected in a grid with hundreds of interconnection points. These do not normally require conversion of a.c. into d.c. and then back to a.c. But wherever the grid is very vast involving very long transmission lines and low powers, such as in India, a back-to-back d.c. link is utilised to maintain stability. Such back-to-back stations for interconnection of two systems at the same frequency are planned for Belgaum (Karnataka-Maharashtra Tie), Kota (Mumbai-Rajasthan Tie), Vindhyachal (U.P.-M.P. Tie), etc. Thus, it is not absolutely necessary that for a back-to-back terminal using a.c.-d.c.-a.c. interchanges the frequencies of the two a.c. systems should be necessarily different.

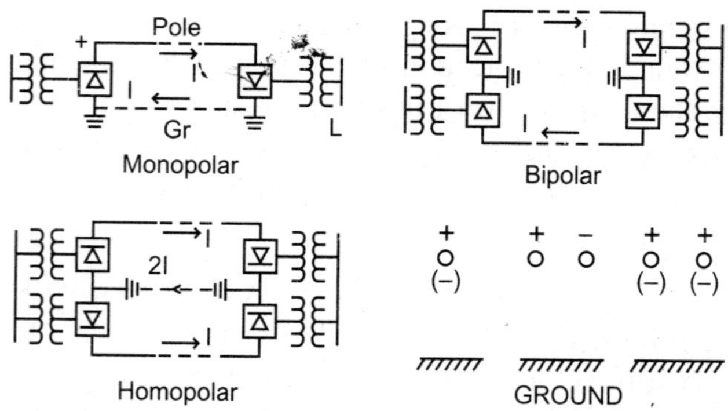

Fig. 6.1: Schematic arrangement of Monopolar, Bipolar and Homopolar D.C. transmission system.

Q.4. What are the different types of d.c. links or transmission schemes available and are used in the world?

Ans. D.C. transmission systems can be classified into three schemes: (*a*) Monopolar; (*b*) Homopolar; and (*c*) Bipolar. Fig. 6.1.

(*a*) In the **Monopolar System**, there is only one conductor or pole on a tower. (The pole can be on a pole, in which case the former is meant to be indicative of the electrical polarity of the conductor with respect to ground, and the latter word is a supporting structure for the conductor). The return current flows in the ground, thereby involving ground-return resistance and other effects. However, since ground return, (by which is also meant sea-water in those cases where a cross-channel link is used), is infinite in extent so that its resistance is zero. The only resistance to consider is that of the electrodes through which the system-current enters the ground (could be at either end) and leaves the ground. The potential gradient at these points is very high and is very harmful to any acquatic life.

[Potential gradient, E = specific resistance $\beta(=1/\sigma)$ × current density δ, since $V = RI = (\beta l/A)I$, or, $E = V/l = \beta(IA) = \beta.\delta$].

In a monopolar arrangement, the overhead conductor or pole can be either positive with respect to ground or negative.

(*b*) **Homopolar:** In the monopolar scheme, if the line develops a fault, the power is interrupted 100%. In the homopolar scheme, there are two conductors or poles of the same polarity with respect to ground on the tower. Ground forms the return for both conductors. Loads are connected across the poles and ground. An advantage of homopolar over the monopolar arrangement is that not only two circuits are available for more reliability against fault to one circuit, but also for very little extra expenses two conductors or poles can be strung on a tower. Since the two poles are at the same voltage and of the same polarity, they are at the same potential so that very little spacing need be given and the tower will be smaller than having two conductors at different polarities as in the bipolar scheme described next.

The poles can either be both positive or negative with respect to ground. Again, problems associated with ground return are present.

(*c*) **Bipolar:** By far the most popular and widely-used configuration for d.c. transmission lines is the bipolar arrangement. Here, there are two poles on the same tower, one of them positive with respect to ground and the other negative. (When we use the term "on the same tower", it is not necessary that the conductors are on the same structure, although this is what is understood normally. Each pole can be strung on its own tower. But this is unnecessary). The voltage between the conductors is 2V, where V = voltage between one pole and ground. There is normally no return-current in the ground for balanced loading of the two circuits in practice, but in case of one pole being out of service, 50% of the power can still be transmitted between the healthy pole and ground return. In 3-phase a.c. systems, this is not possible since all 3 phases are open and must be taken out of service when a system fault occurs.

Q.5. What are the comparative advantages and disadvantages of a.c. and d.c. transmission systems? Compare only a 3-phase a.c. system with a bipolar d.c. link.

Ans. We will consider that the powers to be transmitted are equal in both cases and the line-to-line voltages are equal.

1. For the a.c. system, $P = \sqrt{3} \cdot V \cdot I \cdot \cos\varphi$. For $\cos\varphi = 1$, $I = P/\sqrt{3}V$. For the d.c. system, $I_{dc} = P/V$. Therefore, the current flowing in any conductor or pole of the d.c. system is $I_{dc} = \sqrt{3} \times$ line current in the a.c. system. The cross-section for each conductor for the d.c. bipolar system is $\sqrt{3}$ times larger. However, since there are two conductors in the bipolar line and 3 in the a.c. line, the weight and cost of conductors for the two systems will be:
   d.c./a.c. = $2\sqrt{3}/3$ = 1.15 or 15% greater for the d.c. line. Thus, there is practically very little difference as far as cost and weight of conductors are concerned, but might be serious when the systems are huge.
2. Since insulation required depends upon line-to-ground and line-to-line voltages, for equal line-to-line voltages, the a.c. line requires twice the width of tower as compared to the d.c. line, assuming both to have horizontal configuration of the conductors. The weight of tower is consequently larger in the a.c. system, as well as width of land required, (called the width of Right-of-Way, ROW). Refer to Fig. 6.2.

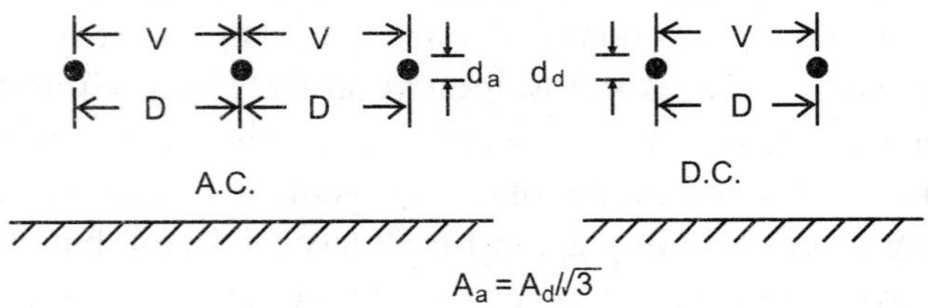

Fig. 6.2: Comparison of S/C 3-phase A.C. line and bipolar D.C. line.

3. High voltage effects are more severe in general with a.c. lines than with d.c. lines. This has to do with Corona Discharges which are present around the conductors. The line-to-ground voltage (line to metallic tower in real cases) is $V/2$ for d.c. while the same for the a.c. line is $V/\sqrt{3}$. Furthermore, the d.c. line has a larger diameter for conductor by the factor $(3)^{1/4} = 1.316$ since the current in the d.c. pole is $\sqrt{3}$ times the current in the a.c. line and requires $\sqrt{3}$ times the area, normally. Therefore, the voltage gradient on the conductor surface is lower for the d.c. line nearly in the ratio $(V/2 \times 1.316)$ to $(V/\sqrt{3}) = 0.38/0.578$ i.e., in the ratio of nearly 2 : 3. Any high voltage effects caused by the dielectric stress on the conductor surface will be lower or less severe for the d.c. line especially the power loss caused by corona on the conductors, interference to radios located in houses or other establishments close to the line, etc. We cannot discuss all of the effects which form a highly-specialised branch of high voltage transmission engineering, but the general picture is that a d.c. transmission line gives less severe high voltage effects than a.c. lines with the same line-to-line voltage and for handling the same power.

We cannot compare the electrostatic effects near the line since the physics of the two types of excitation voltages is different. For example, considering the ground-level field which might affect human beings, animals and vegetation under and near the line, in the d.c. line there is a continuous movement of electric charges from line to ground while for the a.c. line, the charges are localised around the conductors. These are respectively called the Poissonian field and Laplacian field.

Q.6. Draw the schematic of a representative rectifier station and an inverter terminal for a bipolar d.c. link.

Ans. The rectifier station or terminal converts the input 50/60 Hz power-frequency voltage, current and power into d.c., while the inverter terminal does the reverse. They are identical in all electrical aspects and are equivalent, see Fig. 6.3. The same equipment is present at both terminals, except for a difference in what is known as the Firing Angle of the Thyristors or Controlled Rectifiers.

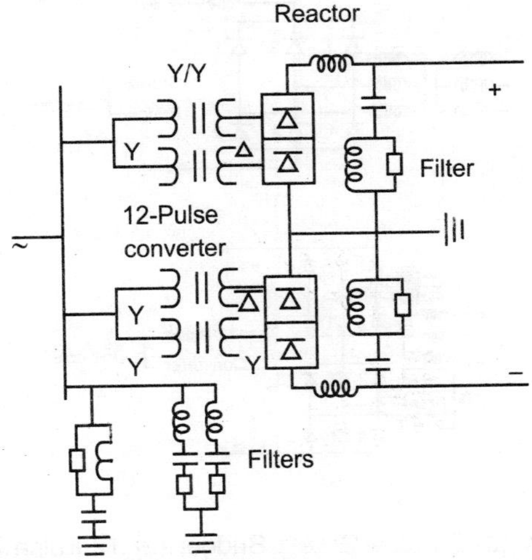

Fig. 6.3: Schematic diagram of rectifier (and inverter) station in bipolar D.C. transmission system.

They consist mainly of the following electrical equipment:

1. 3-phase generators behind converter or rectifier, and loads in front of inverter station;
2. Four 3-phase transformer units making up 12 phases. Two of them are connected in *Y*/Delta and the other two in *Y*/*Y*. See next Chapter for a review of 3-phase networks). These are necessary not only to match the generator voltage (13.2 – 13.8 kV) with the system high voltage level, but also to provide 30° phase difference required for successful operation of the solid-state bridge circuits, as will be described shortly;
3. Smoothing reactors at the entrance to the line at converter station and the load at the inverter station;
4. Various filter circuits on the d.c. side at the converter station to filter out troublesome harmonics in the output voltage.

Q.7. Describe some of the more commonly-used converter circuits in practice. In particular, describe (*a*) 6-Pulse Graetz Bridge Circuit, and (*b*) 12-Pulse Bridge Converter Circuit. Refer to Fig. 6.4.

Ans. (*a*) **Graetz Circuit:** This is also known as 6-Pulse Converter, because it comprises of 6 gate-controlled rectifiers. By arranging to fire two rectifiers in turn with a time-shift of 1/6-th cycle or 60°, each pair of rectifiers conducts for 60° in only one direction. The resulting output waveform of voltage consists of parts of a sine wave over ± 30° on either side of the peak value, if conduction of the rectifiers is arranged such that the rectifier gate fires the device ±30° about the peak. The lowest harmonic in the output voltage is thus 6 × power frequency (300 or 360 Hz). The harmonics are filtered by a heavy reactor or 'choke' connected in series with the line. The negative pole of the bridge is usually grounded, or considered as ground.

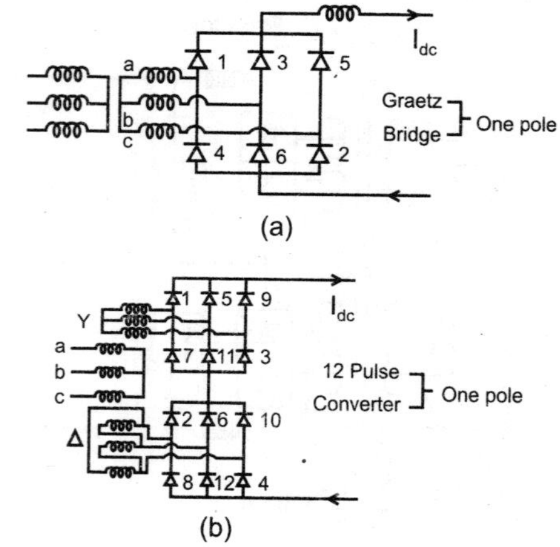

Fig. 6.4: (*a*) 6-Pulse Graetz Bridge; (*b*) 12-Pulse Bridge.

The sequence of firing of the rectifiers is 1–4, 2–5, 3–6. Thus, one period of the a.c. power supply voltage is divided into 6 intervals when one pair is conducting in each interval. The average d.c. voltage is, where α = the beginning of conduction from zero of sine wave in radians:

$$V_{dc} = (3/\pi)\int_{\alpha}^{\alpha+\pi/3} \sqrt{2}\cdot V_{ac}\cdot \sin(\omega t + \pi/3)\cdot d(\omega t), \quad \text{where} \quad \omega = 2\pi f,$$

$$= (3\sqrt{2}/\pi)\int_{\alpha}^{\alpha+\pi/3} V_{ac}\cdot \sin(\theta + \pi/3)d\theta = (3\sqrt{2}/\pi).$$

$$V_{ac}\cdot \cos(\theta + \pi/3)\Big|_{\alpha+\pi/3}^{\alpha}$$

$$= (3\sqrt{2}/\pi)\cdot V_{ac}\,[\cos(\alpha + \pi/3) - \cos(\alpha + 2\pi/3)]$$

Now, using the trigonometric identity:

$$\cos A - \cos B = 2\sin (B - A)/2 \cdot \sin (B + A)/2$$

we obtain:

$$V_{dc} = (3\sqrt{2}/\pi) \cdot V_{ac} \cdot 2 \sin (\alpha + \pi/2) \cdot \sin (\pi/6)$$
$$= (3\sqrt{2}/\pi)\ V_{ac} \cdot \cos \alpha$$

The angle α is known as the "Firing Angle". For $(-\pi/2)$ to $(+ \pi/2)$ for α, $\cos \alpha$ is positive while it is negative from $(\pi/2)$ to $(3\pi/2)$. Thus, by varying the gate-pulse timings to the thyristors, we can obtain d.c. voltage of both polarities, positive or negative, and of varying magnitudes.

When the d.c. voltage is positive, the converter acts like a rectifier. When it is negative, it acts as an inverter. Each pair of valves conducts over an interval of 120°, two shifts of 60° each, half a cycle apart.

Even though the output is d.c. (with ripple which is smoothed out) the a.c. side operates with a power factor. It is clear that when $\alpha = 90°$, i.e., a thyristor fires on the peak of the a.c. voltage, $\cos \alpha = 0$, and the output voltage is zero. The source of power does not supply any real power. Its power factor is zero. On the other hand, when $\alpha = 0$ and $\cos \alpha = 1$, the d.c. output voltage is maximum possible at $V_{dc} = (3\sqrt{2}/\pi)V_{ac}$. This corresponds to maximum power supplied by the a.c. source. The a.c. power supplied to the converter can therefore be written as:

$$P_{ac} = \sqrt{3}\ V_{l-l}\ I_l \cos \varphi = \sqrt{3}\ V_{l-l}\ I_l \cos\alpha$$

A reactive power is also associated with the firing angle α. When $\alpha = 0$, only real power is supplied. On the other hand, when $\alpha = \pi/2$, only reactive power is supplied and consumed.

(*b*) **12-Pulse Converter:** This is a series combination of 2 six-pulse Graetz converters yielding a smoother waveform, as shown in Fig. 6.4(*b*). It requires 12 thyristors in each circuit. The lowest ripple-frequency is 12 × power frequency (600 or 720 Hz) so that smoothing is easier, requiring smaller choke.

Q.8. What is a high phase-order a.c. system?

Ans. High phase order is the name of a system which uses more than 3 phases. The most usual phase order is 6, although there is no restriction on the number of phases that can be used. In a 3-phase system, the three voltages to neutral are, with

$$V = \text{r.m.s. value:}$$
$$V_a = V\underline{|0°}$$
$$V_b = V\underline{|-120°}$$
and $$V_c = V\underline{|+120°}$$

The standard notation for the phasor $1\angle 120°$ is the letter 'a'. Just like $j = 1\angle 90°$ we always use $a = 1\angle 120°$.

Therefore the three a.c. voltages are written as:

$$V_a = V,\ V_b = a^2V, \text{ and } V_c = aV,$$

since $a^2 = (1\angle 120°)^2 = 1\angle 240° = 1\angle -120°$.

The magnitudes of the three line-to-line voltages will be $V_{l-l} = \sqrt{3}V$. In the 6-phase system, the 6 phases are energised by the voltages:

$$V_1 = V\angle 0°, \qquad V_2 = V\angle -60°,$$

$$V_3 = V\angle -120°, \qquad V_4 = V\angle -180° = -V_1,$$

$$V_5 = V\angle -240° = -V_2,$$

and

$$V_6 = V\angle -300°$$

$$= V\angle +60° = -V_3.$$

The line-to-line voltages are:

$$V_{l-l} = V\angle 0° - V\angle -60° = V\angle +60°, \text{ etc.}$$

showing that the line-to-line voltage magnitude is equal to the line-to-neutral voltage. Indeed, they occupy the three sides of an equilateral triangle.

The high phase-order system requires 6 conductors on the same tower while the 3-phase system has 3 conductors. This is the usual arrangement of the two systems. In practice, since 6-phase lines are limited to medium voltage up to 132 kV, it is usual for 3-phase towers to be strung with 2 circuits called Double-Circuit (D/C) lines. Thus, there are 6 conductors on a tower even in 3-phase systems. The same arrangement is used for the 6-phase system also. Therefore, there is essentially no difference in the tower structures, except in size, as we will see.

Let us consider that the same power is to be transmitted at the same power factor. Then, for the 3-phase system: $P = 3V_3I_3\cos\varphi$, and for the 6-phase system: $P = 6V_6I_6\cos\varphi$. (The subscripts 3 and 6 refer to number of phases for identification and not to voltage numbers mentioned previously). Therefore, for equal line-to-neutral voltages in the two systems, the line current in each conductor of the 3-phase system is twice that in the 6-phase system. However, since there are twice the number of conductors in the 6-phase system, the amount of copper or aluminium is equal in both systems. In fact, the conductor sizes will be the same in a double-circuit 3-phase system as in the 6-phase system. In other words, to transmit the same amount of power at the same line-to-neutral voltages, the amount of conductor materials is the same, regardless of the number of phases used. As an engineer, this could have been guessed as being reasonable.

The advantage of the 6-phase system is that since the line-to-line voltage
= $(1/\sqrt{3}) \times$ line-to-line voltage of the 3-phase system,
the insulation clearances are lower and towers can be made smaller.

However, the 6-phase system requires special transformers which convert the 3-phase generated power in conventional generators into 6-phase system. This is not a serious restriction since the design of transformers for rectifiers and inverters involving 3–6 phase or 3–12 phase conversion has been with the industry for developing d.c. transmission.

There are also proposals for increasing the number of phases to 12, but as of date only one 6-phase system is in operation in the eastern United States of America.

Q.9. What are the major high voltage cable transmission systems in the world?

Ans. As discussed earlier in Chapter 5, the charging current of a cable when operated under a.c. is very high. As the cable becomes longer, the charging MVAR or kVAR may amount to a considerable percentage of the load current which it has to carry. Therefore, transmission of a.c. power by cable is limited to short lengths only. Overhead lines are exclusively used for long-distance power transmission by a.c. However, as the system voltage increases to extra high voltages, the power is brought from the generating station to a metropolitan load centre at a receiving station which has to be located outside the city for various reasons, chiefly because of the vehicular and human traffic as also the need for acquiring a large area of land for the station. If the power is to be brought inside the metropolitan area proper, then high voltage or even extra high voltage cables have to be used in underground trenches. This is a major use of e.h.v. cables on land.

By far the most dramatic use of e.h.v. cables is for d.c. transmission, especially for under-sea transmission of power. The Table below gives ten major d.c. cable systems interconnecting an island with the mainland, except Kingsnorth in Great Britain.

Where the flow of current in the sea water disturbs shipping by deflecting compasses, a bipole is necessary. Otherwise, a monopolar system is adequate.

Table: Major E.H.V. D.C. Cable Transmission Systems of the World

No.	System	km	kV	MW Power
1.	Gotland-Swedish Mainland	100	150	260
2.	Konti-Scan (Denmark-Sweden, Cable only)	100	250	500
3.	Cross Channel (Gt. Britain-France)	65	±100	160
		±270		2,000
4.	Vancouver Island-Mainland	35	±260	312
5.	Finland-Sweden	185	350	420
6.	New Zealand (North-South Islands across Cook Strait)	40	±250	600
7.	Kingsnorth (GB)	82	±266	640
8.	Japan (Hokkaido-Honshu)	44	250	300
9.	Skagerrat (Denmark-Norway)	127	300	320
			±250	500
10.	Sardinia - Italy	121	200	200

Q.10. What are meant by S/C, D/C, and M/C a.c. transmission lines?

Ans. These symbols or abbreviations stand for (*a*) Single Circuit; (*b*) Double Circuit; and (*c*) Multi-Circuit transmission lines. Up to now, d.c. transmission lines are limited to single circuit, S/C.

In an S/C line, there is only one circuit on the tower. In bipolar d.c., the tower carries a positive pole and a negative pole. The pole separation is determined by air clearances required for insulation. Referring to Fig. 6.5, an S/C line has 3 conductors of the phases along a horizontal configuration or spread. These are generally preferred for e.h.v. and u.h.v. transmission i.e., 345 kV and above.

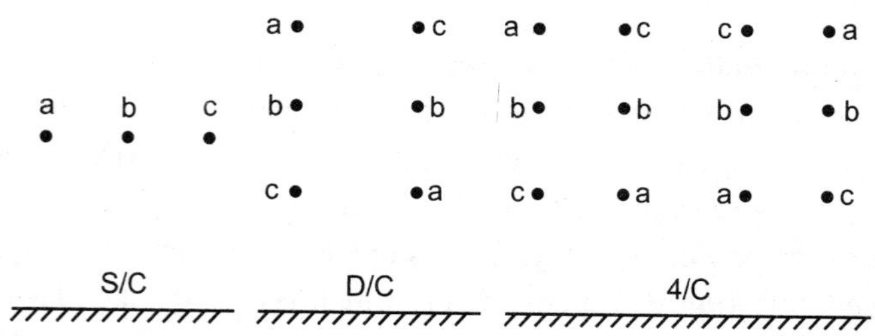

Fig. 6.5: S/C, D/C, 4/C A.C. transmission lines on tower.

A D/C a.c. tower carries 6 conductors, each circuit having its conductors in vertical configuration. The phase order is *A–B–C* and *C–B–A* for the top, centre or middle, and bottom conductors for the two circuits. This scheme is used for voltages of 220 kV and below, but there are some D/C 345 kV and 400 kV lines in the world.

Up to date, the maximum number of circuits planned on a tower is limited to four in an M/C line. It is chiefly used in Germany. We will show later on in Chapter 13 that one 750-kV S/C line can carry as much power as four 400-kV lines for the same distance of transmission. Thus, where 750-kV technology is not available, or, land for four S/C 400-kV lines is also not available, Germany has preferred the planning of a 4-circuit 400-kV tower. The limitation with German technology is not the 750-kV level, but land availability. This arrangement is exceptional and most a.c. lines in the world are either S/C or D/C.

Q.11. What is an insulator string?

Ans. The conductors or poles at high voltages above ground and between themselves have to be supported by towers, usually steel and sometimes wood for 132 kV and below. These towers are at ground potential and the conductors must therefore be insulated from the tower. They are suspended by insulators from the cross-arm of the tower in the vertical direction. Since the conductor can swing in the wind and the clearance from tower is large, the long porcelain or other insulator must be built from several smaller discs strung together properly, see Fig. 6.6.

Each disc is made of high-strength porcelain (or Pyrex glass) with its shape following the contour of the electrostatic lines of force, as dictated by the field between the conductor and the tower structure. At the top and bottom of each disc, galvanised-steel metallic cap and pin are cemented firmly to the porcelain by special cementing processes. The pin of one disc is inserted in a slot in the cap of the disc below it and a cotter pin holds the pin firmly in the cap, but allows movement.

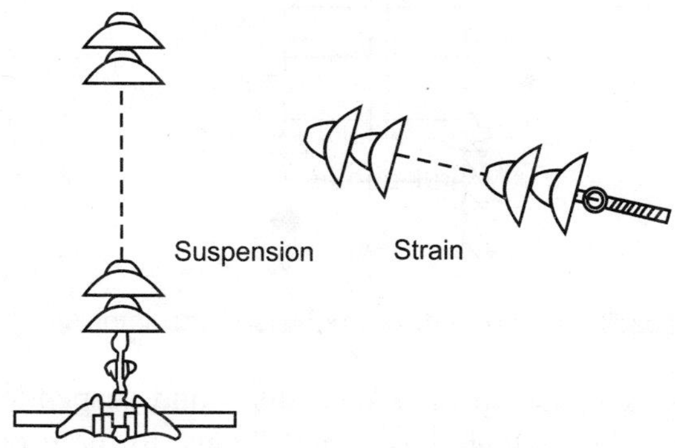

Fig. 6.6: Suspension and strain insulator strings.

While a suspension insulator is used on a tower when the transmission line runs nearly straight between the two adjacent towers, when the line contour has to bend by more than 15°, a pair of "strain" insulators is used on either side of the tower cross-arm, as shown in Fig. 6.6 on the right. The two insulator strings are jumped by a conductor which clears the tower structure by the proper insulation clearance. The jumper can also be suspended from the cross-arm by a suspension insulator string so as to provide some rigidity and safety. The construction of both suspension and strain insulators is identical, the former being vertical and not taking as much strain as a strain insulator which is at high mechanical tension and in horizontal position.

Q.12. How is the voltage distributed among the many discs of a normal insulator string with *n* discs?

Ans. When an insulator string with *n* discs is used to support a d.c. voltage of magnitude *V*, each disc has a voltage $V_d = V/n$ across it. This is because only the resistance network of the *n* discs comes into the picture and since the discs are identical, their resistances are equal.

However, the picture is completely altered under a.c. Here, it is the capacitance network that determines the voltage distribution. Each disc has a capacitance C_I between the cap and pin separated by the porcelain or glass dielectric. A standard disc is designated 5¾" × 10" (14.6 cm × 25.4 cm) which means that it has a diameter of 10" and the cap-pin separation is 5¾". Such a disc has

$$C_I \approx 50 \text{ pF} = 50 \ \mu\mu\text{F} = 50 \times 10^{-12} \text{ farad}$$

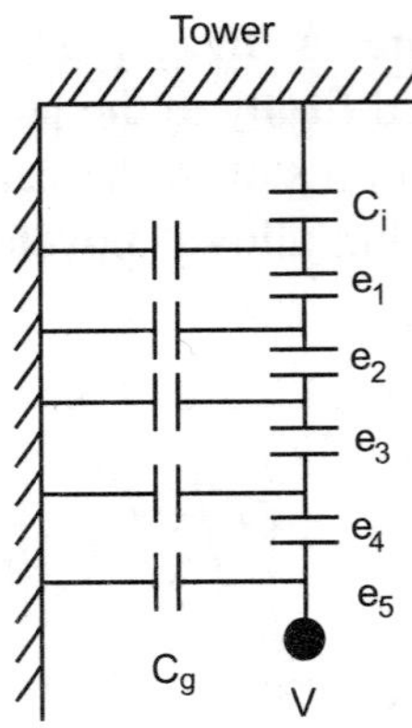

Fig. 6.7: Capacitance network of unshielded insulator string. 5-disc unit.

When a string is made up of n such discs and suspended from a tower, each cap and pin joint at the high voltage has a capacitance to the grounded tower structure. Let this be denoted by C_g. The resulting capacitance network is shown in Fig. 6.7. The voltage distribution becomes highly non-uniform, with the line-end disc (the bottom most in a suspension insulator string) taking a disproportionately large portion of the voltage. For uniform distribution, each disc is stressed equally to V/n. But this is far from true in the actual string. This distribution is calculated as follows:

For the sake of illustrating the procedure, let us first consider a string with 5 discs. Let $C_g/C_I = \alpha$, which is normally about 0.2. The current-flow in any capacitor is $I = \omega Ce$, where $\omega = 2\pi f$ and e = voltage across the capacitor. Consider the voltage to ground (tower structure) of the 5 junctions and denote them as e_1, e_2, e_3, e_4 and $e_5 = V$, the full applied voltage across the string.

At Junction 1: Current flowing from Junction 2 towards Junction 1 is $\omega(e_2 - e_1)\, C_I$. Current flowing out of Junction 1 to ground is $\omega e_1 \cdot (C_I + C_g)$. Equating the two currents (and omitting ω), there is:

$$(e_2 - e_1)C_I = e_1(C_I + C_g)$$

yielding

$$e_2 = (2C_I + C_g)e_1/C_I$$

$$= (2 + \alpha)e_1 \qquad \text{.... (1)}$$

At Junction 2:

$$(e_3 - e_2)C_I = e_2\, C_g + (e_2 - e_1)C_I$$

$$\therefore \quad e_3 = [(2C_I + C_g)e_2 - e_1 C_I]/C_I$$

$$= (2 + \alpha)e_2 - e_1 \qquad \text{... (2)}$$

At Junction 3:

$$(e_4 - e_3)C_I = (e_3 - e_2)C_I + e_3 C_g$$

$$\therefore \quad e_4 = (2 + \alpha)e_3 - e_2 \qquad \text{... (3)}$$

At Junction 4:

$$e_5 = V = (2 + \alpha)\, e_4 - e_3 \qquad \text{... (4)}$$

There are now four equations for the four cap-pin voltages e_1 , e_2, e_3 and e_4. We can substitute the values of e_2, e_3, e_4 successively in terms of e_1 and obtain an equation relating e_1 and V in equation (4). Once e_1 is found in terms of a and V, the other 3 voltages can be easily calculated. Let us do this for $V = 100$ and $\alpha = 0.2$.

Then,

$$\begin{aligned} 100 &= 2.2e_4 - e_3 \\ &= 2.2(2.2e_3 - e_2) - e_3 \\ &= 3.84e_3 - 2.2e_2 \\ &= 3.84(2.2e_2 - e_1) - 2.2e_2 \\ &= 6.248e_2 - 3.84e_1 \\ &= 6.248 \times 2.2e_1 - 3.84e_1 \\ &= 9.9056e_1. \end{aligned}$$

Therefore, we obtain the junction voltages and disc voltages as follows:

Junction Voltages					Disc Voltages		
e_1	=	100/9.9056	=	10.095	e_1	=	10.1
e_2	=	2.2 e_1	=	22.2	$e_2 - e_1$	=	12.1
e_3	=	2.2 $e_2 - e_1$	=	38.766	$e_3 - e_2$	=	16.6
e_4	=	2.2 $e_3 - e_2$	=	63.08	$e_4 - e_3$	=	24.2
e_5	=	2.2 $e_4 - e_3$	=	100, which checks.	$e_5 - e_4$	=	37
					Total	=	**100**

We now make the most important observation that the line-side insulator disc is stressed to 37% of the line-to-ground voltage, whereas the ground-side one is stressed only to 10% of full voltage. This sort of voltage distribution is very wasteful since the discs are identical and are designed to take equal voltage stresses. The voltage distribution is highly uneven. Had the voltage distribution been uniform, each of the 5 discs should be stressed to 20% of full voltage across the string.

In a string with more discs, the voltage distribution is even worse. For the same value of $\alpha = 0.2$, the reader should calculate the voltage distribution for a 12-disc string and obtain the following values:

Voltage to Ground	e_1	e_2	e_3	e_4	e_5	e_6	e_7	e_8	e_9	e_{10}	e_{11}	$e_{12} = V$
	0.45	0.98	1.72	2.8	4.43	6.5	10.9	16.95	26.4	41.2	64.2	100
Disc Vol.	0.45	0.53	0.74	1.08	1.63	2.1	4.4	6.05	9.45	14.8	23	35.8

Notice that the line-side disc is taking 35.8% of full voltage while the ground-side disc is taking practically no voltage (0.45%). If the voltage distribution were to be uniform, the voltage across each disc would have been equal to 100/12 = 8.33%.

As an indication of the number of discs used normally at different voltage levels of transmission in practice, the following Table shows the practice:

System kV	66	132	220	345	400	500	750	±400 kV
Maximum kV	72.6	145	245	362	420	550	765	D.C.
No. of Discs	5	9	(13–20) (Av. 15)	18	22	26	28 (6¾x11")	24 (6¾x12–5/8")

Q.13. What is a "grading ring", and how does it improve the voltage distribution across the discs of an insulator string?

Ans. A close examination for the reason why the line-end insulator disc experiences a very high proportion of voltage will reveal that this disc carries the charging current of all other units including the current through C_g of one cap-and-pin joint to tower. If, therefore, this charging current of other discs can be diverted from the discs closer to the line end, then the currents will be more nearly equal in all the discs, resulting in more equal voltage distribution among the insulator discs.

Such a scheme is carried out by connecting a metallic ring in the form of a large doughnut to the line and extending it to nearly half way along the height of the insulator string, as shown in Fig. 6.8. This is called a grading ring or shield. It offers additional capacitances shown as C_s (for shield) which will supply currents to the discs directly from the line without all the charging current having to flow through the line-end disc.

The resulting voltage distribution for such a shielded insulator string with guard ring is calculated as follows:

At Junction 1: Currents flowing towards the junction $= \omega C_I(e_2 - e_1) + \omega C_s(V - e_1)$.

Current flowing out of the junction $= \omega(C_I + C_g)e_1$.

$\therefore \quad C_I(e_2 - e_1) + C_s(V - e_1) = (C_I + C_g)e_1$, which yields

$$e_2 = [2 + (C_g + C_s)/C_I]e_1 - (C_s/C_I)V.$$

Let, as before, $\alpha = C_g/C_I$, and let $C_s/C_I = \beta$.

Then,
$$e_2 = (2 + \alpha + \beta)e_1 - \beta V. \quad \text{... (1)}$$

At Junction 2: $(e_3 - e_2)C_I + (V - e_2)C_s = C_g\, e_2 + C_I(e_2 - e_1)$, giving

$$e_3 = (2 + \alpha + \beta)e_2 - e_1 - \beta V. \quad \text{... (2)}$$

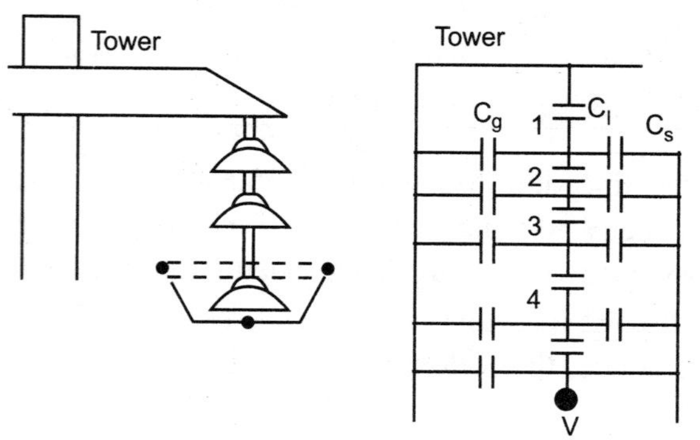

Fig. 6.8: Capacitance network of shielded insulator string.

At Junction 3: $(e_4 - e_3)C_I - (V - e_3)C_s = C_g\, e_3 + C_I\,(e_3 - e_2)$

or, $$e_4 = (2 + \alpha + \beta)e_3 - e_2 - \beta V \quad \text{... (3)}$$

At Junction 4: $(V - e_4)(C_I + C_s) = (e_4 - e_3)C_I + e_4 C_g$

or, $$V(1 + \beta) = (2 + \alpha + \beta)e_4 - e_3 \quad \text{... (4)}$$

If there are n discs, it is easily observed that the general equation will be

$$e_n = (2 + \alpha + \beta)e_{n-1} - e_{n-2} - \beta V.$$

By substituting successively the values of e_4, e_3, and e_2 in terms of V and e_1, we can finally obtain an equation connecting e_1 with V and the constants α and β. The entire algebra can be of course programmed on a Digital Computer where there are also special subroutines for solving such problems in related fields such as Heat Transfer, Surge Transfer, etc.

But we will solve equations (1) through (4) by long hand here by assuming values for $\alpha = 0.2$, $\beta = 0.1$, and $V = 100$, to show the improvement obtained in voltage distribution across discs by providing the shield. The resulting numerical equations are:

1. $e_2 = 2.3e_1 - 0.1 \times 100 = 2.3e_1 - 10$
2. $e_3 = 2.3e_2 - e_1 - 10 = 4.29e_1 - 10$
3. $e_4 = 2.3e_3 - e_2 - 10 = 7.567e_1 - 75.9$; and finally,
4. $100 \times 1.1 = 2.3(7.567e_1 - 75.9) - (4.29e_1 - 33)$,

or, $13.1141e_1 = 251.57$

$\therefore \quad e_1 = 19.48,\ e_2 = 34.8,\ e_3 = 50.97,\ e_4 = 71.5$

and $V = 100$, given.

These are the voltages to the grounded tower structure of the 5 junctions. Consequently, the voltages experienced by the discs will be as follows:

	With Guard Ring	Without Guard Ring
Disc 1:	$e_1 = 19.48$ %	10 %
Disc 2:	$e_2 - e_1 = 15.32$ %	12.1 %
Disc 3:	$e_3 - e_2 = 15.77$ %	16.6 %
Disc 4:	$e_4 - e_3 = 20.9$ %	24.2 %
Disc 5:	$V - e_4 = 28.5$ %	37 %
	Total = 100 %	Total = 100 %

Average is 20% of V across each disc for uniform voltage distribution.

We now observe clearly that the voltage distribution with the guard ring is more uniform or improved with the line-end disc experiencing 28.5% of applied voltage instead of 37% without the grading ring or shield offering a capacitance of $C_s = 10\%\ C_I$ to the cap-and-pin junctions of all discs. The discs at the ground end as we near the tower are also stressed as high as other discs or nearly so.

The reader should work out this problem for the 12-disc insulator with $\alpha = 0.2$ and $\beta = 0.1$, and observe the resulting voltage distribution as follows:

Junction	e_1	e_2	e_3	e_4	e_5	e_6	e_7	e_8	e_9	e_{10}	e_{11}	$e_{12} = V$
Voltage (Approx.)	14	22.5	27.5	31	33.5	35.5	38	39.5	45	54.5	70.5	100 %

Note: Because of high charging current supplied by shield at line potential, ground-side discs have high voltage.

Q.14. What is meant by "String Efficiency" and what are their values for the 5-disc insulator string considered before without the shield and with the shield?

Ans. We observed that some discs are carrying more than the average voltage as determined by e_{av} = Applied voltage/number of discs. In most cases, it is the line-end disc connected to the conductor at high voltage which is stressed more than any other disc in the string. Let this voltage be denoted as e_{max}. Then, it is clear that this disc must be selected so as to withstand e_{max}. Now, even though the discs may experience different voltages which are less than e_{max}, it is not a sound engineering practice or economical to use different types or sizes of discs but use identical discs which can each withstand e_{max}. Therefore, the withstand capability of the string is $(n \times e_{max})$. But the n-disc insulator string is being used in practice for the voltage V, the line-to-tower applied voltage. This waste of electrical strength is defined by a "string efficiency" by the relation:
String Efficiency,

$$\varepsilon = V/(n.e_{max}) = \text{Applied voltage}/n \times (\text{max. vol. of any disc in situ}).$$

The percentage string efficiency is: $\% \ \varepsilon = (V/n\ e_{max}) \times 100\%$. For the 5-disc insulator strings of Q. 12 and 13, we have, with $\alpha = 0.2$ and $\beta = 0.1$:

(*a*) Without guard ring: $e_{max} = 37$ for $n = 5$ and $V = 100$.

$$\therefore \qquad \% \ \varepsilon = (100/5 \times 37) \times 100 = 54\%.$$

The string is used only up to 54% of its possible dielectric withstand voltage.

(*b*) With guard ring: $e_{max} = 28.5$ for $V = 100$.

$$\therefore \qquad \% \ \varepsilon = (100/5 \times 28.5) \times 100 = 70\%.$$

Thus, the utilisation of the electric strength of the 5 discs is now 70%. Each disc can be selected on the basis of being able to withstand 28.5% of the applied voltage when the guard ring is provided, instead of 37% of the voltage. A simple aluminium doughnut or toroidal ring, usually 1 metre in diameter and 10 cm ring will be quite adequate, although the diameter is determined on the basis of the system voltage and its size varies vastly.

In practice, the shield capacitance C_s never extends all the way over all the discs and their cap-and-pin joints, but only part way. The calculation is not very difficult for such partial shielding, since the equations can be set up in a straightforward manner from simple electric circuit theory as discussed earlier.

For the 12-disc insulator-string without guard ring in Question 12, e_{max} = 35.8% and n = 12 with V = 100. The string efficiency is:

$$\% \ \varepsilon = (100/12 \times 35.8) \times 100 = 23.28\%,$$

which is extremely poor.

The reader should calculate the string efficiency with a guard ring or shield for $\alpha = C_g/C_I = 0.2$ and $\beta = C_s/C_I = 0.1$. The answer turns out to be:

% String Efficiency, $\varepsilon = (100/12 \times 29.4) \times 100 = 28.3\%$, which is still not a very great improvement over 23.28%. This is the design problem in selecting an n-disc insulator and a shield.

[Disc Voltages: 14, 8.5, 5, 3.5, 2.5, 2, 2.5, 1.5, 5.5, 9.6, 16, 29.4 %]

❑❑

CHAPTER 7

THREE-PHASE CALCULATIONS IN HIGH VOLTAGE NETWORKS: STEADY-STATE AND TRANSIENT-STATE (REVIEW)

Q.1. What is the historical development of 3-phase a.c. and high voltage power transmission and distribution?

Ans. The phenomenal and revolutionary progress in human welfare through large-scale industrial development is entirely due to the discovery and use of the 3-phase a.c. system in 1880 and since then. Its discoverer was the great Yugoslavian electrical genius, Nicola Tesla. His discovery was preceded by the design of the single-phase transformer by the German engineer Dobrovolsky, which made h.v. transmission possible over long distances as well as interconnection of a system with various voltage levels. Prior to their work, the only known method of generation, transmission from central station to loads in cities, and utilisation of electrical power was by d.c., which is associated with electrical giants like Thomas Edison, Swann and Thomson-Houston (who founded the General Electric Company in the U.S.A.)

But d.c. has very serious limitations as it requires a rotating commutator in the dynamo with copper segments insulated with mica sheets. As the current commutated by the brushes becomes larger, sparking is prohibitive. However, the rotating armature winding generates alternating voltage in itself, one cycle corresponding to two pole pitches. Tesla realised this and designed a 3-phase motor that would work directly with the generated a.c. and had self-starting capability.

Furthermore, the most serious limitation of d.c. is that it cannot be transformed to voltage levels different from the generated value. Such a transformation is possible only through a.c. transformers using a good magnetic core of high permeability. When the 3-phase system, a motor (the 3-phase induction motor) and the transformer became available (through the scientific effort of another electrical genius, Charles Steinmetz, who placed all machine design principles on a scientific and mathematical basis), real progress in power generation, transmission over long distances and induction motors to turn the wheels of large-scale industry became an accomplished reality. It was George Westinghouse in Pennsylvania, U.S.A., who adopted Tesla's ideas in constructing major 3-phase generating station in the world at Niagara Falls and transmitting the power at 33 kV to nearby Buffalo City and then extending the line to Pittsburgh in the south to feed the giant steel mills at higher voltages.

At the present time, 120 years later, single generators have increased in size to 1500 MVA, transformers to similar rating, transmission voltage to 1500 kV, and motors in excess of 50,000 HP each. All this has become possible entirely because of existence of the 3-phase high voltage generation, transmission and distribution system.

Q.2. What are the basics of a 3-phase system?

Ans. Inside a generator, 2 pole-pitches correspond to one full cycle of alternation or 360°of the sine-wave of voltage generated, the shape of the pole shoe controlling the distribution of the flux density over a pole pitch. If the winding over two pole pitches is split into 3 distinct and equal parts, the voltage output of each section is out of phase with the other two by 120°. These three voltages are directly led out of the ends to an external circuit as shown in Fig. 7.1. One terminal of each winding section is connected together with one from each of the other two sections to form a neutral or star-point *n*, and the three phase-voltages are denoted by the phasors $V_a = V\underline{|0°}$, $V_b = V\underline{|-120°}$, and $V_c = V\underline{|+120°}$, where V = r.m.s. value of the generated phase voltage, which is also called the e.m.f. (electromotive force).

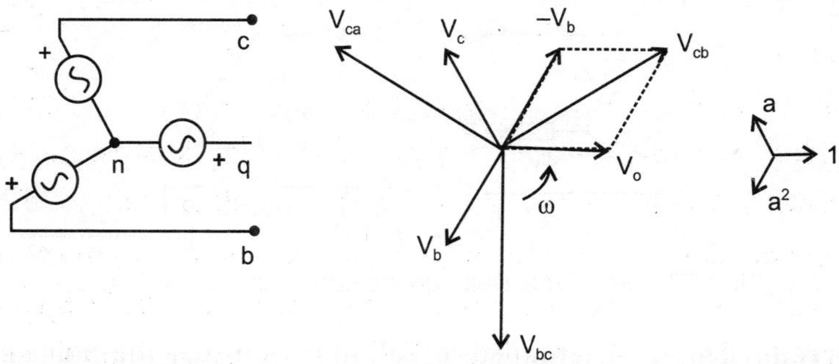

Fig. 7.1: 3-phase source, and its line-to-neutral and line-to-line voltage relations. Unit vectors $1\underline{|0°}$, $a = 1\underline{|120°}$, and $a^2 = 1\underline{|240°} = 1\underline{|-120°}$.

The resulting line-to-line voltages are written as follows:

$$V_{ab} = V_a - V_b = V\underline{|0°} - V\underline{|-120°}$$
$$= V(1 + j\,0) - V(-0.5 - j\,0.866)$$

$$= V(1.5 + j0.866)$$
$$= \{1.5^2 + 0.866^2\}\ V/\tan^{-1}(0.866/1.5)$$
$$= \sqrt{3}\ V\lfloor 30^\circ .$$

Similarly, $V_{bc} = \sqrt{3}\ V\lfloor -90^\circ$,

and $V_{ca} = \sqrt{3}\ V\lfloor -210^\circ = \sqrt{3}\ V\lfloor +150^\circ$

The magnitude of line-to-line voltage is $\sqrt{3}$ times the phase voltage which is the voltage between any line and neutral. By convention, the three line-to-line voltages form a balanced 3-phase voltage system leading the line-to-neutral system by 30°. Transmission lines and loads can be connected to the three lines and the neutral, if necessary.

Q.3. What are the properties of Wye-connected and Delta-connected loads, and the transformation properties of Wye into Delta, and vice versa?

Ans. The system shown in Question 2 and Fig. 7.1 is called a *Y*-connected source with a neutral or star point. Figure 7.2 shows a *Y*-connected system of 3 equal impedances and a Delta-connected system across the same 3 lines, *a*, *b*, *c*. The line-to-line voltages can be written as $V_1\lfloor 0^\circ$, $V_1\lfloor -120^\circ$, and $V_1\lfloor 120^\circ$ for convenience, as long as they have equal magnitudes and are different in phase by 120°. In both systems, the line currents are I_a, I_b, I_c. For balanced loads, the three currents also form a balanced set of three-phase currents of equal magnitude and displaced 120° from each other.

(*a*) In the *Y*-connected system, we can observe the following properties:
Line-to-neutral voltage = Line-to-line voltage/√3;
Phase current in any impedance = line current.
$I_a + I_b + I_c = 0 \cdot I_a \cdot Z_Y = V_{an};\ I_b \cdot Z_Y = V_{bn};\ I_c \cdot Z_Y = V_{cn}$.

(*b*) In the Delta-connected system, there is no neutral point. Phase voltage across any impedance = line-to-line voltage.

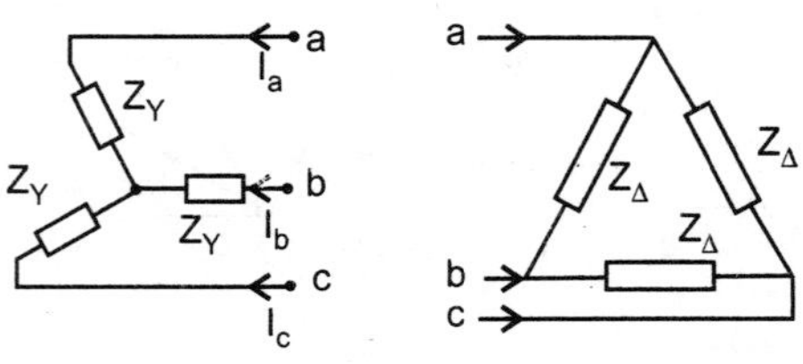

Fig. 7.2: Wye- and Delta-connected loads. $Z_Y = Z_D/3 \cdot L$.

Phase current: This is determined as follows: Consider the point *a*. Current coming towards $a = I_a$.

Current going away from $a = (V_{ac} + V_{ab})/Z_D$.

Taking $V_{ab} = V\lfloor 30^\circ$,

$$V_{bc} = V\lfloor -90^\circ$$

and $V_{ca} = V\angle 150°$,

we have $V_{ac} = -V_{ca} = V\angle -30°$.

$\therefore \quad V_{ac} + V_{ab} = V(0.866 - j\,0.5) + V(0.866 + j\,0.5) = \sqrt{3}\,V\angle 0°$.

Thus, $I_a = \sqrt{3}\,V\angle 0° / Z_D$.

But, V/Z_D = current flowing in the Delta-connected phase-impedance and is called the phase current. Therefore, for the Delta:

line current = $\sqrt{3}$ × phase current.

The total power supplied by source or consumed by load is:

$P_{3\text{-ph}}$ = 3 × phase voltage × phase current × phase power factor.

(*a*) In the *Y*-connected system:

$P_{3\text{-ph}}$ = 3 × (1/$\sqrt{3}$) × (line-to-line voltage) × (line current) × (phase p.f.)

$= \sqrt{3}\ V_{l-l}\, I_l \cos \varphi_{ph}$.

(*b*) In the Delta-connected system:

$P_{3\text{-ph}}$ = 3 × (phase voltage) × (phase current) × (phase power factor)

= 3 × (line-to-line voltage) × (1/$\sqrt{3}$) × (line current) × (phase p.f.)

$= \sqrt{3}\ V_{l-l}\, I_l \cos \varphi_{ph}$.

Note that the power factor is always to be reckoned as that of the phase impedance. The expressions for power for both systems are the same.

Delta-Wye Transformation of Impedances

When a Delta-connected system of 3 impedances is to be transformed into an equivalent *Y*-connected system, it means that the line quantities in both systems remain unchanged. Such a transformation does not disturb the voltage and current conditions external to the load terminals.

In Z_D, voltage = line-to-line voltage, and current = line current/$\sqrt{3}$.

In Z_Y, voltage = line-to-neutral voltage = line-to-line voltage/$\sqrt{3}$;

$\therefore$ Z_D/Z_Y = [(line-to-voltage)/(line current/$\sqrt{3}$)]/ [(line-to-line voltage)/$\sqrt{3}$/ (line current)]

= 3.

Thus, if three impedances Z_D are connected in Delta, an equivalent *Y*-connected set of impedances Z_Y will have the value $Z_Y = Z_D/3$. Conversely, if 3 equal impedances Z_Y are connected in *Y* or star, they are equivalent to three Delta-connected impedances $Z_D = 3Z_Y$ connected in Delta or Loop.

Example: A 3-phase balanced system of Delta-connected impedances across 132 kV lines consume a total power of 100 MW at 0.8 lagging power factor (MVA = 125). Calculate the following:

(*a*) The line currents I_a, I_b and I_c. Also, the reactive power Q consumed.

(*b*) The phase currents through each impedance.

(*c*) An equivalent Y-connected system of impedances drawing the same power at the same power factor across the same lines.

Answer or Solution:

(*a*)
$$P = \sqrt{3}\, V_1\, I_l \cos\varphi, \text{ so that } 100 \times 10^3 = \sqrt{3} \times 132 \times 0.8 \times I_l,$$

given
$$I_l = 10^5/(\sqrt{3} \times 132 \times 0.8) = 0.5467 \times 10^3 \text{ A} = 0.5467 \text{ kA}$$

$$Q = \sqrt{3}\, V_l\, I_l \sin\varphi = P \tan\varphi = 100 \times 0.6/0.8 = 75 \text{ MVAR}$$

$$W = P + jQ = 125\angle 37^\circ = 100 + j\,75 \text{ MVA}$$

(Note that for a lagging power-factor load, the reactive power occurs with a positive sign, by convention.)

(*b*) Phase current = $I_l/\sqrt{3}$ = 0.3157 kA.

(*c*) Phase impedance $Z_D = \{132 \text{ kV}/0.3157 \text{ kA}\}\angle \cos^{-1} 0.8 = 418.2\angle 36.87^\circ$ ohm.

Since the power factor is lagging, Z_D has inductive part and the internal angle of the impedance is positive.

$$Z_Y = Z_D/3 = 139.4\angle 36.87^\circ$$

As a check for Z_Y: Line-to-neutral voltage = $132/\sqrt{3}$ kV = 76.21 kV.

Phase current = line current = 0.5467 kA

$$\therefore \quad Z_Y = (76.2 \text{ kV}/0.5467 \text{ kA})\angle \cos^{-1} 0.8 = 139.4\angle 36.87^\circ$$

Q.4. A single-phase transformer has a load impedance on the secondry side which draws a power of 100/3 MW at 0.8 p.f. lag across 76.2 kV. It is energised on the primary side by a voltage of 7.62 kV = $13.2/\sqrt{3}$.

Calculate:

(*a*) The currents flowing in the two windings;

(*b*) The impedance of load as seen from the primary-side terminals.

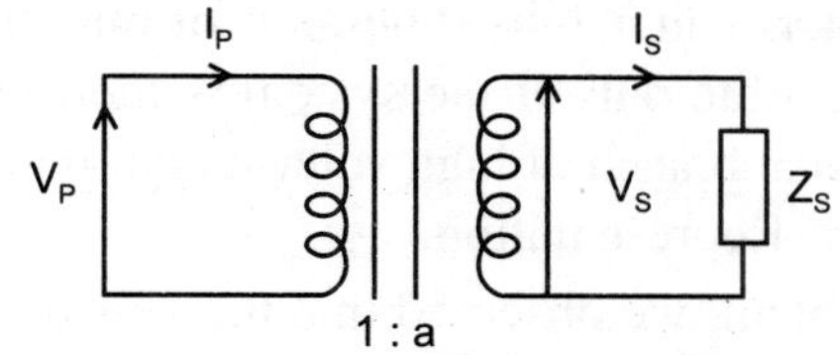

Fig. 7.3: Transformer with turns ratio 1 : a.

Ans. See Fig. 7.3.

$P = VI \cos \varphi;\ I = P/(V \cos \varphi) = W/V$

(*a*) $W = 100/(3 \times 0.8) = 125/3 = 41.67$ MVA

$\therefore I_s = (41.67 \times 10^3 \text{ kVA})/(13.2/\sqrt{3} \text{ kV}) = 0.5467 \text{ kA} = 546.7$ amp

Turns ratio = 76.2 kV/7.62 kV = 10/1

$I_P = 10\, I_S = 5.467$ kA

(*b*) The load impedance on the secondary side

$= Z_L = Z_S = V_S/I_S = \{132 \text{ kV} \underline{|\cos^{-1} 0.8} / \sqrt{3} \times 0.5467 \text{ kA}\}$

$= 139.4 \underline{|36.87°}$

$= 139.4(0.8 + j\ 0.6)$

$= 111.52 + j\, 83.64$ ohm

Impedance of load looking from the primary terminals:

$Z_P = V_P/I_P$

$= 13.2 \text{ kV}/(\sqrt{3} \times 5.467 \text{ kA})$

$= 1.394\ \Omega$ at 0.8 p.f lag or 36.87°.

$Z_L/Z_P = 100$ = square of turns ratio or voltage ratio, being higher on the higher-voltage side, because of higher voltage and lower current than the secondary side.

In general, if the turns ratio is written as $V_S/V_P = 1/a$, then the following relationships are valid for the two sides of the transformer:

$V_S/V_P = a/1;$

$I_S/I_P = 1/a;$

$Z_S/Z_P = a^2/1;$

$Z_P/Z_S = 1/a^2$. In the problem at hand: $a = 10$.

$\therefore$ $V_S/V_P = 10;$

$I_S/I_P = 0.1;$

$Z_S/Z_P = 100.$

Q.5. What are "base quantities" and how are they used in a high voltage network?

Ans. A high voltage a.c. network is a vast interconnection of generating stations, transmission lines and loads. They are all interconnected by transformers at very different voltage levels. It is impossible to deal with such vast interconnections to solve for operating conditions by using the actual numerical values of voltage, current, loads and impedances.

Modern Digital Computers can handle thousands of inter-connected points (called Buses) in one stage. This is possible only if the system is represented properly. This is in itself a very challenging and vital branch of high voltage high-power system analysis called Load-Flow Study and System Representation.

In the previous problem, we observed that the voltage, current and impedance values on the two sides of the transformer vary over wide range in numerical values, but the one quantity which remains constant on both sides is the power (real power P, reactive power Q, and the volt-ampere power $W = P + j\,Q$). The voltage, current, and impedance values on the primary side adjust themselves to the given load and the secondary voltage across the load (as determined by the transformer turns ratio).

We define "base" quantities as follows, in a preliminary manner:

1. Base kVA = rated load kVA at its power-factor angle;
2. Base kV = rated load kV;
3. Base current = Base kVA/Base kV, ampere;
4. Base impedance (in ohms)
   = Base kV/Base current in kA
   = Base kV/(Base MVA/Base kV)
   = (Base kV)^2/Base MVA
   = (Base voltage)^2/Base volt-amps

Thus, any two quantities can be chosen as bases and all other quantities can be derived from these. Here, we have written current and impedance bases in terms of base voltage and base power. This is the usual way of choosing bases, but we could equally well have chosen other combinations. But only two quantities have to be chosen and the others will follow automatically out of these two chosen base quantities.

In the previous example, let us choose:

$$\begin{aligned}\text{Base kVA} &= \text{Base real power/Power factor}\\ &= (100/3) \times 10^3/0.8, \text{kVA}\\ &= 41.67 \text{ MVA}\end{aligned}$$

Base voltage, $V_{bS} = 132/\sqrt{3}$ kV, on the high voltage secondary side.

Then, base current $I_{bS} = 41.67 \times 10^3$ kVA/$(132/\sqrt{3})$ kV

$= 546.78$ amps $= 0.5468$ kA

Note immediately that the base current cannot be chosen arbitrarily, but will depend upon the previous two chosen bases. This is crucial to this game.

Now, base impedance on the secondary side,

$$\begin{aligned}Z_{bS} &= V_{bS}/I_{bS}\\ &= (132/\sqrt{3})/0.5468\\ &= 139.4 \text{ ohms}\end{aligned}$$

Also,

$$\begin{aligned}Z_{bS} &= (\text{Base kV})^2/\text{Base MVA}\\ &= (132/\sqrt{3})^2/41.67\\ &= 139.4 \text{ ohms}\end{aligned}$$

The above quantities refer to the high voltage secondary side. Now, consider the primary or low-voltage side with given turns ratio of 1/10. The base power must be equal on both sides of the transformer.

$\therefore$ Base kVA = $W_{bP} = W_{bS} = 41.67 \times 10^3$ kVA, which is 41.67 MVA.

Base kV = $V_{bP} = V_{bS}/10 = 13.2/\sqrt{3}$ kV = 7.62 kV.

Consequently, base current,

$$I_{bP} = 41.67 \times 10^3/7.62 = 5468 \text{ amps} = 5.468 \text{ kA}$$

Note that $I_{bP} = 10\, I_{bS}$, which could have been used directly. Base impedance on primary low-voltage side,

$$\begin{aligned} Z_{bP} &= V_{bP}/I_{bP} \\ &= 7.62 \text{ kV}/5.468 \text{ kA} \\ &= 1.394 \text{ ohm} \end{aligned}$$

Also,

$$\begin{aligned} Z_{bP} &= Z_{bS}/10^2 \\ &= 139.4/100 \\ &= 1.394 \text{ ohm, which could have been used directly.} \end{aligned}$$

Therefore, the base values on the primary side depend solely on the chosen base values on the secondary side and the turns ratio. If therefore there are hundreds of transformers in an interconnected system, and the turns ratios of all transformers are known, we are permitted to choose the base kVA for the system in an arbitrary manner based on convenience, and the base kV and kVA in only one part of the system or network. The base kV in every other part of the network is fixed by the turns ratios of transformers, while the base kVA remains the same in every part of the system. This also determines or fixes the base currents and impedances in all parts of the network. They cannot be chosen arbitrarily in the same manner as choosing the base kVA or MVA and the base kV, after selecting these two in some part of the system.

Q.6. What is meant by "per unit" values of power, voltage, current and impedance?

Ans. We observed in the simple two-winding transformer problem that the numbers to be handled differed widely in magnitude, and some of them are higher and others are lower by some magnitudes. It is very difficult to handle such variations when there are hundreds of transformers at different voltage levels ranging from say 11 kV to 750 kV in the interconnected system. The immensity of numerical values is incredible and it is quite unnecessary too if we observe the following properties coming out of the use of "per unit" values.

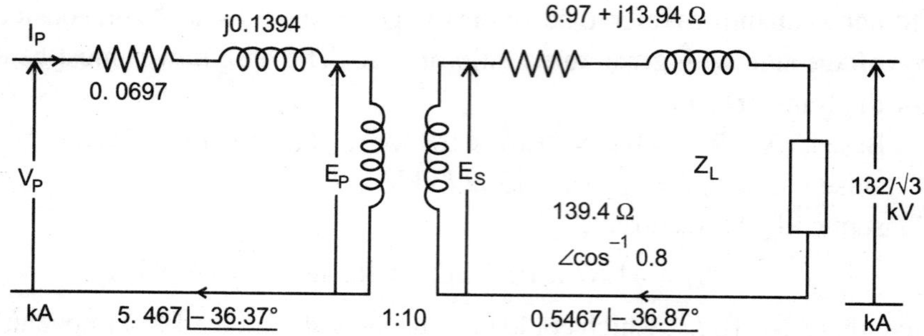

Fig. 7.4: Equivalent circuit of transformer for Question 6. Actual values of all quantities.

Let us consider the previous transformer problem but include representative voltage losses by introducing series impedances on both sides, to be more realistic. The practical equivalent circuit is redrawn in Fig. 7.4.

There are four voltages shown which are usually called:

V_S = terminal voltage of secondary, or the load voltage (V_L also);

E_S = induced voltage in the secondary caused by the core flux;

E_P = induced voltage in the primary due to the core flux = E_S/a;

V_P = voltage applied by the source of power to the primary, which is usually held constant by the large system behind the transformer, irrespective of load changes on this transformer.

From elementary circuit calculations, we obtain, for Fig. 7.4:

$$E_S = V_S + (r_S + j\,x_S)\,I_S\,(\cos\varphi - j\sin\varphi)$$
$$= (76.2 + j\,0) + (6.97 + j\,13.94) \times 0.5467\,(0.8 - j\,0.6)$$
$$= 83.82 + j\,3.81 \text{ kV}$$

$$\therefore \quad E_P = 0.1\,E_S = 8.382 + j\,0.381 \text{ kV}$$
$$I_S = 0.5467\,(0.8 - j\,0.6) \text{ kA}; \; I_P = 10\,I$$
$$I_S = 4.736 - j\,3.28 \text{ kA}$$
$$\therefore \quad V_P = 8.382 + j\,0.381 + (0.0697 + j\,0.1394) \times 5.467\,(0.8 - j\,0.6)$$
$$= 9.144 + j\,0.762 = 9.1757\angle 4.76^\circ \text{ kV}$$

Voltage drop in series impedance of secondary side = 7.62 + j 3.81 kV;

Voltage drop in series impedance of primary side = 0.762 + j 0.381 kV.

We observe that in these few calculations, we had to handle numbers of widely differing magnitudes. Now, we can ease our calculations considerably by adopting the per-unit system which is as follows:

We first choose two base quantities. For now, let us choose 132/√3 (= 76.2) kV and the power of 41.67 MVA on the secondary side, which so happen to be the rated load voltage and power. But this restriction will be removed in the next question which is more general and universal procedure, and which is the usual procedure adopted in using per-unit values in calculations and system representations.

Then, the per-unit value of load voltage = actual voltage/base voltage = 1. Since we are taking the load voltage as the reference phasor, we write this voltage as: P.U. value of load voltage, $V_S = 1\underline{|0°}$.

The per-unit value of power = actual power/base power = 1.

From the base voltage and base power, we obtain the base current as

$$\begin{aligned} I_{bS} &= \text{base load/base voltage} \\ &= 41.67 \text{ MVA}/76.2 \text{ kV} \\ &= 0.5467 \text{ kA} \end{aligned}$$

∴ P.U. value of secondary current,

$$\begin{aligned} I_{bS} &= 0.5467(0.8 - j\,0.6)/0.5467 \\ &= 0.8 - j\,0.6 \\ &= 1\underline{|-36.87°} \end{aligned}$$

Base impedance on secondary side,

$$\begin{aligned} Z_{bS} &= \text{base kV/base kA} \\ &= 76.2/0.5467 \\ &= 139.4 \text{ ohms} \end{aligned}$$

We now convert the series impedance into per-unit values as follows:

P.U. value of $r_S + j\,x_S = (6.97 + j\,13.94)/139.4 = 0.05 + j\,0.1$ p.u.

On the primary side, the base quantities depend on the turns ratio.

These are:

$$\begin{aligned} V_{bP} &= 0.1 \times 76.2 \text{ kV} \\ W_{bP} &= 41.67 \text{ MVA} \\ a &= 10 \\ I_{bS} &= 5.467 \text{ kA} \end{aligned}$$

and $\quad Z_{bP} = Z_{bS}/a^2 = 1.394$ ohms

The corresponding per-unit values of all quantities referred to the above base quantities will be:

$$I_P = 5.467\underline{|-36.87°}/5.467 = 1\underline{|-36.87°} \text{ p.u.}$$

series impedance, $r_P + j\,x_P = (0.0697 + j\,0.1394)/1.394 = 0.05 + j\,0.1$ p.u.

The equivalent circuit for the entire loaded transformer with the series impedances included is shown in Fig. 7.5. The transformer is now shown as having a turns ratio of 1:1 and not the actual value of turns ratio of 1 : 10. Then, we can combine the series impedances into one impedance of value $(0.1 + j\,0.2)$. The current flowing is $(0.8 - j\,0.6)$ p.u., while the terminal voltage is $V_S = 1\underline{|0°}$ p.u. The applied primary voltage is:

$$V_P = 1\angle 0^\circ + (0.1 + j\,0.2)(0.8 - j\,0.6)$$
$$= 1.2 + j\,0.1$$
$$= 1.204\angle 4.75^\circ \text{ p.u.}$$

The power input is:

$$W_P = V_P \cdot I_P{}^*$$
$$= (1.2 + j\,0.1)(0.8 + j\,0.6)$$
$$= 1.204\angle 41.63^\circ \text{ p.u.}$$
$$= 0.9 + j\,0.8 \text{ MVA}$$
$$= 1.204\angle 41.63^\circ \text{ p.u.}$$

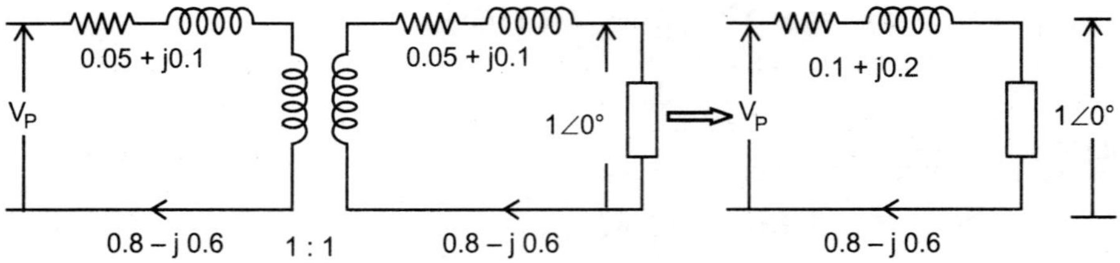

Fig. 7.5: Equivalent circuit of transformer. Per unit values on bases of $132/\sqrt{3}$ kV and 41.67 MVA (rated values).

In order to convert the per-unit values of primary voltage and input power to actual values, we only have to multiply the p.u. values by the chosen base values at the beginning,

thus: actual $V_P = 1.204 \times 76.2$

$= 91.76$ kV leading V_S by the angle 4.75°

and the input power = $(0.9 + j\ 0.8) \times 41.67 = 37.5$ MW $+ j$ 33.33 MVAR

The load power = $(0.8 + j\ 0.6)$ p.u. = 33.33 MW $+ j$ 25 MVA

∴ Losses in the series impedance amount to $(0.1 + j\,0.2)$ p.u.

$= 4.17$ MW$+ j$ 8.33 MVAR

So, % efficiency, ε = 100 × output MW power/ input MW power

$= 100 \times 33.33/37.5 = 89\%$

$$\% \text{ regulation} = \frac{\text{Rise in voltage when load is thrown off}}{\text{Primary input voltage (which is constant)}} \times 100$$

$= 100 \times (1.204 - 1.0)/1.204 = 16.94\%$

Q.7. Repeat the previous problem by choosing any other set of base quantities. What will be the equivalent circuit, efficiency and regulation on bases of 100 kV and 50 MVA on the secondary side?

Ans. In a vast interconnected system, it is the more usual practice to choose very convenient values for the base quantities instead of restricting to the rated values at any one particular part of the network. Whatever the choice of base values, the final result will turn out to be the same for the actual magnitudes of all quantities, as will be seen in this Question.

Secondary side: Base kV = 100 kV

Base MVA = 50 MVA

∴ P.U. value of load voltage,

V_S = actual voltage/base voltage

= 76.2/100

= $0.762\angle 0°$

P.U. load = $41.67(0.8 + j\,0.6)/50$

= $0.667 + j\,0.5$ p.u.

P.U. current = $(0.667 - j\,0.5)/0.762$

= $0.875 - j\,0.656$

= $1.09375\angle -36.87°$

Note that the base current = 50 MVA/100 kV

= 0.5 kA

Actual current = 0.5467 kA at 0.8 p.f. lag

= $0.5467\angle -36.87°$.

∴ The per-unit value of current is also found from

$0.5467\angle -36.87°/0.5 = 1.09375\angle -36.87°$.

Base impedance = kV^2/MVA

= $100^2/50$

= 200 ohms

∴ P.U. value of series impedance

= $(6.97 + j\,13.94)/200$

= $0.03485 + j\,0.0697$

Primary side: Base kV = 10 kV as determined by actual turns ratio.

Base MVA = 50 MVA

This stays constant throughout the network.

∴ Base impedance = $10^2/50 = 2$ ohm. (Also, $Z_{bP} = Z_{bS}/a^2 = 200/100$).

∴ P.U. value of series impedance

= $(0.0697 + j\,0.1394)/2$

= $0.03485 + j\,0.0697$

[Note one important property here: The P.U. values of series impedances on the secondary and primary sides of transformers are equal, even though the voltages, currents and actual impedance values may differ widely].

The resulting equivalent circuit on base quantities of 100 kV and 50 MVA chosen on the secondary side now becomes the circuit shown in Fig. 7.6. The primary applied voltage in p.u. based on the chosen bases is:

$$V_P = 0.762 + j\,0 + (0.0697 + j\,0.1394) \times (0.875 - j\,0.656) = 0.9176\angle 4.77°$$

Since the base voltage is 10 kV, actual voltage = 9.176 kV, and leads the load voltage by 4.77°, which is the same as were obtained on bases of 132/√3 kV and 41.67 MVA chosen previously in Question 6.

%reg, $r = 100 \times (0.9176 - 0.762)/0.9176 = 16.96\%$

%ε : Load on sec. side = $0.667 + j\,0.5$ p.u.

Input power = input volt × complex conjugate of input current

$$= (0.9144 + j\,0.0763)(0.875 + j\,0.656)$$
$$= 0.75 + j\,0.6668.$$

Also,

$$0.9176\angle 4.77° \times 1.0937\angle 36.87°$$
$$= 1.00365/41.64°$$
$$= 0.75 + j\,0.6668 \text{ p.u.}$$

In this calculation, the real power consumed by load = 0.667 p.u., and the real power drawn from the source on the primary side = 0.75 p.u.

∴ % efficiency, ε = 100 × 0.667/0.75 = 89%, which is the same as in Question 6.

Q.8. Give a summary of the procedure for solving electric networks with transformers for a general network by using per-unit method.

Ans. 1. Draw the entire circuit or network showing the power and voltage sources, series and shunt impedances, turns ratio of all transformers, the loads with their power factors, and the input voltage. Usually this is drawn for one phase only in a balanced 3-phase system and is known as a single-line diagram;

2. Select any two quantities related to each other in one part of the network on which all remaining quantities depend. These form the base quantities. The most usual base quantities one would select are the base kV and base MVA. These choices need not have anything to do with rated values of any equipment given in the problem, but are based on convenience.

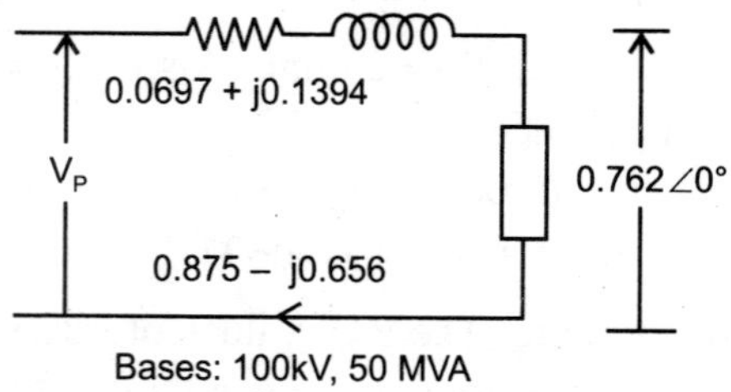

Fig. 7.6: Equivalent circuit of transformer in Figs. 7.4 and 7.5. Base values of 100 kV and 50 MVA.

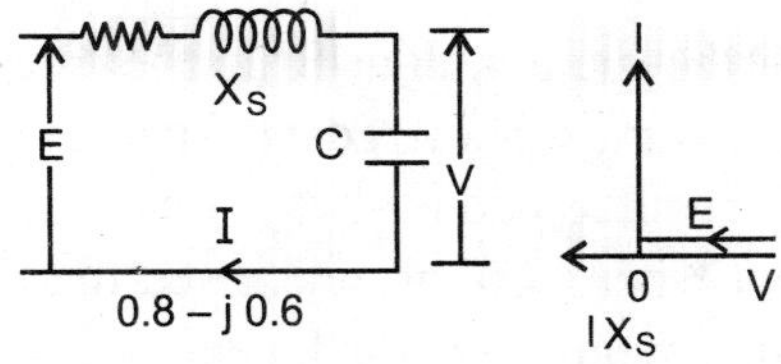

Fig. 7.7: Ferranti effect on an open cable or line.

3. Divide all voltages, powers, currents and impedances by these base quantities on their respective sides in the network and obtain their per-unit (p.u.) values. If there are transformers, fixing the base quantities on one side will automatically determine the base quantities in all other parts of the connected network, as governed by the turns ratios of all transformers.
4. Redraw the network replacing all components by their per-unit values. Phase angles must be preserved as in the original circuit or network. Where the network is extremely large, it is usual to neglect all series resistances but only preserve inductive and capacitive reactances. This amounts to omitting all series losses but only include voltage drops since resistive voltage drops are very small and add at right angles to the inductive voltage drops. But each problem must be examined on its merits.
5. Solve the single-line equivalent circuit to determine any quantities which are to be evaluated, such as voltages at required points of interest, power losses, efficiencies, regulation, source input power, etc., all in per-unit values.
6. Convert to actual values by multiplying the p.u. values by the base values assumed initially for the problem at every part of the network. This calculation does not apply to efficiency and regulation.

Q.9. What is meant by "Ferranti Effect" cables and overhead in a.c. transmission lines?

Ans. Consider a source with e.m.f. E connected to a load through a length of cable or a transmission line. The cable or line represents a capacitance which is shown as concentrated at the load end across the voltage $V\angle 0°$. It also has a series inductance or reactance shown lumped as X_S, see Fig. 7.7. Let the cable or line be unloaded to understand the phenomenon.

Referring to the phasor diagram, the phasor

OV = the voltage at the far end,

OI = the charging current which leads V by 90°,

since it is a pure capacitive current. The voltage drop in X_S will lead the current by 90° and is shown as OX. The source voltage is the phasor sum of V and IX_S. We observe that IX_S falls on the same line as load voltage V but is directed along the opposite direction.

∴ The source voltage $E = V - IX_S$, and its magnitude is less than the load-end voltage V. In other words, when an unloaded a.c. cable or overhead transmission line with capacitance to ground is energised by a source of e.m.f. E, the voltage at the far end is higher in magnitude than the source voltage. This rise in voltage is known as Ferranti Effect, which was first observed in the 1880s by the great Italian-British electrical engineer who laid the first underground cables in London and other large industrial cities in Great Britain. In d.c. cables, the far-end voltage from the source-end is always lower than the source voltage by the drop in the series resistance. But when Ferranti energised the first a.c. cable from a generating station and posted his assistant engineer at the far end or load end, the assistant was 'shocked' to observe that the load-end voltage was much higher than expected, from his experience of d.c. The story goes that this was prior to the invention of the telephone, so that he sent a workman on bicycle about 10 km away to report this curious and unexpected phenomenon and asked that the source voltage be lowered. The workman had to travel back and forth several times on his bicycle before finally the voltage at the load end was reduced to normal rated value in order not to damage the insulation. But then, when the cable took on load, and the source voltage was held at the previous no-load value, the far-end voltage dropped. So, the bicycle trips had to be repeated once again, etc.

Q.10. A source with voltage E_S is feeding a load at voltage $E_r\angle 0°$ through a series impedance $Z_S\angle\Theta$. Assume that the series impedance is purely inductive, neglecting resistance, so that $Z_S = j\,X_S$. The load has a power factor cos φ. What is the expression for the real power supplied by source in terms of E_S, E_r, X_S and the phase angle δ between E_S and E_r? Take E_r as reference vector: $E_r\angle 0°$ and $E_S\angle\delta$.

Ans. See Fig. 7.8 for circuit and phasor diagrams.

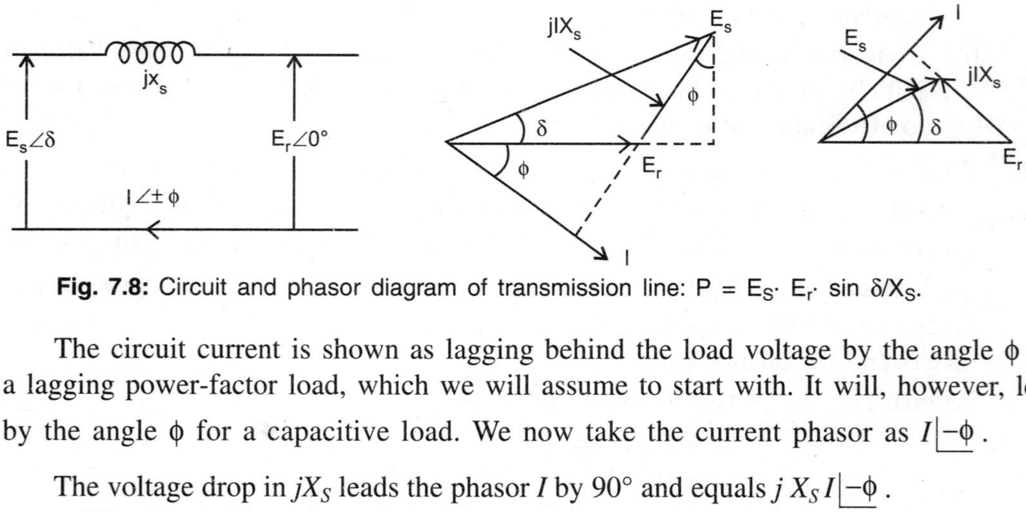

Fig. 7.8: Circuit and phasor diagram of transmission line: P = $E_S \cdot E_r \cdot \sin\delta/X_S$.

The circuit current is shown as lagging behind the load voltage by the angle ϕ for a lagging power-factor load, which we will assume to start with. It will, however, lead by the angle ϕ for a capacitive load. We now take the current phasor as $I\angle -\phi$.

The voltage drop in jX_S leads the phasor I by 90° and equals $j\,X_S\,I\angle -\phi$.

∴ $E_S \cos\delta = E_r + I\,X_S \sin\phi$, and $E_S \sin\delta = I\,X_S \cos\varphi$.

The real power consumed by the load is:

$P = E_r\,I \cos\phi$. But $I\cos\phi = E_S \sin\delta/X_S$.

∴ $P = E_S \cdot E_r \cdot \sin\delta/\,X_S$.

This is one of the most important relations in all of power transmission by a.c. It is called the "tie-line" equation.

Q.11. What is the significance of the "tie-line" equation, $P = E_S \cdot E_r \cdot \sin \delta / X_S$?

Ans. Note that if $\sin \delta = 0$, i.e., if there is no phase difference between sending-end voltage E_S and receiving-end voltage E_r, no real power P can be transmitted. We also observe that if the magnitude of E_S equals E_r, or even lower than E_r, real power can be transmitted as long as there exists a phase difference δ between E_S and E_r. This is not possible with d.c. transmission. The angle δ is known as the "power angle". For a leading power-factor load, current I leads E_r and the phasor diagram is shown in Fig. 7.8 for this case also.

Q.12. What is the meaning of "power-factor improvement" and "power-factor correction"?

Ans. Normally, most industrial drives use the 3-phase induction motor whose power factor at full load is around 0.8–0.9 lag. On the other hand, pure lighting load operates between 0.95–1.0 power factor, fluorescent lighting having chokes or ballasts or inductors in them, and incandescent lamps are purely resistive. Capacitive loads operate at leading power factors. A pure capacitance has 0 p.f. lead, while a pure inductor has 0 p.f. lag.

When a large industrial establishment or a city contains only induction-motor drives and lighting, the combined power factor is lagging, and the source has to supply the total current at lagging power factor. However, if the power factor is unity, the source current for the same kW or MW load is the minimum possible value. The losses in transmission as well as the MVA of the source generators and transformers and the cost of the entire system will reduce.

Therefore, every effort is made by power-supply companies to bring the resulting power factor at the load terminals as close to unity as possible (at a substation in a major establishment). This is carried out normally by connecting high-energy capacitor banks in parallel with the loads or at the substation which will draw a leading current. If the amount of capacitive current equals the inductive component of load current, the line current is minimised and the system operates at unity power factor. This is called "power-factor improvement" or "correction". A numerical example will clarify this very basic and important aspect of power transmission from the source end to the load end.

Q.13. A large industry has the following loads connected across 3-phase 50-Hz 11 kV lines:

Induction motors: 10,000 HP output, 0.8 p.f. lag and 80% efficiency;

Lighting and other loads: 5,000 kW at 0.95 p.f. lag.

Calculate:

(*a*) The input kW to the motor load and the line current drawn;

(*b*) The combined kW and kVAR of motor and lighting load and the line current drawn;

(*c*) The power factor of the combined load;

(*d*) The capacitive reactive power required to improve the overall power factor of the system to unity;

(*e*) The value of each capacitor in μF to be connected in Y to effect the power-factor improvement or correction to unity, see Fig. 7.9.

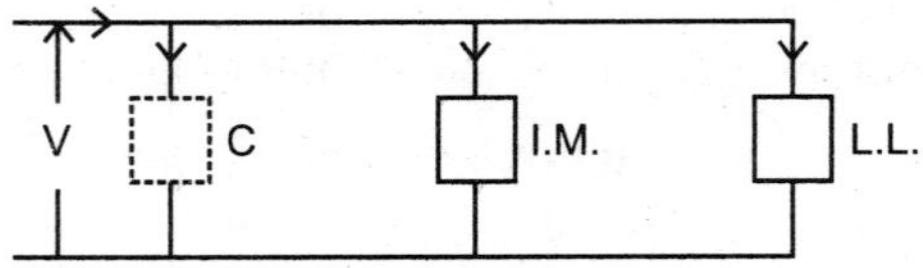

Fig. 7.9: Circuit diagram.

Ans. (*a*) 1 HP = 746 watt = 0.746 kW.

$\therefore$ Output of motor load = 7,460 kW

Since efficiency, ε_m = 0.8 = 80%

input = 7,460/0.8 = 9,325 kW

P.F. = 0.8 lag

$\therefore$ kVA = 9325/0.8

= 11,656 kVA

Also, reactive power drawn by motor,

$$Q_m = 11{,}656 \times 0.6$$
$$= 6{,}994 \text{ kVAR, lag}$$

Also, $Q_m = P_m \tan \phi$

$$= 9{,}325 \times 0.6/0.8$$
$$= 6{,}994 \text{ kVAR, inductive}$$

Total power drawn by motor load,

$$W_m = P_m + j\,Q_m = 9{,}325 + j\,6{,}994 \text{ kVA}$$

(*b*) Lighting and other loads,

$$P_1 = 5{,}000 \text{ kW at } 0.95 \text{ lag}$$

$\therefore$ $\cos \phi = 0.95$

and $\sin \phi = 0.312$

so that $\tan \phi = 0.329$ lag

$\therefore$ Reactive power = 1,645 kVAR

Real power and reactive power of combined load:

$$W = (9325 + 5000) + j\,(6994 + 1645)$$
$$= 14{,}325 + j\,8{,}639$$
$$= 16{,}728 \underline{|32^\circ} \text{ kVA}$$

Line current for combined load:

$$I = 16{,}728/11\sqrt{3}$$
$$= 878 \text{ amperes}$$

Line current for motor: $I_m = (9325 - j\,6994)/11\sqrt{3}$

$$= 489.4 - j\,367 \text{ amps}$$

$$= 751.9\underline{|-36.87^\circ}$$

Line current for lighting load:

$$I_l = (5000 - j\,1645)/11\sqrt{3}$$

$$= 262.4 - j\,86.3 \text{ amps}$$

$$= 276.2\underline{|-18.2^\circ}\,.$$

(*c*) Power factor of combined load = cos 32° = 0.857 lag.

(*d*) In order to improve the power factor to unity, the capacitance to be connected must supply 8,639 kVAR (leading). Under this correction, the power source will have to supply 14,325 kW or kVA (at 1.0 p.f.) instead of 16,728 kVA at 0.857 p.f. lag. The line current is also reduced from 878 amperes to (489.4 + 262.4) = 752 amps.

(*e*) Now, if the 3 capacitors in the phases are connected in *Y*, then each capacitor will draw the line current, which is $I_C = 8639/11\sqrt{3} = 453.4$ amps, which is also the sum of (367 + 86) amps, the lagging current of the combined loads.

$$\therefore \quad 453.4 = 2\pi \times 50 \times 11{,}000\, C\sqrt{3} \Rightarrow C = 227\ \mu\text{F}$$

Note that except for this part, there was no need for us to concern ourselves whether the motor and lighting loads were connected in *Y* or Delta. As a matter of interest, the reader should calculate the value of each capacitor if the three capacitors were connected in Delta. (Ans. $C_D = 227/3 = 75.7\ \mu$F each, whose impedance is 3 times that of $C_Y = 227\ \mu$F).

Q.14. How does the use of Synchronous Motors in a large industrial load improve the overall power factor of the system?

Ans. A 3-phase induction motor always operates at a lagging power factor because the stator winding carrying 3-phase voltages or currents has to generate the magnetic field or induce one in the rotor winding, and this is a lagging power factor operation. On the other hand, a synchronous motor has not only a 3-phase a.c. winding on the stator but also d.c. field windings or poles on the rotor. The operation of the motor for continuous torque production requires that the fields "lock" themselves together. Therefore, the rotor speed always equals the speed of the rotating field of the stator generated by the 3-phase a.c. winding. This speed is called the "synchronous speed" and is given by the relation $n_S = 120{\cdot}f/P$, revs/minute, where f = frequency of a.c. side voltage and P = number of poles on the machine as determined by the construction of the stator winding. This motor has the important property that if the mechanical load on the shaft is fixed, by controlling the d.c. excitation current fed to the rotor field winding the magnitude and power factor of the current drawn on the a.c. side can be changed. This is not possible in an induction motor.

Normal value of excitation current in the d.c. field is known as that current which makes the a.c. side operate at unity power factor, so that the stator current drawn from the supply lines is the minimum required for the shaft load. If the field is under-excited, namely, the d.c. field current is less than for normal excitation, then the stator of the motor draws a current at a lagging power factor. Conversely, if the motor is over-excited,the a.c. side operates at a leading power factor and in this sense supplies capacitive reactive power to the connected system. [See books on Electrical Machine Theory for Operation of Synchronous Motors, such as for example the author's book: "Electromechanical Energy Conversion with Dynamics of Machines" (Ist Ed. Wiley Eastern 1988, IInd Ed. New Age International 1998, New Delhi)].

In a large industrial load such as considered before with 10,000 HP of induction motors operating at 0.8 p.f. lag, it would be desirable to run the mechanical loads partly by using synchronous motors operating at leading power factor. This usually eliminates the need for switched capacitor banks in order to change their values during load changes.

Q.15. In Question 13, we arrange the loads as follows:

5,000 HP of induction motors at 0.8 lagging p.f. and 80% efficiency;

5,000 HP of synchronous motors at 80% efficiency;

5,000 kW of lighting load at 0.95 lagging p.f.

What should be the power factor of the synchronous motor if the overall power factor of the combined load should be unity?

Ans. (*a*) Power drawn by induction motor, $W_{IM} = 4{,}662 + j\,3{,}497$ kVA;

(*b*) Power drawn by lighting load, $W_L = 5{,}000 + j\,1{,}645$ kVA;

(*c*) Power drawn by synch. motor, $W_{SM} = 4{,}662 + j\,Q_S$ kVA.

For unity power-factor operation overall, the system kVAR must be zero.

$\therefore\ Q_S + 3{,}497 + 1{,}645 = 0 \Rightarrow Q_S = -5{,}142$ kVAR

Therefore, the synchronous motor is now drawing:

$$\begin{aligned} W_{SM} &= P_{SM} + j\,Q_S \\ &= 4{,}662 - j\,5{,}142 \text{ kVA} \\ &= 6{,}940\underline{|-47.8^\circ} \text{ kVA} \end{aligned}$$

$$\therefore \qquad \cos 47.8^\circ = 0.67$$

The synchronous motor has to be worked in an over-excited condition at 0.67 leading p.f. to improve the overall power factor of the combined industrial load to unity. By varying the d.c. excitation current easily, its kVAR can be controlled in order to supply the leading VARS at all load conditions.

We note that the synchronous motor is taking up its part of the shaft kW (HP) mechanical load in addition to supplying the leading reactive power kVAR.

Q.16. What is meant by a Synchronous Condenser and where is it used?

Ans. As we saw before, a synchronous motor can operate at any power factor from 0 p.f. lag to 0 p.f. lead through 1.0 p.f. by simply varying its d.c. field excitation. This property is utilised in order to provide lagging or leading VARS only at 0 p.f. by connecting the motor in the network but without a shaft load at the end of a long transmission line from a generating station for control of receiving-end voltage. In the early days of power transmission, since the motor was used specifically to supply only capacitive reactive power and so performed the same function as a "condenser" or capacitor, the unloaded synchronous motor became known as a Synchronous Condenser. However, since it can also be used to draw lagging reactive power, it is known as Synchronous Phase Modifier.

The only losses present in an unloaded motor are its electrical heating losses, shaft friction losses, and the windage losses caused by the cutting of the atmosphere by its rotor in which it rotates. Therefore, for very large sizes (greater than 5,000 kVA), where the additional expense permits, the motor is enclosed in an explosion-proof casing which is filled with hydrogen, usually at 2 atmospheres pressure in order that no air may leak inside to form an explosive mixture. Hydrogen is not only much lighter than air so that windage losses of the rotor are reduced, but also it is a much better heat carrier so that cooling becomes easier than in air. Before filling hydrogen, the casing is first filled with CO_2 for safety.

When the system has no load or very light load, as during night time, the machine is required to draw lagging VARS since the only current flowing in the line is the charging capacitive current of the line. But when the system is loaded and the power factor of the load becomes lagging, the motor or condenser supplies capacitive or leading VARS to compensate the inductive portion of the line current. In order to operate the two ends, the generating and receiving ends of the transmission line at fixed pre-determined voltages, voltage regulators are provided at both ends. These regulators operate on the d.c. field excitation of the generators at the sending end and of the synchronous condenser at the load end. In modern times with transmission lines becoming very long and voltages on the increase, the charging current of the line is prohibitively large for rotating synchronous condensers to handle the voltage-regulation problem. With the advent of high voltage high-powered Thyristors, static reactors are used which can be switched extremely fast to provide variable lagging reactive power to control the receiving-end voltage at light loads. Switched capacitors take care of load condition subsequently. As such these systems are called Static VAR Systems. Synchronous condensers are not usually used for transmission systems beyond 220 kV nor in a vast grid-connected system requiring fast voltage control for power transfers from one place to another. We discuss these in Chapter 13.

THREE-PHASE A.C. SYSTEM IN TRANSIENT STATE

Q.17. What are the transient conditions or states in which a 3-phase a.c. system can operate or be subjected to?

Ans. These are too numerous and varied to list all of them here. But the most important ones that give grave concern and on which the design of the system is based, are the following:

(*a*) Short circuits;

(*b*) Lightning strokes;

(*c*) Switching operations;

(*d*) Load swings and stability of the system; and

(*e*) Voltage control.

Here, we will discuss only the short-circuit problem, and postpone lightning and switching surge problems to later Chapters. The last two topics are highly specialised as subjects of their own and cannot be covered in a book on High Voltage, although some principles of voltage control will be given in a later Chapter.

Q.18. What are the different types of short circuits that can occur on a 3-phase system?

Ans. Refer to Fig. 7.10. In this discussion, we will always assume that the neutral of the source generator is connected to ground or "earthed". This represents the most severe conditions as far as currents and voltages are concerned during short circuits. A high voltage system rarely operates without its neutral being grounded or earthed, although it can be handled equally well with an "isolated" neutral.

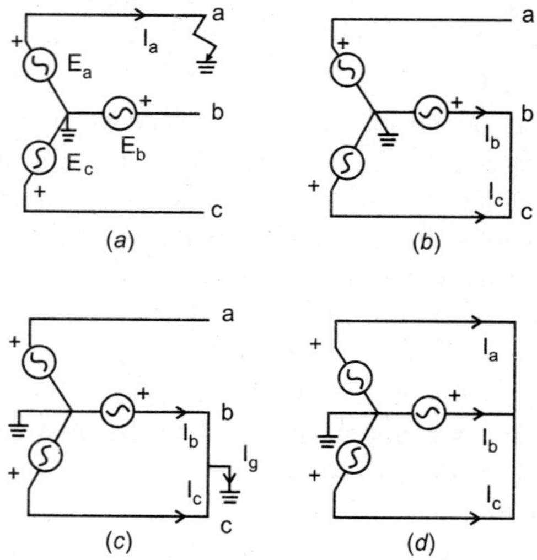

Fig. 7.10: Different types of short circuits on 3-phase transmission system.
(*a*) Single-line-to-ground S.C.;
(*b*) Line-to-line short circuit not involving ground;
(*c*) Double-line-to-ground fault;
(*d*) 3-phase symmetrical short circuit.

The most important types of short circuits are as follows:

1. **Single Line-to-Ground Short Circuit:** About 85% of all short circuits or "faults" occurring in practice are single line-to-ground short circuits. Here, one line or conductor develops a fault due to various circumstances and touches the ground. Such conditions are normally due to: (*a*) Squirrels and mice running inside the warm generator pits where the 6 terminals of the synchronous generator are exposed. This is a direct short circuit at the machine terminals, and the only impedance that controls the magnitude of current is the generator reactance (impedance). (*b*) One conductor

of a 3-phase transmission line breaks and falls on the ground anywhere along the length of the long line. (*c*) One insulator string flashes over to tower. This can take place when a severe lightning stroke contacts the line conductor and the insulator flashes over. The condition is removed when the line is opened automatically due to the flow of "power-follow" current, whereby the conducting path created by the lightning is followed by the power-frequency voltage feeding more current. The condition is also caused by "de-capping" of the insulator discs where due to faulty cement used in manufacture, or due to any severity of the environment in which the insulator is used, a cap on an insulator disc loosens and the string falls down thereby making the line conductor touch the tower or ground.

In all such cases of fault involving ground, the voltage of the faulted phase is zero with respect to ground i.e., the ground assumes the potential of the concerned phase until the fault is cleared. Normally, the voltage to ground of all 3-phases is (line-to-line voltage)/$\sqrt{3}$. But when the phase is grounded, the voltage of the healthy phases to ground becomes the line-to-line voltage. This is called "Dynamic Rise" in voltage. Any equipment connected between line and ground, such as most importantly the lightning arrester, will be subject to this dynamic voltage rise. The voltage rating of a lightning arrester is controlled by this dynamic voltage rise during a single-line-to-ground fault, even though it is connected between line and ground and is subject to the line-to-ground voltage or phase voltage during normal operating conditions.

2. **Line-to-Line Short Circuit:** Here, two of the 3 line conductors touch each other or are "shorted". The currents in the shorted phases are very high, but less than in a single-line-to-ground short circuit. The voltage driving the current in the faulted circuit is the line-to-line voltage and the current is governed by twice the impedance of any one phase, mostly reactance since resistances can usually be neglected as being small.
3. **Double-Line-to-Ground Short Circuit:** Here, the two lines are not only shorted but they also touch the ground. Thus, a ground loop is involved.
4. **Three-Phase Short Circuit:** This is a very severe condition on the system particularly if it occurs at the terminals of the source generator. Such short circuits are rare but must be guarded against in any event. It is clear that the machine voltages are behaving in the same symmetrical manner as when carrying normal 3-phase balanced load, but the currents during the fault are more severe. Thus, this condition is still balanced unlike the three previous short-circuit conditions enumerated earlier where the system is operating in an "unbalanced" state.

Understanding the conditions of the system that are taking place under these short circuits is important from the following points of view:

(*a*) Estimating over-currents in order to set protective relays to operate the circuit breakers to open the faulted phase or the entire system;

(*b*) Estimate voltage rises in order to design and select lightning arresters;

(*c*) Design synchronous machines to withstand the short circuit currents;

(*d*) Design windings in generators and transformers to withstand the severe mechanical forces which are caused by over currents, since mechanical forces are proportional to the square of the magnitude of current;

(*e*) Design protection and communication equipment;

(*f*) Estimate ground-current effects. This is particularly directed towards voltage rise of low-voltage telephone and other circuits normally handled by human operators;

(*g*) Design circuit breakers; And many other problems created by short circuits and system faults. We are not discussing stability of the power system under fault conditions but only restrict to high voltage aspects.

Q.19. What are 3-phase Symmetrical Components, and how are they used in analysing transients in a.c. systems?

Ans. In the year 1915, the great Canadian electrical genius, Charles L. Fortescue, working in the Westinghouse Research Laboratories in the U.S.A. gave the power industry his famous Symmetrical Components for handling transients in all types of electrical systems, in particular for analysing short circuits in 3-phase a.c. systems. According to this concept or method, if there are three voltages or currents of any magnitude and phase angles, they can be resolved into their symmetrical components as follows:

1. **Zero-Sequence Components:** These are denoted as V_{a0}, V_{b0}, V_{c0} which are equal in magnitude and are in phase, as shown in Fig. 7.11.

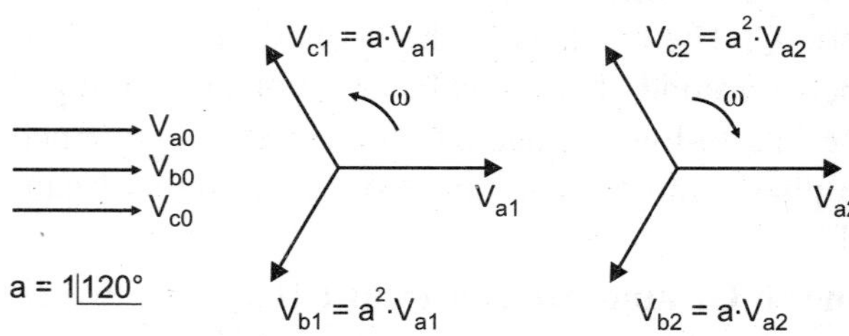

Fig. 7.11: Zero-, positive- and negative-sequence systems. Phasors.

2. **Positive-Sequence Components:** These are denoted as V_{a1}, V_{b1}, V_{c1} which are also equal in magnitude and form three phasors with phase difference of 120° from each other. When a system is balanced, these are the only three driving voltages present. They are rotating in the usual positive direction of increasing angle, namely, in the counter-clockwise direction.

3. **Negative-Sequence Components:** These are also three equal phasors inclined at 120° to each other but having a direction of rotation in the opposite direction to the positive-sequence set of voltages. These are denoted as V_{a2}, V_{b2}, V_{c2} in Fig. 7.11. Similarly for currents.

Let at any location in the system, the three phase-voltages have any magnitudes and phase angles, denoted by the phasors V_a, V_b, V_c. For example, let there be a single line-to-ground fault on phase a and the remaining two phases be healthy. Then, $V_a = 0$, $V_b = V\underline{|-120°}$ and $V_c = V\underline{|-120°}$. For other types of faults or operating conditions, we can always find these voltages, or currents. Then, Fortescue gave the following rules for resolving the three voltages into their symmetrical components:

1. Zero-sequence voltage of phase *a* is: $V_{a0} = (V_a + V_b + V_c)/3$. The zero-sequence voltages of phases *b* and *c* are equal to V_{a0}.

   Thus;

$$V_{a0} = V_{b0} = V_{c0} = (V_a + V_b + V_c)/3$$

   They are phasors; they have angles.

2. The positive-sequence component of voltage of phase a is:

$$V_{a1} = (1/3)(V_a + a{\cdot}V_b + a^2 \cdot V_c),$$

where *a* is the unit vector $a = 1\underline{|120°}$. Since V_{b1} is displaced 240° from the component V_{a1}, it is written as $V_{b1} = a^2 \cdot V_{a1}$, and similarly, $V_{c1} = a \cdot V_{a1}$.

The unit vector $a = 1\underline{|120°}$ denotes a rotation in the positive direction by 120° without change in magnitude of a vector or phasor which it multiplies.

Now, the unit vector $a = 1\underline{|120°}$ has the following properties:

(*a*) $a = 1\underline{|120°} = e^{j120} = e^{j2\pi/3} = \cos 120° + j \sin 120° = -0.5 + j\,0.866$

(*b*) $a^2 = 1\underline{|240°} = \cos 240° + j \sin 240° = -0.5 - j\,0.866$

(*c*) $1 + a + a^2 = 0$; $a + a^2 = -1$; $a^3 = 1$; $a - a^2 = j\sqrt{3}$; etc.

(*d*) The three unit vectors $1 = 1\underline{|0°}$, $a = 1\underline{|120°}$, $a^2 = 1\underline{|240°}$ are the three cube roots of unity.

Using these properties, the three positive-sequence components are written as follows:

$$\begin{aligned} V_{a1} &= (1/3)(V_a + a\,.V_b + a^2.V_c); \\ V_{b1} &= a^2 \cdot V_{a1} = (1/3)\,(a^2 \cdot V_a + V_b + a \cdot V_c)\,; \\ V_{c1} &= a \cdot V_{a1} = (1/3)\,(a \cdot V_a + a^2 \cdot V_b + V_c). \end{aligned}$$

The reader must remember these equations.

3. The Negative-sequence components of voltages of phases *a*, *b*, *c* are:

$$\begin{aligned} V_{a2} &= (1/3)\,(V_a + a^2 \cdot V_b + a \cdot V_c)\,; \\ V_{b2} &= a \cdot V_{a2} = (1/3)\,(a \cdot V_a + V_b + a^2 \cdot V_c); \\ V_{c2} &= a^2 \cdot V_{a2} = (1/3)\,(a^2 \cdot V_a + a \cdot V_b + V_c). \end{aligned}$$

From the above equations, it is evident that there are a total of 9 symmetrical components, 3 for each phase. But in reality we only have to determine the zero, positive-, and negative-sequence components of phase *a* only as follows: From the known values of V_a, V_b, V_c at the fault or any other location, we obtain:

$$V_{a0} = (1/3)(V_a + V_b + V_c);$$

then, $V_{b0} = V_{c0} = V_{a0}$.

Next, $V_{a1} = (1/3)(V_a + a.V_b + a^2 \cdot Vc)$,

giving $V_{b1} = a^2 \cdot V_{a1}$, and $V_{c1} = a \cdot V_{a1}$.

Finally, $V_{a2} = (1/3)(V_a + a^2 \cdot V_b + a \cdot V_c)$,

yielding $V_{b2} = a \cdot V_{a2}$, and $V_{c2} = a^2 \cdot V_{a2}$.

The above operations are known as "resolution" of three given voltages V_a, V_b, V_c into their symmetrical components.

The inverse relations to determine phase quantities from their symmetrical components follows easily according to the following rules:

$$V_{a0} + V_{a1} + V_{a2} = (1/3) \cdot 3V_a + (1/3)(1 + a + a^2)V_b + (1/3)(1 + a^2 + a)\, V_c = V_a,$$

since $(1 + a + a^2) = 0$

Similarly, $V_{b0} + V_{b1} + V_{b2} = V_b$

and $V_{c0} + V_{c1} + V_{c2} = V_c$

If the reader is familiar with matrix notation and matrix algebra, it becomes easier to handle these equations in one stage. But this will not be pursued further.

Example: Consider that phase a is experiencing a ground fault. Then, at the fault location, $V_a = 0$, $V_b = V\underline{|-120°} = a^2.V$, and $V_c = a.V$. If the system is unloaded to start with when the fault occurred, the currents are: $I_b = I_c = 0$. The problem is to determine the current flowing in the fault. The actual procedure will be explained in the next question. But the sequence components of voltages and currents will be found from the previously-given equations as follows:

$$\begin{aligned}
V_{a0} &= (1/3)\,(V_a + V_b + V_c) \\
&= V \cdot (1/3)\,(0 + a^2 + a) \\
&= -\,(1/3)\,V \\
V_{a0} &= V_{b0} = V_{c0} \\
V_{a1} &= (1/3)\,(V_a + a \cdot V_b + a^2 \cdot V_c) \\
&= V \cdot (1/3)\,(0 + a^3 + a^3) \\
&= (2/3)V \\
V_{b1} &= a^2 \cdot V_{a1} \\
V_{c1} &= a \cdot V_{a1} \\
V_{a2} &= (1/3)(V_a + a^2 \cdot V_b + a \cdot V_c) \\
&= V \cdot (1/3)(0 + a + a^2) \\
&= -\,(1/3)V \\
V_{b2} &= a \cdot V_{a2} \\
V_{c2} &= a^2 \cdot V_{a2}
\end{aligned}$$

The currents can also be resolved into symmetrical components thus: Since $I_b = I_c = 0$, and only I_a is finite (and is to be evaluated), it is easy to see that $I_{a0} = I_{a1} = I_{a2} = I_a/3$. Thus, it is established that the three sequence components are equal, and this is the essential part of solving the problem of finding current I_a flowing in the fault from the source voltage and what are known as "sequence impedances". This is the next step. But prior to this, we establish what are known as "sequence networks".

Q.20. What are "sequence networks" and what is their utility in fault calculations?

Ans. We have observed how the voltages V_a, V_b, V_c at the fault can be resolved into symmetrical components. These voltages give rise to currents through impedances up to the fault. They are known as zero-, positive-, and negative-sequence impedances denoted by Z_0, Z_1 and Z_2, respectively. We are therefore in a position to construct three networks called the "sequence" networks in each of which there is a driving voltage, an impedance, and a terminal voltage, as shown in Fig. 7.12.

We also observe that the source generator which feeds into the fault generates the three phase-voltages which form a positive-sequence set of three voltages. It does not generate zero-sequence or negative-sequence voltages. Thus, the e.m.f.s of the source are: 0 for zero-sequence emf, E for positive-sequence emf, and again 0 for the negative-sequence emf, which are shown in Fig. 7.12. In general, the three sequence-impedances are known and we will briefly describe how they are experimentally determined in Question 21 next. We now proceed to describe how these sequence networks, with their e.m.f.s, impedances and terminal voltages are used for solving short-circuit problems.

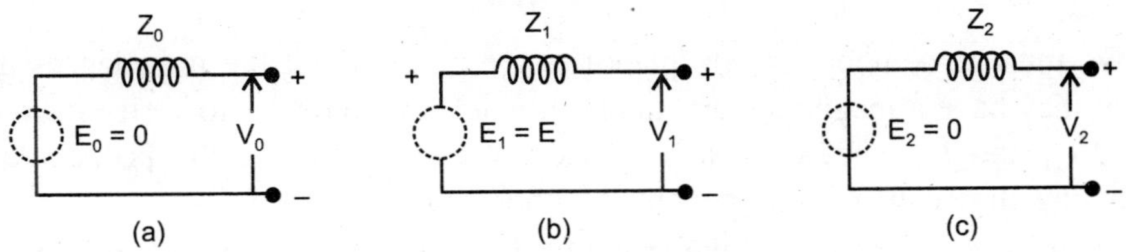

Fig. 7.12: Zero-, positive- and negative-sequence networks. Only the positive-sequence network has driving voltage or e.m.f. E_1.

1. **Single line-to-ground fault:** We have established the fact that for this case, when $I_b = I_c = 0$, the three sequence components of I_a are equal,

i.e., $$I_{a0} = I_{a1} = I_{a2}$$

This immediately establishes the property that the three sequence networks are in series, as shown in Fig. 7.13.

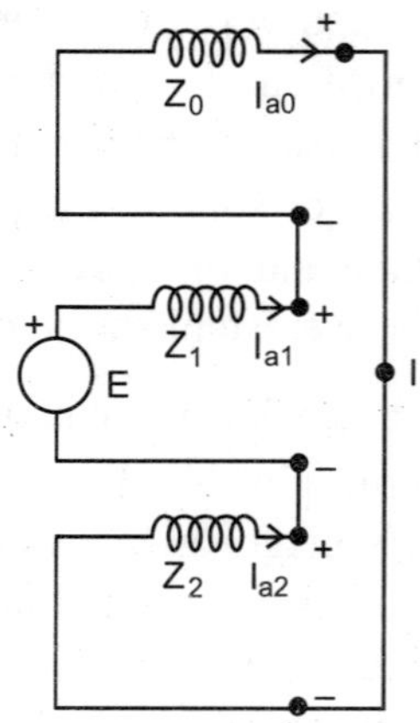

Fig. 7.13: Sequence-network interconnection during single-line-to-ground fault. All three sequence networks in series.

Since the only driving e.m.f. is E_1, the positive-sequence generated voltage of the source, we obtain for the currents:

$$I_{a0} = I_{a1} = I_{a2} = E/(Z_0 + Z_1 + Z_2)$$

Since E, Z_0, Z_1, Z_2 are known up to the fault, the three sequence components of current are known. The fault current is then:

$$I_a = I_{a0} + I_{a1} + I_{a2} = 3E/(Z_0 + Z_1 + Z_2)$$

We must remember that all quantities are phasors and the impedances are complex so that they have a magnitude and angle. It is left to the reader to verify from the equality of I_{a0}, I_{a1} and I_{a2} whether I_b and I_c become zero, by using the proper equations for combining the sequence components to obtain phase values.

We have now established the ground rules of the game, namely, for any type of fault or operaing condition, find out some relation between the voltages or currents which enable us to interconnect the three sequence networks properly. Then from the source positive-sequence e.m.f. and the sequence impedances, determine the sequence currents. Having done this, combine the sequence quantities thus found to obtain actual phase quantities.

2. As the next example, consider a phase-to-phase short circuit on an initially unloaded system as shown in Fig. 7.10(*b*). The known conditions at the fault are:

$$I_a = 0,\ I_b + I_c = 0, \text{ or, } I_b = -I_c$$

We do not know any other relation between other quantities, but this is sufficient in order to determine how the three sequence networks should be connected.

For this case, the three sequence currents are:

$$I_{a0} = (I_a + I_b + I_c)/3 = 0$$
$$I_{a1} = (I_a + a\,.\,I_b + a^2 \cdot I_c)/3$$

$$= I_b(a - a^2)/3$$
$$I_{a2} = (I_a + a^2 \cdot I_b + a \cdot I_c)/3$$
$$= I_b(a^2 - a)/3 = -I_{a1}$$

Since $I_{a0} = 0$ and the zero-sequence e.m.f. is also zero, this network is closed upon itself and does not take part in the interconnection with other two networks. However, the condition $I_{a1} = -I_{a2}$ shows that the positive-and negative-sequence networks are connected in parallel. The resulting connection is shown in Fig. 7.14, and the reader should check carefully if with such an interconnection, the two currents are equal and opposite. The currents flowing will be $I_{a1} = -I_{a2} = E/(Z_1 + Z_2)$.

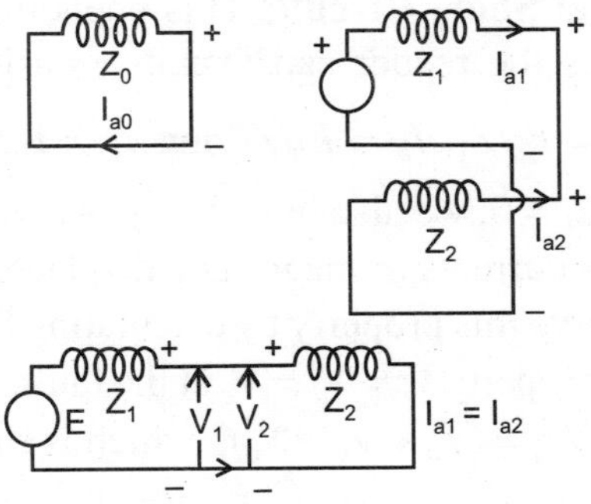

Fig. 7.14: Line-to-Line short circuit. Positive- and negative-sequence networks in parallel.

We can also find the three sequence voltages in order to obtain the property that the positive- and negative-sequence networks are to be connected in parallel, thus: From Fig. 7.10(*b*) we have the three voltages V_a, V_b and V_c with $V_b = V_c$, since the two lines are connected at the fault.

$$\therefore \quad V_{a0} = (V_a + 2V_b)$$
$$V_{a1} = [V_a + (a + a^2)\, V_b]/3$$
$$= (V_a - V_b)/3$$
$$\text{and} \quad V_{a2} = [V_a + (a^2 + a)\, V_b]/3 = V_{a1}$$

Thus, the two networks have equal terminal voltages and must be connected with their positive terminals together as well as their negative ones. Hence, they are in parallel. An equivalent description is that since $I_{a1} = -I_{a2}$, the two networks are also in series opposition.

Having found the three sequence components of current, the resulting phase currents are: $I_a = I_{a0} + I_{a1} + I_{a2} = 0$;

$$I_b = I_{a0} + a^2 \cdot I_{a1} + a \cdot I_{a2}$$
$$= (a^2 - a) \cdot E/(Z_1 + Z_2)$$
$$= -j\sqrt{3}\, E/(Z_1 + Z_2)$$

which lags 90° behind $[E/(Z_1 + Z_2)]$,

since $a^2 - a = (-0.5 - j\ 0.866) - (-0.5 + j\ 0.866) = -j\sqrt{3}$.

Similarly,

$$I_c = I_{a0} + a \cdot I_{a1} + a^2 I_{a2}$$
$$= (a - a^2) \cdot E/(Z_1 + Z_2) = -I_b$$

The three currents in the three phases are now found out by the above procedure. We observe the important property that for a line-to-line short circuit not involving ground, only the positive- and negative-sequence networks are involved. The zero-sequence network is not involved. It is of course a salient property that the zero-sequence network comes into the picture only for a fault condition involving ground.

3. **3-Phase Symmetrical Short Circuit:** This phenomenon involves only the positive-sequence network, as the reader can visualise easily. Thus, the 3 currents are:

$$I_a = E_a/Z_1,\ I_b = E_b/Z_1 \text{ and } I_c = E_c/Z_1$$

Since $E_a + E_b + E_c = 0$, we also have $I_a + I_b + I_c = 0$, and they also form a positive-sequence set of equal currents in magnitude displaced 120° from each other.

We can also observe this property by calculating the three sequence components of the voltage from the property $V_a = V_b = V_c$ at the fault, since the terminals are connected together at the fault. $V_{a0} = (V_a + V_b + V_c)$, which is finite.

$$V_{a1} = (V_a + a \cdot V_b + a^2 \cdot V_c)/3$$
$$= (1 + a + a^2)\ V_a/3 = 0 = V_{a2}$$

The current flowing is $I_{a1} = E/Z_1$. Since the zero- and negative-sequence circuits have no driving e.m.f.s and the terminals are shorted, there are no currents in these two sequence networks. The network dispositions are shown in Fig. 7.15.

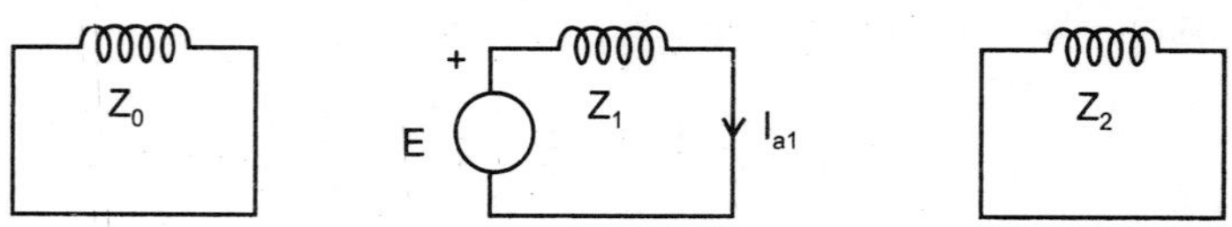

Fig. 7.15: 3-phase symmetrical short circuit.
All 3 sequence networks are independent.

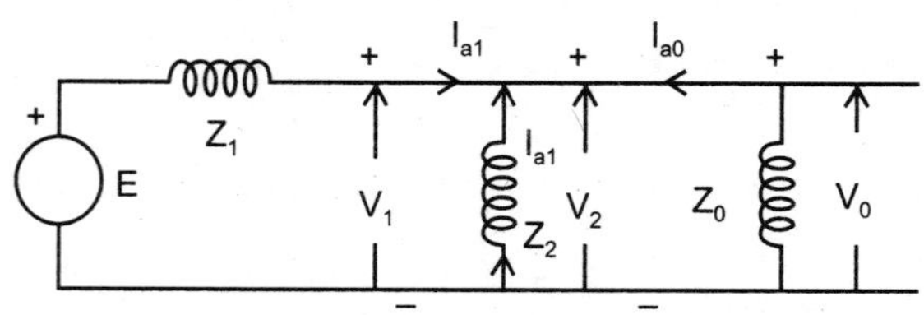

Fig. 7.16: Double-line-to-ground fault. All 3 sequence networks are in parallel.

4. Finally, consider a **Double Line-to-Ground Fault** as shown in Fig. 7.10(*c*). For this case, the known conditions are: At the fault,

$$I_a = 0$$
$$I_b + I_c = I_g$$
$$V_b = V_c = 0,$$

since they are connected to ground at the fault.

$$\therefore \quad V_{a0} = (V_a + V_b + V_c)/3 = V_a/3$$
$$V_{a1} = (V_a + a \cdot V_b + a^2 \cdot V_c)/3 = V_a/3$$

Similarly, $V_{a2} = V_a/3$

Since the terminal voltages of all the three sequence networks are equal, they must be connected in parallel, as shown in Fig. 7.16. The only driving e.m.f. is the positive-sequence voltage of the source generator. Note the proper connection of polarities of the three networks yielding $I_{a0} + I_{a1} + I_{a2} = I_a = 0$. Here, we have Z_0 and Z_2 in parallel, and this parallel combination is in series with Z_1. Therefore, the three sequence currents are given as follows:

$$I_{a1} = E/[Z_1 + Z_0 \cdot Z_2/(Z_0 + Z_2)] = E \cdot (Z_0 + Z_2)/(Z_0 Z_1 + Z_1 Z_2 + Z_2 Z_0);$$
$$I_{a0} = I_{a1} \cdot Z_2/(Z_0 + Z_2) = -E \cdot Z_2/(Z_0 Z_1 + Z_1 Z_2 + Z_2 Z_0) \text{ ; and}$$
$$I_{a2} = I_{a1} \cdot Z_0/(Z_0 + Z_2) = -E \cdot Z_0/(Z_0 Z_1 + Z_1 Z_2 + Z_2 Z_0).$$

From these sequence components, the phase currents are obtained as follows:

$$I_a = I_{a0} + I_{a1} + I_{a2} = 0$$

It is left to the reader to calculate the currents I_b and I_c to obtain the following values:

$$I_b = I_{a0} + a^2 \cdot I_{a1} + a \cdot I_{a2}$$
$$= E \cdot \sqrt{3}\,(Z_2 \angle{-210^\circ} - j Z_0)/(Z_0 Z_1 + Z_1 Z_2 + Z_2 Z_0) \text{ ; and}$$
$$I_c = I_{ao} + a \cdot I_{a1} + a^2 \cdot I_{a2}$$
$$= E \cdot \sqrt{3} \cdot (j Z_0 + Z_2 \angle 150^\circ)/(Z_0 Z_1 + Z_1 Z_2 + Z_2 Z_0)$$

The current flowing in the ground and generator neutral is, with

$$a^2 + a - 2 = (a^2 + a + 1) - 3 = -3:$$
$$I_g = I_b + I_c$$
$$= E \cdot [Z_0(a^2 - a) + Z_2(a^2 - 1)$$
$$+ Z_0(a - a^2) + Z_2(a - 1)]/(Z_0Z_1 + Z_1Z_2 + Z_2Z_0)$$
$$= -3\,E \cdot Z_2/(Z_0Z_1 + Z_1Z_2 + Z_2Z_0) = 3\,I_{a0}.$$

This high value of ground current can be used for operating a relay which would open a circuit breaker and clear this fault.

The three sequence voltages are:

$$\begin{aligned} V_{a0} &= V_{a1} = V_{a2} = E - I_{a1} \cdot Z_1 \\ &= E - E \cdot Z_1(Z_0 + Z_2)/(Z_0 Z_1 + Z_1 Z_2 + Z_2 Z_0) \\ &= E \cdot Z_2 Z_0 / (Z_0 Z_1 + Z_1 Z_2 + Z_2 Z_0) \\ &= -Z_0 I_{a0} \end{aligned}$$

Consequently, the voltage V_a of the healthy or unfaulted phase to ground is:

$$V_a = V_{a0} + V_{a1} + V_{a2} = -Z_0 I_g$$

This gives: $Z_0 = -V_a / I_g$

This is a very famous property of this type of short circuit, namely, that the ground current is determined entirely from the healthy-phase voltage and the zero-sequence impedance. It also gives us a method of experimentally determining the zero-sequence impedance which we will elaborate in Question 22.

Q.21. In a Line-to-Line fault not involving ground, what are the expressions for the following voltages? (*a*) Healthy phase '*a*' to neutral? (*b*) Voltage between healthy phase *a* and the common of the shorted phases *b* and *c*?

Ans. We obtained the following sequence currents and voltages for the line-to-line fault on phases *b* and *c* with phase *a* open, and the common of phases *b* and *c* isolated from neutral or ground, [see Fig. 7.10(*b*)]:

$$\begin{aligned} I_a &= 0 \\ I_b &= -I_c \\ I_{a0} &= 0 \\ I_{a1} &= I_{a2} \\ V_{a0} &= 0 \\ V_{a1} &= V_{a2} \\ &= (V_a - V_b)/3 \\ V_b &= V_c \\ V_{a1} &= V_{a2} = E \cdot Z_2/(Z_1 + Z_2) \end{aligned}$$

which is obtained by voltage division from Fig. 7.14,

and $I_b = -j\sqrt{3}\, E/(Z_1 + Z_2)$

(*a*) ∴ Voltage of healthy phase *a* to neutral:

$$V_a = V_{a0} + V_{a1} + V_{a2} = E \cdot 2Z_2/(Z_1 + Z_2)$$

(*b*) Voltage of common point of phases *b* and *c* to neutral will be:

$$\begin{aligned} V_b &= V_c = V_{a0} + a^2 \cdot V_{a1} + a \cdot V_{a2} \\ &= V_{a0} + (a^2 + a) V_{a1} = V_{a0} - V_{a1} \\ &= -E Z_2/(Z_1 + Z_2) \end{aligned}$$

∴ Voltage between healthy phase *a* and common of phases *b* and *c* will be:

$$V_a - V_b = E \cdot 2 Z_2/(Z_1 + Z_2) + E \cdot Z_2/(Z_1 + Z_2) = E \cdot 3Z_2/(Z_1 + Z_2).$$

Q.22. What are the experimental methods of determining the three sequence impedances Z_0, Z_1 and Z_2 arising out of the previous analysis?

Ans. (*a*) **Zero-Sequence Impedance, Z_0:** In the double-line-to-ground fault we saw that the current flowing in the ground loop was $I_g = -V_a/Z_0$. (Question 20).

$$\text{Thus,} \qquad |Z_0| = |V_a/I_g| = \frac{\text{magnitude of healthy phase voltage to ground}}{\text{magnitude of current flowing in ground-loop}}.$$

If therefore, an ammeter is connected in the ground loop and a voltmeter is connected between the healthy unfaulted phase *a* and ground, the ratio of resulting voltage to ground to the current in the ground loop is the magnitude of the zero-sequence impedance up to the fault location, Z_0. This is shown in Fig. 7.17. A wattmeter is also connected with its current coil in the ground loop in series with the ammeter, and its voltage or potential coil between healthy phase and ground in parallel with the voltmeter. Its reading divided by the product of voltage and current will give the phase angle of the zero-sequence impedance from which the resistance and reactance components can be determined.

It is to be borne in mind that all numerical values of impedances in any system must be converted to per-unit values on chosen base quantities.

(*b*) **Negative-Sequence Impedance, Z_2:** In the line-to-line fault not involving ground, we obtained the following results, (see Question 20):

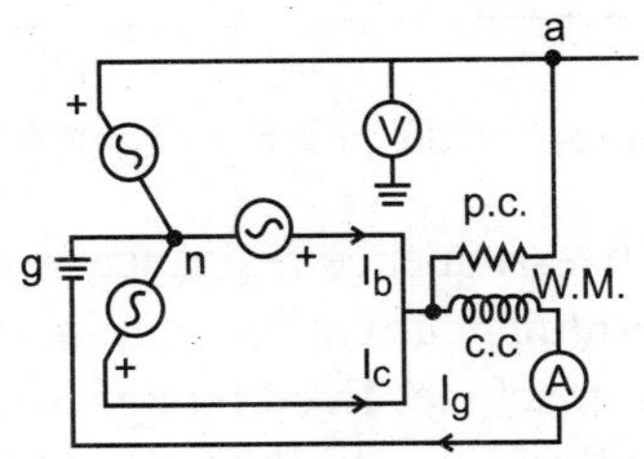

Fig. 7.17: Test arrangement for determining zero-sequence impedance, Z_0.

$$I_b = -j\sqrt{3} \cdot E/(Z_1 + Z_2), \text{ and } I_b = -I_c$$

Voltage between healthy phase *a* and common of faulted phases *b* and *c* is:

$$V_a - V_b = 3E \cdot Z_2/(Z_1 + Z_2) = j\sqrt{3}\ Z_2 \cdot I_b$$

$$\therefore \qquad Z_2 = (j/\sqrt{3}) \cdot (V_a - V_b)/I_b$$

$$\therefore \qquad |Z_2| = \frac{1}{\sqrt{3}} \left| \frac{\text{Voltage between healthy phase and common}}{\text{current in the fault}} \right|.$$

Another expression is obtained as follows: Voltage of healthy phase and neutral is (see Question 21)

$$V_a = 2\ E{\cdot}Z_2/(Z_1 + Z_2) = j(2/\sqrt{3})\ Z_2\ I_b$$

$$\therefore \quad |Z_2| = \left(\frac{\sqrt{3}}{2}\right) \cdot |V_a/I_b| = \frac{\sqrt{3}}{2} \left| \frac{\text{healthy phase voltage to neutral}}{\text{current in the fault}} \right|.$$

The resulting test circuits for determining Z_2, the negative-sequence impedance, are shown in Figs. 7.18(*a*) and (*b*), along with voltmeter, ammeter and wattmeter connections.

(*c*) **Positive-Sequence Impedance, Z_1:** In this test, the three lines are connected together through an ammeter in one of them, and the source generator voltage is read from its excitation characteristics. Since the lines are short-circuited, there is no way of connecting a voltmeter, and so the open-circuit characteristics of the synchronous generator must be used. The ratio of generator e.m.f. and the current in any one of the phases will give the positive-sequence impedance up to the fault, Z_1. The resistance must be measured separately by an ohm meter, since a wattmeter cannot be connected in the circuit.

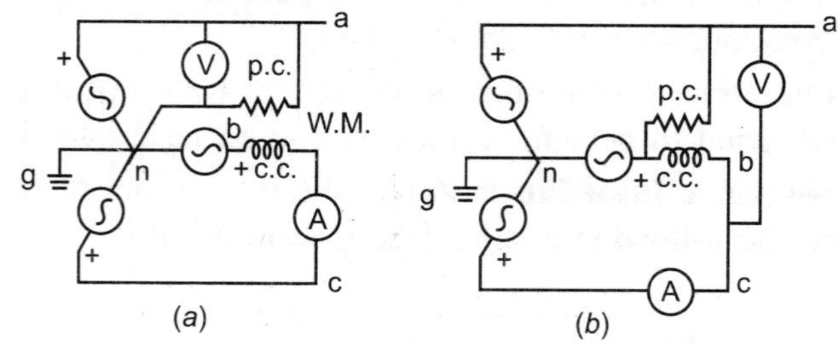

Fig. 7.18: Test arrangements for determining negative-sequence impedance, Z_2.

Normally, the positive-sequence reactance is nearly 1 per unit for a synchronous generator so that the resulting current flowing in a 3-phase symmetrical fault is nearly the rated current of the machine. Even though we are measuring the impedance, the resistance is quite negligible and all of the measured impedance is equal to the reactance. This test is not dangerous to the system even if carried out at full voltage, as is the case with the other tests for determining Z_0 and Z_2. Their per-unit values are normally less than 0.2 per unit, meaning that when normal voltage is applied the current flowing is 5 times the rated current. However, in the symmetrical short-circuit, the voltage of the generator must be gradually increased after connecting the three lines together externally, by raising the excitation field current. A sudden short-circuit applied on a synchronous generator brings in very high initial components of current which are known as "sub-transient" and "transient" stages before the steady-state short-circuit current is established. This is a topic in Synchronous Machine Theory and cannot be discussed here. [See the author's book "Electromechanical Energy Conversion with Dynamics of Machines," Wiley Eastern (Ist Ed.)1988, New Age International (IInd Ed.) 1998].

Q.23. What are the approximate values of various reactances of generators, transformers and lines used in a high voltage network?

Ans. Although the performance of a system in the steady-state and transient-state must take into account the proper impedance values, it is quite adequate to concentrate only on the reactive part and include resistances wherever necessary if they are of much concern. The resistances affect power losses and damping or decay of transients, but for determining the worst conditions or the initial values, they can be omitted. Therefore, in this question, only reactances will be discussed.

All quantities are in their per-unit values with 1 p.u. reactance referring to the reactance which permits rated current (1 p.u.) to flow at the rated or 1 p.u. voltage of the proper type. They are also based on the machine or equipment rating as base quantities, except for transmission-line conductors for which the reactance values are the actual values in ohms.

The Synchronous Machine has many reactances other than merely the zero, positive-, and negative-sequence reactances because of 3 windings and the fact that the rotor is a rotating member which controls the magnetic paths taken by the fluxes. For static equipment, such as transformers and lines, the positive- and negative-sequence reactances are equal. But the zero-sequence reactance of a 3-phase transformer will be different according as the 3 phase-windings are wound on 3 limbs (core-type) or 5 limbs (shell type).

1. **Synchronous Generator and Motor Reactances:**

(*a*) *Synchronous Reactance*: This is also the positive-sequence reactance and is the response of the 3-phase stator winding or armature when it is excited by positive-sequence voltage which the machine normally generates due to its d.c. field excitation. It normally consists of two components in the cylindrical-rotor machine and three in the salient-pole machine. These are: X_{aS} and X_l for the round-rotor generator or motor, and X_{ad}, X_{aq} and X_l for the salient-pole machine. The synchronous reactance equals the sum of the armature-reaction reactance and the leakage reactance, X_l, as described below.

(*i*) For a **cylindrical-rotor machine**, the air-gap is uniform and the resulting reactance stays constant at $X_S = X_{aS} + X_l = 0.9 + 0.1 = 1.0$ p.u.;

(*ii*) For a **salient-pole machine**, the air-gap is non-uniform. But the performance of the machine is analysed by treating it as composed of two machines called the "direct-axis" machine and "quadrature-axis" machine. These two correspond respectively to polar-axis when the air-gap length is minimum, and to inter-polar axis when the air-gap length is maximum and generally the reactance is minimum. The direct-axis synchronous reactance is equal to that of the cylindrical-rotor machine, but the quadrature-axis reactance is lower. Average values are:

$$X_{ad} = 0.9 \text{ p.u.}$$
$$X_1 = 0.1 \text{ p.u.}$$
$$X_S = 1.0 \text{ p.u.}$$
$$X_d = X_{ad} + X_1 = 1.0 \text{ p.u.}$$

and $X_{aq} = 0.6$ p.u., giving $X_q = 0.7$ p.u.

(*b*) *Transient Reactances*: X', X'_d, X'_q. When a short-circuit occurs on the machine terminals or when sudden load changes occur, because of the inductance of the field winding on the rotor, it does not allow the flux to change instantaneously to the required new value. In order to keep the flux-linkage constant through it, a current is induced in it which reacts on the armature flux, reducing the flux linked by the armature winding on the stator. This lowers the effective inductance and reactance of the stator winding. The resulting reactance is called its Transient Reactance. It lasts only until the current in the field winding has decayed as governed by the field time constant $\tau_f = L_{ff}/R_{ff}$.

Since there is no field winding in the quadrature axis, $X'_q = X_q$, the quadrature-axis synchronous reactance. Average values are:

Cylindrical-rotor machine: $X' = 0.3$ p.u.;

Salient-pole machine: $X'_d = 0.3$ p.u., $X'_q = 0.7$ p.u.

(*c*) *Sub-Transient Reactances*: X''_d, X''_q: In salient-pole machines, pole-face windings are used in order to damp hunting of the rotor during transient conditions, especially during sudden load changes and for generating an induction-motor type torque for starting a synchronous motor and bringing the rotor nearly to synchronous speed before switching on the d.c. field excitation current. These pole-face windings are also known as 'damper' windings or by the French word 'amortisseur' windings. When the rotor is rotating at its constant synchronous speed, there are no currents induced in these damper windings by the rotating field of the stator winding. But any relative speed between the rotor and these amortisseur windings will give rise to currents in them. Since they are located on the pole surface and are very close to the stator surface, the air-gap is minimum and the flux generated by the induced currents hits back or reacts on the armature flux. Thereby, the armature threads less flux and its reactance is reduced during this "sub-transient" stage, until the currents in the pole-face windings decay to zero. This reactance is lower than the "transient" reactance which is caused by currents induced in the main field winding located deeper in the rotor.

Average values are: $X''_d = 0.15$ p.u., and $X''_q = 0.15$ p.u. These reactances are quite low and consequently bring in very heavy short-circuit currents.

(*d*) *Negative-Sequence Reactance*, X_2: During this state, the stator winding is energised externally by 3-phase balanced system of currents which generate a rotating magnetic field at synchronous speed but rotating in the opposite direction to the rotor, or in the negative sequence. It is therefore evident that the flux generated by the armature sweeps past the rotor at twice the synchronous speed. Thus, the only winding which is affected is the pole-face or amortisseur winding, the phenomenon being too rapid for the flux to penetrate deeper into the field windings. Since the flux sweeps past alternately across the direct axis and quadrature axis, the resulting reactance is the average of X_d'' and X_q''. Thus, $X_2 = (X_d'' + X_q'')/2$, which is approximately 0.15 p.u.

(*e*) *Zero-Sequence Reactance, X_0*: By definition, this is the reactance presented by the armature winding when the 3 phases are excited by 3 equal voltages or currents in phase with each other: $V_{a0} = V_{b0} = V_{c0}$. Therefore, during a test to determine this reactance, the three phases are connected in series (or parallel) if all 6 terminals are available, and excited by a single voltage. Since the flux is generally high, a low voltage must be applied, and can induce a high voltage in the d.c. field winding by transformer action. The rotor need not be rotated. The field winding is usually short-circuited during this test.

It is clear that no rotating field is generated and the winding acts like a single-phase winding. The only return path for the flux generated in the armature steel stampings is through leakage paths in air. Therefore, the zero-sequence reactance is nearly equal to the leakage reactance.

Thus, $X_0 \approx X_1 = 0.1$ p.u. for normal machines.

2. **Transformers:** In all studies involving transformers as circuit elements, it is usual to omit the magnetising current required to generate the core flux. This also means neglecting the core losses, namely the hysteresis and eddy-current losses. Therefore, the only reactance of concern is the series leakage reactance. This is intentionally kept low in order not to cause a large voltage drop. Although this is a desirable condition from the steady-state operational point of view, when the transformer suffers a short-circuit at its terminals at full voltage, the only reactance which limits the current is this leakage reactance.

The leakage reactance of large transformers is usually less than 0.1 p.u. The positive- and negative-sequence reactances used in system-study representation are nearly equal to 0.1 p.u. on transformer base rating.

The zero-sequence reactance for a 3-limb core-type transformer is also equal to this value, since if the three phases are excited by equal voltages which are all in phase with each other, the core flux in the 3 limbs are equal and flow in the same direction. There is no return path in the steel core; the only return path is in the oil, air and other non-magnetic materials inside the tank. Therefore, the reactance is equal to the leakage reactance or even lower, since the length of air paths for the flux is longer than for normal leakage flux which takes place in the space between the windings.

However, if the 3 phase-windings are wound on 5 limbs, as in the shell-type transformers, the two extra limbs serve as return paths for the zero-sequence flux and the resulting reactance value increases to the magnetising reactance which is extremely high, in the order of over 20 p.u. in good magnetic steels. This means that the magnetising current drawn is less than 5% of the full rated current. Thus, the zero-sequence circuit offers nearly an open circuit.

3. **Transmission Lines:** The three sequence reactances of lines are unequal. The method of calculating series inductance will be discussed in detail in the next Chapter 8. They depend entirely upon the sizes of conductor and the dimensions of the line, such as line height above ground, separation of phases, and bundling of conductors. Average values of actual reactances are as follows:

	400 kV	500 kV	750 kV	1000 kV	1200 kV
X_S :	0.33	0.3	0.27	0.23	0.23 ohm/km at 50 Hz.

The above reactances are positive-sequence values applicable to steady-state operation under normal conditions. They are also equal to the negative-sequence reactances. However, the zero-sequence reactance is higher than both these by about 50%, as will be shown in Question 8.13 in the next Chapter.

❑❑

CHAPTER 8

INDUCTANCE, CAPACITANCE AND VOLTAGE GRADIENTS OF OVERHEAD TRANSMISSION-LINE CONDUCTORS

Q.1. What are the expressions for: (*a*) Magnetic Field Strength; (*b*) Flux Density; and (*c*) Flux flowing up to a distance x from a round conductor carrying a current I?

Ans. (*a*) Ampere's Circuital law states that the line integral of H, the magnetic field strength over a closed curve C is equal to the current enclosed by the curve. Consider a circle of radius x with its centre coinciding with the centre of the conductor. The conductor radius is r and carries a current I.

Then, because of symmetry, the magnetic field strength H is equal on the periphery of the circle of radius x, as shown in Fig. 8.1.

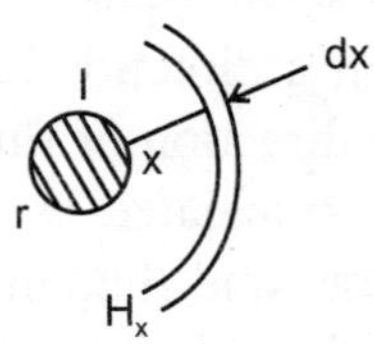

Fig. 8.1: Magnetic field around a round conductor of radius r carrying a current I.

Ampere's Circuital law applied along the circle gives:

$$2\pi x \cdot H = I, \text{ or, } H = I/2\pi x \text{ ampere/metre}$$

(*b*) The flux density at x is $B = \mu_0 \mu_r H$, tesla or weber/m^2,
where $\mu_0 = 4\pi.10^{-7}$ henry/metre = permeability of free space, and μ_r = relative permeability of medium in which the flux density is to be evaluated = 1 for non-magnetic materials such as air, and >100 normally for ferrous materials.

(*c*) Considering an elementary width dx at distance x from the conductor centre, the flux $d\phi$ flowing in dx and 1 metre length of conductor is:

$$d\phi = B \cdot dx = (\mu_0\,\mu_r/2\pi) \cdot \text{I} \cdot (1/x) \cdot dx \text{ weber}$$

∴ Total flux flowing up to the distance x is

$$\phi = \int_r^x d\phi = \frac{\mu_0\mu_r}{2\pi} I \int_r^x (1/x)/dx = \frac{\mu_0\mu_r}{2\pi} I \cdot \ln(x/r) \text{ weber}$$

Q.2. What is the inductance of two long straight parallel conductors, of radii r, per unit length and separated by distance d, and carrying currents $+I$ and $-I$?

Ans. Referring to Fig. 8.2, we observe that one conductor forms the 'go' and the other the 'return' of a single-phase a.c. line or the two poles of a d.c. line. Consider a distance x from the centre of the left conductor. We consider two flux lines shown as 1 and 2. Along the line 1 the line integral of magnetic field strength on a closed path enclosing the two conductors is zero, since the total current is zero. Therefore, we only have to concentrate on the flux flowing in between the two conductors from r to $(d - r)$ from the left conductor.

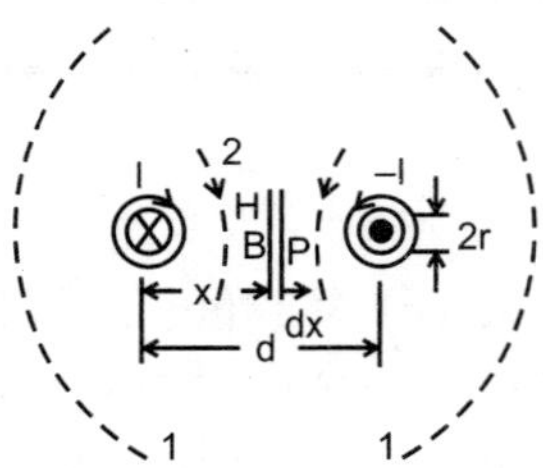

Fig. 8.2: Inductance calculation of two long straight parallel conductors. Radii = r, separation = d, currents + I and –I.

According to Faraday's conception, the direction of flux line (which is really the direction in which a small magnetic compass needle will align itself) is given by the following rule, which is also called the right-hand corkscrew rule of Fleming:

If the screw is turned in such a direction as to advance in the direction of flow of current in the conductor, then the direction of flux is given by the direction of rotation of the screw. This rule can also be translated as the simple rule that if the conductor is grabbed by the fingers of the right hand (hopefully without shock!) such that the thumb points to the direction of current flow, then the flux lines follow the direction of fingers.

If therefore a current is flowing into the paper as shown by the cross (or tail of an arrow) the flux lines are directed clockwise. If the current is coming out of the paper shown by a dot in the right conductor (the head of a pointed arrow), the flux lines flow counter-clockwise or anti-clockwise. Therefore, at the point P at distance x from the left conductor, the flux created by both conductors flow in the same direction, namely from top to bottom, clockwise around the left conductor and counter-clockwise around the right conductor. The two flux-densities at x caused by the currents $\pm\, I$ are:

$$B_L = (\mu_0\,\mu_r/2\pi)\cdot I/x$$

and $$B_R = (\mu_0\,\mu_r/2\pi)\cdot I/(d-x) \text{ tesla}$$

∴ The total flux density at P is:

$$B = B_L + B_R$$
$$= (\mu_0\,\mu_r/2\pi)\cdot I\cdot[(1/x + 1/(d-x)]$$

Taking a small width dx at x whose length is 1 metre into the paper, (along the length of the line), the differential flux $d\phi$ flowing in the area $(1\cdot dx)$ is:

$$d\phi = B\cdot dx = (\mu_0\,\mu_r/2\pi)\cdot I\cdot[(1/x + 1/(d-x)]\cdot dx$$

Since x varies from r on the surface of the left conductor to $(d-r)$ on the surface of the right conductor, the total flux flowing in the space between the two conductors will be:

$$\phi = \int_r^{(d-r)} B\cdot dx = (\mu_0\mu_r/2\pi)\cdot I\int_r^{(d-r)}[(1/x + 1/(d-x)]\cdot dx$$
$$= (\mu_0\mu_r/\pi)\cdot I\cdot \ln[(d-r)/r] \text{ weber}$$

The resulting inductance is

$$L_2 = \phi/I = (\mu_0\mu_r/\pi)\cdot \ln[(d-r)/r] \text{ henry/metre}$$

(The two conductors form only 1 turn of a loop in which the total flux flows or links. The definition of inductance = flux-linkage per ampere = flux × number of turns in a coil or loop/current). Therefore, we can consider each conductor to have one-half of this inductance, so that the inductance of each conductor per metre is:

$$L = (\mu_0\,\mu_r/2\pi)\cdot \ln[(d-r)/r]$$
$$= 2\times 10^{-7}\,\mu_r\cdot \ln[(d-r)/r] \text{ henry/metre}$$

since $$\mu_0 = 4\pi\times 10^{-7} \text{ henry/metre}$$

Example: Two conductors of diameter 31.7 mm (Moose ACSR) are separated in air by 10 metres. Calculate the self inductance per km of each conductor.

Note that $d = 10$ metres and $r = 15.85$ mm $= 0.01585$ m. Therefore, in most if not all calculations with overhead conductors, $d - r \approx d$. $\mu_r = 1$ for air.

∴ $$L = 2\times 10^{-7}\ln(10/0.01585)$$
$$= 1.29\times 10^{-6} \text{ H/m}$$
$$= 1.29 \text{ mH/km}$$

Q.3. An overhead conductor of diameter d and radius r is located at height H above the ground plane. The ground can be considered as a flux line. What is the self inductance of the conductor per unit length?

Ans. In Fig. 8.2, the currents in the conductors are equal and flow in opposite directions. The magnetic field is symmetrical about the mid-plane and the flux line is vertical. If therefore we turn the diagram by 90° and consider the two conductors located one above the other vertically separated by the distance $2H$, the mid-plane at distance H from the conductors is equivalent to the ground plane. This gives us the most famous Law of Images of James Clerk Maxwell.

Maxwell's Law or Method of Images states that when a conductor is located at a height H above a ground plane which is considered as a flux line in the field, then the ground can be dispensed with if a second imaginary conductor called the Image Conductor is placed at a distance H below ground surface and a current $-I$ is allowed to flow in it. This set of overhead conductor and its image will yield the same field distribution above the ground plane as the overhead conductor and the ground plane geometry.

Accordingly, the self-inductance or the inductance of a conductor caused by its own current will be simply:

$$L = (\mu_0\mu_r/2\pi) \cdot \ln[(2H - r)/r] \approx 2 \times 10^{-7} \cdot \ln(2H/r) \text{ henry/metre}$$

when $r << 2H$ and $\mu_r = 1$ for air.

In the example considered before, for a Moose conductor at 5 metres above ground, (conductor to image distance = 10 m), the self-inductance would be;

$$L_S = 1.29 \text{ mH/km}$$

In a 400 kV line, the 3 phase-conductors are nearly 20 metres above the ground surface. Consequently, for the single Moose ACSR conductor,

$$L_S = 2 \times 10^{-7} \cdot \ln(40/0.01585) \times 10^3 = 15.7 \times 10^{-4} \text{ H/km} = 1.57 \text{ mH/km}$$

Q.4. If a conductor of diameter d and radius $r = d/2$ is strung H metres above the ground plane, and another conductor is located at an aerial distance A from it at the same height H, what is the mutual inductance between the two conductors?

Ans. We first replace the ground surface by placing an image conductor at distance H below the ground surface, as shown in Fig. 8.3.

Now, the inductance of conductor 2 is the result of the flux linked by it due to the current flowing in conductor 1 above ground and its image below ground. There are two components to this flux. Consider the current I in the given conductor and ignore the current $-I$ in the image conductor to begin with. A magnetic field is generated by I in all of space. But since conductor is located at a distance A from conductor 1, only those flux lines or lines of force beyond A will link conductor 2. Let a distance y be chosen from 1 which we will extend to infinity since we have to include all of space. Therefore, the flux linkage of conductor 2 due to the flux generated by the conductor 1 will be, see Question 2:

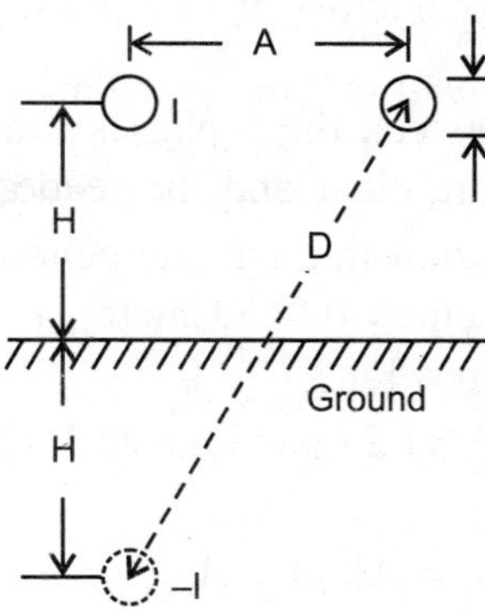

Fig. 8.3: Mutual inductance calculation between two conductors above ground plane in a transmission line.

$$\phi_{12} = (\mu_0\mu_r/2\pi)\, I\int_A^Y (1/x)\, dx$$
$$= (\mu_0\mu_r/2\pi) \cdot I \ln (y/A)$$

Next, ignore the current I in the overhead conductor 1 and consider the current $-I$ in the image conductor. The flux linked by conductor 2 due to this current is, with D = distance of conductor 2 from the image of conductor 1:

$$\phi_{1'2} = -(\mu_0\, \mu_r/2\pi) \cdot I\int_D^Y (1/x)\, dx$$
$$= -(\mu_0\mu_r/2\pi) \cdot I \ln (Y/D)$$
$$= (\mu_0\mu_r/2\pi) \cdot I \ln(D/y)$$

∴ Total flux linked by conductor 2 due to both currents flowing simultaneously is:

$$\phi = \phi_{12} + \phi_{1'2}$$
$$= (\mu_0\mu_r/2\pi) \cdot I\, [(\ln(D/y) + \ln(y/A)]$$
$$= (\mu_0\mu_r/2\pi) \cdot I \ln(D/A)$$

The resulting inductance, called the Mutual Inductance between conductors 1 and 2, both located above the same ground plane is:

$$L_m = (\mu_0\mu_r/2\pi) \cdot \ln(D/A)$$
$$= 2 \times 10^{-7} \ln(D/A) \text{ H/m}$$

A 3-phase S/C a.c. line has 3 conductors, either single or bundled, at the same heights in a horizontal configuration or in D/C vertical configuration of phases at different heights above ground. No matter how they are disposed, we only have to properly evaluate the geometrical distances:

D = distance from image of one conductor of a phase to an overhead conductor of a different phase or location, and

A = aerial distance between conductors of two phases or circuits, which will usually be given.

The mutual inductance between the 3 phases can then be evaluated. An example is provided to make the procedure clear and the reader must check all calculations.

Q.5. A 3-phase 50-Hz a.c. transmission line has the conductor disposition shown in Fig. 8.4. The conductor radius is 1.59 cm = 0.0159 metre, H = 15 m, A = 11 m. What are the self- and mutual inductances per km?

Ans. Number the conductors 1, 2, 3, and their images 1′, 2′, 3′. Then, we have the following aerial and image distances:

$$A_{12} = 11 \text{ m} = A_{23}$$
$$A_{13} = 22 \text{ m}$$
$$D_{11'} = 2H = 30 \text{ m}$$
$$D_{12'} = \sqrt{}(4H^2 + A^2)$$
$$= \sqrt{}(900 + 21) = 32 \text{ m} = D_{23'}$$
$$D_{11'} = D_{22'} = D_{33'} = 2H = 30 \text{ m}$$
$$D_{13'} = \sqrt{}(4H^2 + 4A^2)$$
$$= (900 + 484)$$
$$= 37.2 \text{ m}$$

Self Inductances:

$$L_{11} = L_{22} = L_{33}$$
$$= 2 \times 10^{-7} \ln(30/0.0159) \cdot 10^3$$
$$= 1.51 \text{ mH/km}$$

Reactances at 50 Hz: $X_{11} = X_{22} = X_{33}$

$$= 2\pi f L_S = 0.474 \text{ ohm/km}$$

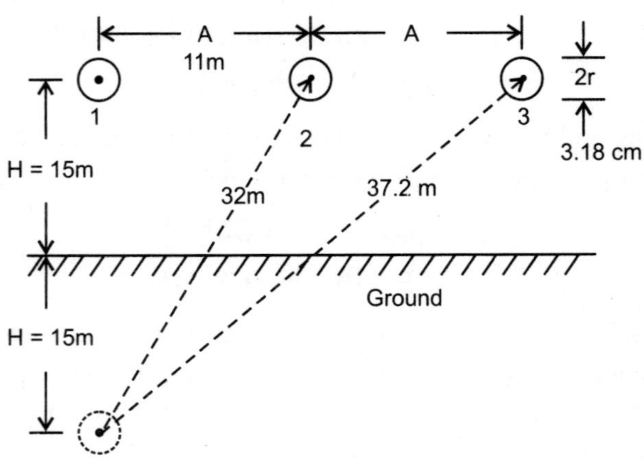

Fig. 8.4: Self-and mutual inductance calculation of S/C line with horizontal configuration.

Mutual Inductances:

$$L_{12} = L_{23}$$
$$= 2 \times 10^{-7} \ln(32/11) \times 10^3 = 0.213 \text{ mH/km}$$
$$L_{13} = 2 \times 10^{-7} \ln(37.2/22) \times 10^3 = 0.105 \text{ mH/km}$$

Reactances:

$$X_{12} = X_{23} = 0.067 \text{ ohm/km}$$
$$X_{13} = 0.033 \text{ ohm/km}$$

These self- and mutual inductances and their reactances at 50 Hz can be conveniently written out or expressed in the following matrix forms:

$$[L] = \begin{array}{c} \\ 1 \\ 2 \\ 3 \end{array}\begin{array}{c} \begin{array}{ccc} 1 & 2 & 3 \end{array} \\ \begin{bmatrix} L_{11} & L_{12} & L_{13} \\ L_{21} & L_{22} & L_{23} \\ L_{31} & L_{32} & L_{33} \end{bmatrix} \end{array}$$

$$= \begin{bmatrix} 1.51 & 0.213 & 0.105 \\ 0.213 & 1.51 & 0.213 \\ 0.105 & 0.213 & 1.51 \end{bmatrix} \text{ mH/km}$$

$$[X] = \begin{bmatrix} 0.474 & 0.067 & 0.033 \\ 0.067 & 0.474 & 0.067 \\ 0.033 & 0.067 & 0.474 \end{bmatrix} \text{ ohm/km at 50 Hz}$$

Q.6. What is meant by a transposed line and what would be the average values of inductance per unit length of the line? What are their values for the line in Question 5 before?

Ans. A transmission line is normally several hundred kilometres in length between the generating station and the receiving station. If the 3 phase-conductors are strung and occupy the geometries shown in Fig. 8.4 in a horizontal configuration, or in a vertical configuration as in a D/C line, the inductances will be unsymmetrical from end to end. Unbalanced conditions will consequently arise chiefly the voltage drops along them. Such a line is known as an "untransposed" line.

In order to balance the inductance and reactance values from end to end, it is desirable to make each phase-conductor occupy all three positions by interchanging their positions for one-third of the total distance. Such a line is called a "fully transposed" line or simply a transposed line.

For a transposed line, the average inductance of the 3 phases, both self and mutual, will become equal:

$$L_{11} = L_{22} = L_{33} = L_S$$

$$L_{12} = L_{13} = L_{23} = L_m$$

The average inductances for the line considered before will then be:

Self:

$$L_s = (1.51 + 1.51 + 1.51)/3$$
$$= 1.51 \text{ mH/km}$$
$$X_s = 0.474 \text{ ohm/km at 50 Hz}$$

Mutual:

$$L_m = (0.213 + 0.105 + 0.213 + 0.213 + 0.105 + 0.213)/6$$
$$= (4 \times 0.213 + 2 \times 0.105)/6 = 0.177 \text{ mH/km}$$
$$X_m = 2\pi \times 50 \times 0.177 \times 10^{-3} = 0.056 \text{ ohm/km at 50 Hz}$$

The matrices of line constants for the transposed line is then:

$$[L_t] = \begin{bmatrix} L_s & L_m & L_m \\ L_m & L_s & L_m \\ L_m & L_m & L_s \end{bmatrix} = \begin{bmatrix} 1.51 & 0.177 & 0.177 \\ 0.177 & 1.51 & 0.177 \\ 0.177 & 0.177 & 1.51 \end{bmatrix} \text{ mH/km}$$

$$[X_t] = \begin{bmatrix} 0.474 & 0.056 & 0.056 \\ 0.056 & 0.474 & 0.056 \\ 0.056 & 0.056 & 0.474 \end{bmatrix} \text{ ohm/km at 50 Hz}$$

Q.7. What are Maxwell's Potential Coefficients?

Ans. In evaluating the self- and mutual-inductances, we encountered the terms: $\ln(2H/r)$ and $\ln(D/A)$. These are known as "Maxwell's Potential Coefficients", the self coefficient and the mutual coefficient, respectively, and denoted by P_s and P_m. For several conductors present on a tower, a double-subscript notation has to be used,

thus $P_{mm} = \ln(2H_m/r_m)$ and $P_{mn} = \ln(D_{mn}/A_{mn})$, $m \neq n$.

Here, H_m = height of conductor m above ground,

r_m = radius of conductor m,

d_{mn} = distance between conductor m and the image of conductor n below ground, $m \neq n$,

and A_{mn} = aerial distance between conductor m and conductor n above ground, $m \neq n$.

The Maxwell's Potential Coefficients are dimensionless.

Then, the self- and mutual inductances per metre length are:

$$L_s = 2 \times 10^{-7} \ln(2H_m/r_m)$$
$$= 2 \times 10^{-7} P_{mm}$$

and

$$L_m = 2 \times 10^{-7} \ln(D_{mn}/A_{mn})$$
$$= 2 \times 10^{-7} P_{mn},\ m \neq n$$

The values of m, n take on as many values as there are conductors on a tower.

It is convenient to evaluate all distances and express them in matrix form. For the problem in Question 5, we can write:

$$[D] = \begin{array}{c} \\ 1 \\ 2 \\ 3 \end{array}\begin{bmatrix} 1 & 2 & 3 \\ H_{11} & D_{12} & D_{13} \\ D_{21} & H_{22} & D_{23} \\ D_{31} & D_{32} & H_{33} \end{bmatrix} = \begin{bmatrix} 30 & 32 & 37.2 \\ 32 & 30 & 32 \\ 37.2 & 32 & 30 \end{bmatrix} \text{ meters}$$

$$[A] = \begin{bmatrix} r_{11} & A_{12} & A_{13} \\ A_{21} & r_{22} & A_{23} \\ A_{31} & A_{32} & r_{33} \end{bmatrix} = \begin{bmatrix} 0.0159 & 11 & 22 \\ 11 & 0.0159, & 11 \\ 22 & 11 & 0.0159 \end{bmatrix} \text{ m}$$

The resulting 3 × 3 Potential-Coefficient matrix will be:

$$[P] = \begin{bmatrix} P_{11} & P_{12} & P_{13} \\ P_{21} & P_{22} & P_{23} \\ P_{31} & P_{32} & P_{33} \end{bmatrix}$$

$$= \begin{bmatrix} \ln(H_{11}/r_{11}) & \ln(D_{12}/A_{12}) & \ln(D_{13}/A_{13}) \\ \ln(D_{21}/A_{21}) & \ln(H_{22}/r_{22}) & \ln(D_{23}/A_{23}) \\ \ln(D_{31}/A_{32}) & \ln(D_{32}/A_{32}) & \ln(H_{33}/r_{33}) \end{bmatrix}$$

$$= \begin{bmatrix} 7.54, & 1.068, & 0.525 \\ 1.068, & 7.54, & 1.068 \\ 0.525, & 1.068, & 7.54 \end{bmatrix}$$

In these expressions, H_{mm} = vertical distance between a phase-conductor above ground and its image below ground = $2H_m$.

The inductance matrix is: $[L] = 2 \times 10^{-4}\ [P]$ henry/km = $0.2\ [P]$ mH/km. With larger number of conductors on a tower, it should not be much of a problem to programme the entire calculation on a Digital Computer.

The reason these coefficients P_s and P_m are called "Potential Coefficients" is that they were first formulated by Maxwell in order to determine the capacitances of the conductors, which are the ratios of relevant charges to their potentials. They have the same values as have been derived for finding the inductances.

Q.8. A D/C 220 kV line has its 6 phase-conductors disposed as shown in Fig. 8.5. Calculate the 6 × 6 inductance matrix for the line if the conductors are 0.028 m in diameter (r = 0.014 m = 1.4 cm).

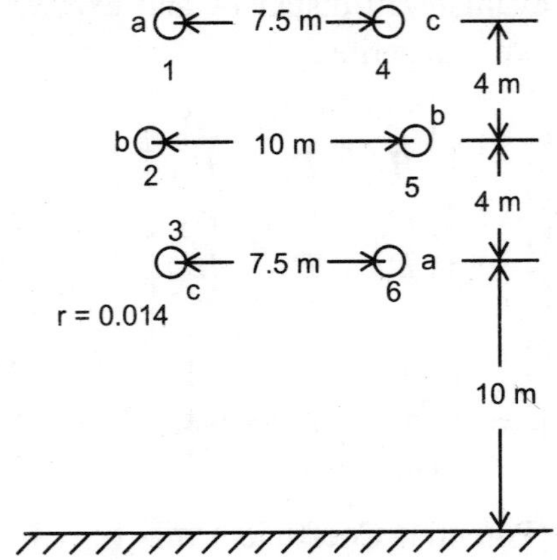

Fig. 8.5: Calculation of 6 × 6 inductance matrix of D/C 220-kV line.

Ans. We note that the middle conductors are spaced under the top and bottom conductor pairs. This is usually necessary if the conductors are not to touch during wind-induced oscillations. Note also the conductor-numbering and phase dispositions of the two circuits.

Step 1: Calculate the aerial distances in metres. The conductor radii are also included in the matrix of aerial distances.

The 6 × 6 matrix has been partitioned into four 3 × 3 matrices to differentiate the two circuits. Note the symmetry in the numbers. It is expected of course, that the reader will check all these values.

$$[A] = \begin{array}{c} \\ 1 \\ 2 \\ 3 \\ 4 \\ 5 \\ 6 \end{array}\begin{array}{c} \begin{array}{cccccc} 1 & 2 & 3 & 4 & 5 & 6 \end{array} \\ \left[\begin{array}{cc} \begin{bmatrix} 0.014 & 4.19 & 8 \\ 4.19 & 0.014 & 4.19 \\ 8 & 4.19 & 0.014 \end{bmatrix} & \begin{bmatrix} 7.5 & 9.62 & 10.97 \\ 9.62 & 10 & 9.62 \\ 10.97 & 9.62 & 7.5 \end{bmatrix} \\ \begin{bmatrix} 7.5 & 9.62 & 10.97 \\ 9.62 & 10 & 9.62 \\ 10.97 & 9.62 & 7.5 \end{bmatrix} & \begin{bmatrix} 0.014 & 4.19 & 8 \\ 4.19 & 0.014 & 4.19 \\ 8 & 4.19 & 0.014 \end{bmatrix} \end{array}\right] \end{array} \text{ metres}$$

Step 2: Calculate the distances from one aerial conductor to the image of itself and another phase-conductor in turn. This gives the matrix $[D]$.

$$[D] = \begin{array}{c} \\ 1 \\ 2 \\ 3 \\ 4 \\ 5 \\ 6 \end{array}\begin{array}{c} \begin{array}{cccccc} 1 & 2 & 3 & 4 & 5 & 6 \end{array} \\ \left[\begin{array}{cc} \begin{bmatrix} 36 & 32.02 & 28 \\ 32.02 & 28 & 24.03 \\ 28 & 24.03 & 20 \end{bmatrix} & \begin{bmatrix} 36.77 & 33.17 & 29 \\ 33.17 & 29.7 & 25.54 \\ 29 & 25.54 & 21.36 \end{bmatrix} \\ \begin{bmatrix} 36.77 & 33.17 & 29 \\ 33.17 & 29.7 & 25.54 \\ 29 & 25.54 & 21.36 \end{bmatrix} & \begin{bmatrix} 36 & 32.02 & 28 \\ 32.02 & 28 & 24.03 \\ 28 & 24.03 & 36 \end{bmatrix} \end{array}\right] \end{array} \text{ metres}$$

Step 3: Calculate the Maxwell's Potential Coefficient matrix.

$$[P]_{ut} = \begin{matrix} & 1 & 2 & 3 & 4 & 5 & 6 \\ 1 & 7.85 & 2.03 & 1.25 & 1.59 & 1.238 & 0.97 \\ 2 & 2.03 & 7.6 & 1.75 & 1.238 & 1.09 & 0.98 \\ 3 & 1.25 & 1.75 & 7.26 & 0.97 & 0.98 & 1.047 \\ 4 & 1.59 & 1.238 & 0.97 & 7.85 & 2.03 & 1.25 \\ 5 & 1.238 & 1.09 & 0.98 & 2.03 & 7.6 & 1.75 \\ 6 & 0.97 & 0.98 & 1.047 & 1.25 & 1.75 & 7.26 \end{matrix}$$

The resulting inductance matrix for the D/C line is $[L] = 0.2\ [P]$ mH/km.

$$\begin{matrix}[L]_{ut} \\ \text{(H/km)}\end{matrix} = \begin{matrix} & 1 & 2 & 3 & 4 & 5 & 6 \\ 1 & 1.57 & 0.406 & 0.25 & 0.318 & 0.248 & 0.194 \\ 2 & 0.406 & 1.52 & 0.35 & 1.248 & 0.218 & 0.196 \\ 3 & 0.25 & 0.35 & 1.45 & 0.194 & 0.198 & 0.21 \\ 4 & 0.318 & 1.248 & 0.194 & 1.57 & 0.406 & 0.25 \\ 5 & 1.248 & 1.218 & 0.196 & 0.406 & 1.52 & 0.35 \\ 6 & 0.194 & 0.198 & 0.21 & 0.25 & 0.35 & 1.45 \end{matrix}$$

For a completely transposed line, the conductors of each circuit occupy all three positions: the top, middle and the bottom for one-third of the length of line. Their self-inductances will all become equal from end to end with the average values:

$$L_{11} = L_{22} = L_{33} = L_{44} = L_{55} = L_{66} = (1.57 + 1.52 + 1.45)/3 = 1.51 \text{ mH/km}$$

Mutual inductances between conductors belonging to the same circuit:

$$L_{12} = L_{13} = L_{23} = L_{45} = L_{46} = L_{56} = (0.406 + 0.25 + 0.35)/3 = 0.335 \text{ mH/km}$$

Mutual inductances between conductors of one circuit and conductors of the other circuit:

$$L_{14} = L_{15} = L_{16} = L_{24} = L_{25} = L_{26} = L_{34} = L_{35} = L_{36} = (0.318 + 0.248 + 0.194 + 0.248 + 0.218 + 0.196 + 0.194 + 0.198 + 0.21)/9 = 0.225 \text{ mH/km}$$

The resulting inductance matrix for a completely transposed D/C line will then become, for this example:

$$\begin{matrix}[L]_{ut} \\ \text{(MH/km)}\end{matrix} = \begin{matrix} & 1 & 2 & 3 & 4 & 5 & 6 \\ 1 & 1.51 & 0.335 & 0.335 & 0.225 & 0.225 & 0.225 \\ 2 & 0.335 & 1.51 & 0.335 & 0.225 & 0.225 & 0.225 \\ 3 & 0.335 & 0.335 & 1.51 & 0.225 & 0.225 & 0.225 \\ 4 & 0.225 & 0.225 & 0.225 & 1.51 & 0.335 & 0.335 \\ 5 & 0.225 & 0.225 & 0.225 & 0.335 & 1.51 & 0.335 \\ 6 & 0.225 & 0.225 & 0.225 & 0.335 & 0.35 & 1.51 \end{matrix}$$

The reader can easily see that the Maxwell's Potential Coefficient matrix for the completely transposed line (or any type of line) will be:

$$[P] = 5\ [L].$$ (The reader should check all values given in this Question).

Q.9. What is the definition of "Geometric Mean Radius" or "Equivalent Radius" of a bundle conductor with N sub-conductors each with radius r, and neighbouring conductors separated by the bundle distance B, and with all conductors arranged on a circle with radius R? See Fig. 8.6.

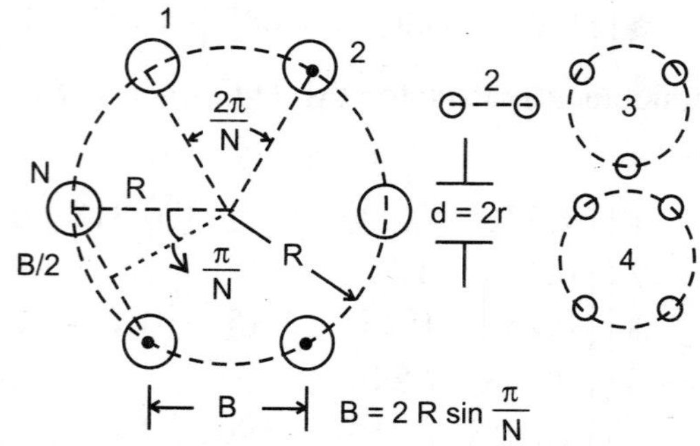

Fig. 8.6: Bundle-conductor configuration for calculation of "Geometric Mean Radius" or "Equivalent Radius", R_{eq}.

Ans. If there are N sub-conductors equidistant from each other, they will fall on a circle with radius R. Now, two neighbouring sub-conductors subtend an angle $2\pi/N$ at the centre and by dropping a perpendicular from the centre on the chord, it will bisect the base B since the triangle is isosceles.

$$\therefore \quad B/2 = R \cdot \sin(\pi/N), \Rightarrow R = B/[2 \cdot \sin(\pi/N)].$$

This is known as the "Bundle Radius", R.

The equivalent radius of the bundle is defined as follows:

$R_{eq} = R_{gm}$ = {Product of [distances from one sub-conductor to all of the remaining $(N - 1)$ companions] × [conductor radius]}$^{1/N}$.

Before developing a general formula, we will derive for several usual values of $N = 2, 3, 4, 6$ and then generalise for any N. The equivalent bundle radius is also called the Geometric Mean Radius since it is the geometric mean of the N values.

$n = 2$:
$$B = 2R$$
$$\therefore \quad R_{eq} = (r \cdot B)^{1/2} = (2rR)^{1/2}$$

$n = 3$:
$$B = 2R \cdot \sin(\pi/3) = \sqrt{3}\,R$$
$$\therefore \quad R_{eq} = (r \cdot B \cdot B\ )^{1/3}$$
$$= (3\,rR^2)^{1/3}$$

$n = 4$:
$$B = 2R \cdot \sin(\pi/4) = \sqrt{2}\,R$$
$$\therefore \quad R_{eq} = (r \cdot B \cdot B \cdot \sqrt{2}\,B)^{1/4}$$
$$= (4 \cdot r \cdot R^3)^{1/4}$$

Note the emergence of a formula:

$$R_{eq} = (N \cdot r \cdot R^{N-1})^{1/N}$$

$N = 6$: $\quad B = R$

$$\therefore \quad R_{eq} = [r \cdot 2R \cdot \sin(\pi/6) \times 2R \cdot \sin(\pi/3) \times 2R \times 2R \cdot \sin(\pi/3) \times 2R \cdot \sin(\pi/6)\ ]^{1/6}$$

$$= (0.25 \times 0.75 \times 32 \times r \times R^5)^{1/6} = (6rR^5)^{1/6}$$

Thus, in terms of the sub-conductor radius r and the bundle radius

$$R = B/\{2 \cdot \sin(\pi/N)\}$$

the general expression for the equivalent bundle radius or geometric mean radius of the bundle is:

$$R_{eq} = R_{gm} = (N \cdot r \cdot R^{N-1})^{1/N}$$

Q.10. What is the utility of the "geometric mean radius" of the bundle conductor?

Ans. In calculating the self potential coefficients of Maxwell, $P_s = \ln(2H/r)$ for a single conductor, we used the actual radius of the conductor. But for high voltage and extra-high voltage or ultra-high voltage a.c. transmission lines or d.c. transmission lines, where each phase or pole consists of a N-conductor bundle, the value of r is to be replaced by $R_{eq} = (N \cdot r \cdot R^{N-1})^{1/N}$. No proof is given here. (Please consult the author's book, "Extra High Voltage A.C. Transmission Engineering", 2nd Ed., 1990, New Age International (P) Ltd., New Delhi, Chapter 3).

Q.11. A 400-kV S/C horizontal configuration transmission line has a two-bundle Moose conductor at 45 cm spacing. The diameter of Moose conductor is 3.18 cm. The conductor heights are 15 m above ground and the phase spacing is A = 11 m, see Fig. 8.4. What are the matrices of: (*a*) Maxwell's Potential Coefficients [P], (*b*) Inductance matrix $[L]_{ut}$ for the untransposed line, and (*c*) Inductance matrix $[L]_t$ for the completely transposed line. Take line length as 400 km.

Ans.

$$r = 0.0159 \text{ m}$$

$$B = 0.45 \text{ m}$$

$$\therefore \quad R_{eq} = (0.0159 \times 0.45)^{1/2} = 0.0846 \text{ m}$$

$$H = 15 \text{ m}$$

$$A = 11 \text{ m}$$

The aerial- and image-distance matrices can be first written down.

$$[A] = \begin{bmatrix} 0.0846 & 11 & 22 \\ 11 & 0.0846 & 11 \\ 22 & 11 & 0.0846 \end{bmatrix} \text{metres}$$

$$[D] = \begin{bmatrix} 30 & 32 & 37.2 \\ 32 & 30 & 32 \\ 37.2 & 32 & 30 \end{bmatrix} \text{metres}$$

Maxwell's Potential Coefficients: $P_s = \ln (2H/R_{eq})$, $P_m = \ln (D/A)$.

$$[P]_{ut} = \begin{bmatrix} 5.87 & 1.068 & 0.525 \\ 1.068 & 5.87 & 1.068 \\ 1.068 & 1.068 & 5.87 \end{bmatrix};$$

$$[P]_t = \begin{bmatrix} 5.87 & 0.89 & 0.89 \\ 0.89 & 5.87 & 0.89 \\ 0.89 & 0.89 & 5.87 \end{bmatrix}$$

Inductances: $[L] = 0.2\ [P]$ mH/km $= 80\ [P]$ mH/400 km.

$$[L]_{ut} = \begin{bmatrix} 470 & 85.44 & 42 \\ 85.44 & 470 & 85.44 \\ 42 & 85.44 & 470 \end{bmatrix}$$

$$\underset{(\text{mH}/400\text{km})}{[L]_t} = \begin{bmatrix} 470 & 71.2 & 71.2 \\ 71.2 & 470 & 71.2 \\ 71.2 & 71.2 & 470 \end{bmatrix}$$

Q.12. What is meant by the self-inductance of a round conductor due to internal flux linkage? What is the expression for this inductance?

Ans. In the analysis considered hitherto, we have been using the surface of the conductor as the lower limit for the distance and calculated the flux linkage beyond this distance. The lower limit for integration in Question 2 is taken as r, the radius of the conductor.

But when the conductor carries the current I, and it is assumed to be uniformly distributed over the cross-sectional area, the current density is $\delta = I/\pi r^2$. The current generates a flux inside the area of the conductor as shown in Figure 8.7. There is, thus, an inductance associated with this internal flux. This is calculated as follows:

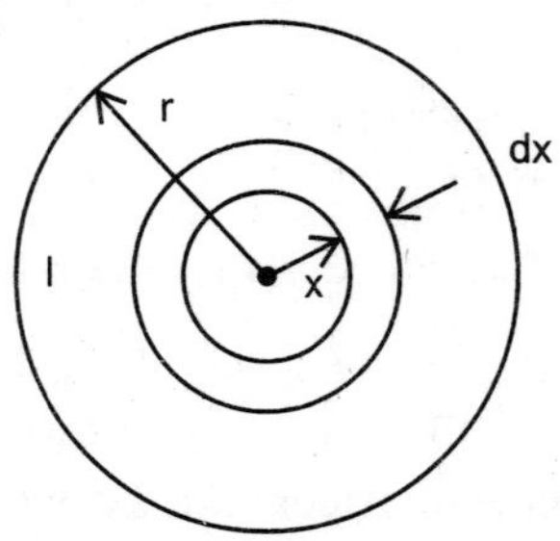

Fig. 8.7: Inductance of round conductor due to internal flux linkage. Uniform current distribution.

Consider a circular tube of radius x and radial thickness dx. The current flowing inside this tube is:

$$I_x = I(\pi x^2/\pi r^2) = I \cdot x^2/r^2$$

The field strength H_x at radius x is, by applying Ampere's Circuital law:

$$2\pi x \cdot H_x = i_x,$$

or,

$$H_x = (I/2\pi r^2)x$$

The flux density $B_x = \mu_0\mu_r H_x = (\mu_0\mu_r/2\pi) \cdot I \cdot x/r^2$

In a width dx (and 1 metre into the paper), the flux is:

$$d\phi_x = B_x \cdot dx = (\mu_0\mu_r/2\pi r^2) \cdot I \cdot x\ dx$$

Now, this flux links only the portion of conductor inside the radius x. Thus, the flux linkage up to x is:

$$d\phi_x = (x^2/r^2) \cdot d\phi_x$$

$$= (\mu_0\mu_r/2\pi) \cdot I\ (x^3/r^4) \cdot dx$$

The total flux linkage of the entire conductor due to the internal flux will be, for $x = 0$ to r:

$$\phi = \int_0^r d\phi_x$$

$$= (\mu_0\mu_r/2\pi)\ r^{-4} \int_0^r x^3 \cdot dx$$

$$= \mu_0\mu_r\ I/8\pi$$

Consequently, the inductance per metre length of conductor due to the internal flux will be: $L_I = \phi/I = \mu_0\ \mu_r/8\pi$, henry/metre. With $\mu_0 = 4\pi \times 10^{-7}$ H/m and $\mu_r = 1$ for non-magnetic material, we obtain the very interesting result that $L_I = 0.5 \times 10^{-7}$ H/m, irrespective of the size r of the conductor. This is also 0.05 mH/km for any round conductor. This value has to be added to the self-inductance we calculated earlier due to external flux from r to ∞. The total self-inductance of a conductor of radius r at height H above a ground plane will then be $L_S = 0.05 + 0.2\ \ln(2H/r)$ mH/km. In Question 2, we obtained $L_s = 1.29$ mH/km. The additional contribution to the self-inductance due to the internal flux linkage will make it a total of 1.34 mH/km, which is an increase of about 5%. In all practical problems, L_I can be neglected.

Q.13. What are the expressions for the zero-, positive- and negative-sequence reactances of a 3-phase transmission line which is perfectly transposed from end to end?

Ans. The sequence reactances govern the response of the system to driving forces, in particular the sequence currents flowing due to impressed sequence voltages, and vice versa. Therefore, in order to find the sequence reactances, we must obtain the relation between voltages and currents. It is easiest to carry this out in matrix form. The final results will be, for a completely transposed line:

$$X_0 = 2\pi f\ (L_s + 2L_m),$$

yielding

$$L_0 = L_s + 2L_m;$$

and

$$X_1 = X_2 = 2\pi f(L_s - L_m),$$

yielding

$$L_1 = L_2 = L_s - L_m$$

The matrix relations are as follows:

In terms of the phase voltages and currents, we have the relations:

$$\begin{bmatrix} V_a \\ V_b \\ V_c \end{bmatrix} = \begin{bmatrix} X_s & X_m & X_m \\ X_m & X_s & X_m \\ X_m & X_m & X_s \end{bmatrix} \begin{bmatrix} I_a \\ I_b \\ I_c \end{bmatrix}.$$

In terms of symmetrical components, with $a = 1\underline{|210°}$, and $a^2 = 1\underline{|240°}$:

$$\begin{bmatrix} V_{a0} \\ V_{a1} \\ V_{a2} \end{bmatrix} = 1/3 \begin{bmatrix} 1 & 1 & 1 \\ 1 & a & a^2 \\ 1 & a^2 & a \end{bmatrix} \begin{bmatrix} V_a \\ V_b \\ V_c \end{bmatrix};$$

$$\begin{bmatrix} I_{a0} \\ I_{a1} \\ I_{a2} \end{bmatrix} = 1/3 \begin{bmatrix} 1 & 1 & 1 \\ 1 & a & a^2 \\ 1 & a^2 & a \end{bmatrix} \begin{bmatrix} I_a \\ I_b \\ I_c \end{bmatrix}$$

Now, the inverse transformation matrix, which converts the sequence quantities into phase quantities are, which the reader should calculate and keep in mind:

$$\begin{bmatrix} V_a \\ V_b \\ V_c \end{bmatrix} = \begin{bmatrix} 1 & 1 & 1 \\ 1 & a^2 & a \\ 1 & a & a^2 \end{bmatrix} \begin{bmatrix} V_{a0} \\ V_{a1} \\ V_{a2} \end{bmatrix}; \begin{bmatrix} I_a \\ I_b \\ I_c \end{bmatrix} = \begin{bmatrix} 1 & 1 & 1 \\ 1 & a^2 & a \\ 1 & a & a^2 \end{bmatrix} \begin{bmatrix} I_{a0} \\ I_{a1} \\ I_{a2} \end{bmatrix}.$$

It can be easily checked by the reader that the transformation matrices are inverses of each other by performing the following multiplication:

$$\begin{bmatrix} 1 & 1 & 1 \\ 1 & a^2 & a \\ 1 & a & a^2 \end{bmatrix} \times 1/3 \begin{bmatrix} 1 & 1 & 1 \\ 1 & a & a^2 \\ 1 & a^2 & a \end{bmatrix} = \begin{bmatrix} 1 & 0 & 0 \\ 0 & 1 & 0 \\ 0 & 0 & 1 \end{bmatrix} = [U] \text{ or } [I], \text{ the unit matrix.}$$

One property of the unit matrix $[U]$ is that if it multiplies any matrix $[A]$ it leaves the value of $[A]$ unchanged. We use this property in evaluating the relation between sequence voltages, currents and impedances as follows: Now,

$$\begin{bmatrix} V_{a0} \\ V_{a1} \\ V_{a2} \end{bmatrix} = 1/3 \begin{bmatrix} 1 & 1 & 1 \\ 1 & a & a^2 \\ 1 & a^2 & a \end{bmatrix} \begin{bmatrix} X_s & X_m & X_m \\ X_m & X_s & X_m \\ X_m & X_m & X_m \end{bmatrix} \begin{bmatrix} 1 & 1 & 1 \\ 1 & a^2 & a \\ 1 & a & a^2 \end{bmatrix}$$

$$\times 1/3 \begin{bmatrix} 1 & 1 & 1 \\ 1 & a & a^2 \\ 1 & a^2 & a \end{bmatrix} \begin{bmatrix} I_a \\ I_b \\ I_c \end{bmatrix}$$

$$= \begin{bmatrix} X_s + 2X_m & 0 & 0 \\ 0 & X_s - X_m & 0 \\ 0 & 0 & X_s - X_m \end{bmatrix} \begin{bmatrix} I_{a0} \\ I_{a1} \\ I_{a2} \end{bmatrix}$$

$$= \begin{bmatrix} X_0 & 0 & 0 \\ 0 & X_1 & 0 \\ 0 & 0 & X_2 \end{bmatrix} \begin{bmatrix} I_{a0} \\ I_{a1} \\ I_{a2} \end{bmatrix}.$$

For the transposed line of Question 6, $X_s = 0.474$ ohm/km,

and $X_m = 0.056$ ohm/km

$\therefore$ $X_0 = 0.586$

and $X_1 = X_2 = 0.418$ ohm/km

The corresponding inductances are:

$L_0 = 1.86$ mH/km

and $L_1 = L_2 = 1.33$ mH/km

Q.14. Two long parallel cylindrical conductors of radius r each are separated by a centre-to-centre distance $2H$. They carry equal and opposite charges of $+Q$ and $-Q$ per unit length, forming a single-phase line or a bipolar d.c. line. What is the capacitance between them?

Ans. Refer to Fig. 8.8:

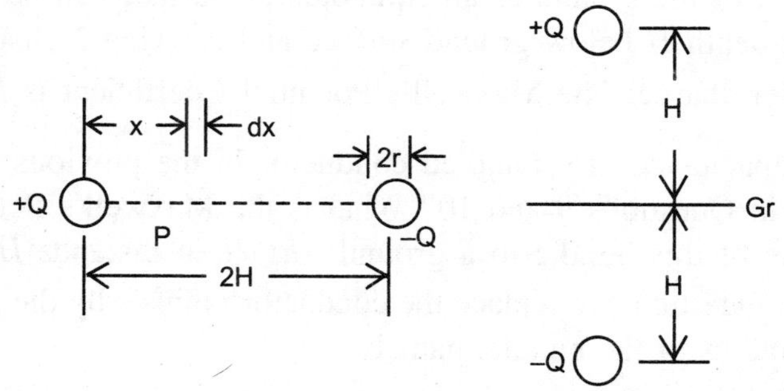

Fig. 8.8: Capacitance calculation of two parallel conductors, and conductor above plane.

Referring back to Question 8 in Chapter 5, application of Gauss's Law yielded the electric stress at a distance x from a charge to be:

$$E_x = [Q/(2\pi\varepsilon_0\varepsilon_r)]/x$$

By definition, the electric stress is the force exerted on a unit test charge. At a point P along the line joining the two charges, at a distance x from the centre of the left conductor, a unit positive test charge will be repelled by $+Q$ and attracted by $-Q$. Therefore, the two forces are in line and are additive. The resulting electric field stress at P is, with $\varepsilon_r = 1$ in air:

$$E_x = (Q/2\pi\varepsilon_0)\ [1/x + 1/(2H - x)] \text{ volt/metre}$$

The potential difference between the conductors is then:

$$V = \int_r^{2H-r} E_x \cdot dx = (Q/2\pi\varepsilon_0) \cdot \int_r^{2H-r} [1/x + 1/(2H-x)]\, dx$$

$$= (Q/\pi\varepsilon_0) \ln [(2H - r)/r]$$

The resulting capacitance is:

$$C_{cc} = Q/V$$

$$= \pi\varepsilon_0 / \ln[(2H - r)/r] \text{ farad/metre}$$

By symmetry, the mid-plane between the conductors is an equipotential plane and it is easy to see that its potential is zero.

The capacitance of each conductor to this mid-plane will be twice the capacitance between the conductors, C_{cc}, since there are two capacitances in series.

∴ Capacitance of one conductor of radius r and an infinite plane at a distance H is:

$$C_{cp} = 2 \cdot C_{cc}$$

$$= 2\pi\varepsilon_0 / \ln[(2H - r)/r] \text{ farad/metre}$$

In all overhead lines,

$$H \rangle\rangle\ r, \text{ and } C_{cp} = 2\pi\varepsilon_0/\ln(2H/r) \text{ farad/metre}$$

By turning the geometry by 90°, again we obtain Maxwell's Method of Images and Maxwell's Potential Coefficient. If an overhead conductor is at a height H above ground plane, the effect of the ground as an equipotential surface can be replaced by an image conductor at a depth H below ground surface and carrying a charge equal but opposite to the conductor charge. The Maxwell's Potential Coefficient is $P = \ln(2H/r)$, $2H \rangle\rangle r$.

Q.15. What is the capacitance of a bundled conductor in the previous Question with N sub-conductors as in Questions 9 and 10? What is the Maxwell's Potential Coefficient and the capacitance of this bundle to a ground surface at distance H?

Ans. In the previous Question, we replace the conductor radius r by the geometric mean radius or equivalent radius of the bundle, namely,

$$R_{eq} = R_{gm} = (N \cdot r \cdot R^{N-1})^{1/N}$$

Maxwell's Potential Coefficient is:

$$P = \ln(2H/R_{eq}), 2H \rangle\rangle R_{eq}.$$

The self-capacitance to ground is:

$$C_s = 2\pi\ \varepsilon_0/\ln(2H/R_{eq}) \text{ farad/metre}$$

$$= (10^{-9}/18) \cdot \ln(2H/R_{eq}) \text{ farad/metre,}$$

since $\varepsilon_0 = (1000/36\pi) \times 10^{-12}$ farad/metre

Q.16. What is the value of the product $1/\sqrt{L_s \times C_s}$? Explain its great significance in high voltage work.

Ans. From Question 3, we had

$$L_s = (\mu_0/2\pi)\ln(2H/r)$$

and we have just seen that

$$C_s = 2\pi\varepsilon_0/\ln(2H/r)$$

The values of μ_0 and ε_0 are:

$$\mu_0 = 4\pi \times 10^{-7}$$

and $$\varepsilon_0 = (1000/36\pi) \times 10^{-12}$$

Their units are henry/m and farad/m, respectively.

$\therefore$ $$L_s = 2 \times 10^{-7} \ln(2H/r)$$

and $$C_s = (10^{-9}/18)/\ln(2H/r)$$

This gives:

$$L_s \cdot C_s = 2 \times 10^{-7} \times 10^{-9}/18$$

$$= 10^{-16}/9 = (10^{-8}/3)^2$$

$\therefore$ $$1/\sqrt{L_s \times C_s} = 3 \times 10^8 \text{ metres/second}$$

= velocity of light or electro-magnetic radiation in free space.

In high voltage work, this is the velocity of electro-magnetic wave propagation in all media for which $\mu_r = 1$ and $\varepsilon_r = 1$. This relation also yields the most famous relation called the "light-velocity theory" for the relation between inductance per unit length and capacitance per unit length. [The reader should prove that the unit of $(L_s \cdot C_s)$ = sec^2/m^2.]

Hint: henry = flux-linkage/ampere = volt-sec/ampere;

farad = coulomb/volt = amp-sec/volt

$\therefore$ $L_s \cdot C_s$ = (henry/m) × (farad/m) = (sec/m)^2

Note $L(dI/dt) = V$].

Q.17. A second conductor is located at a height H_1 above ground plane and at an aerial distance A from the first conductor at height H_2, as shown in Fig. 8.9. What is the mutual capacitance between conductors per unit length?

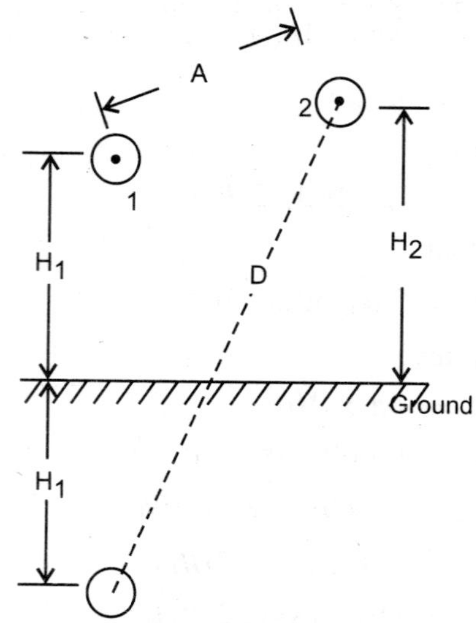

Fig. 8.9: Mutual capacitance calculation of two conductors above ground plane.

Ans. Let the first conductor be energised at a voltage V and carry a charge $+Q$ per unit length. The ground plane is replaced by an image charge $-Q$ at depth H_1 below ground. Referring to Question 14, one can write the potential difference between ground and the surface of any conductor as:

$$V_{12} = (Q/2\pi\varepsilon_0)\ \ln(\text{distance from negative charge/distance from positive charge}).$$

In this case, the potential acquired by the second conductor will be:

$$V_{12} = (Q/2\pi\varepsilon_0) \cdot \ln(D/A)$$
$$= (Q/2\pi\varepsilon_0)P_{12}$$

The mutual capacitance between the two conductors is, consequently:

$$C_m = Q/V_{12} = 2\pi\varepsilon_0/\ln(D/A)$$

Note that D = the usual distance from one conductor in air to the image of the other conductor below ground, and P_{12} = Maxwell's Potential Coefficient.

Q.18. A 400-kV 50 Hz 3-phase S/C horizontal configuration a.c. line has the geometry shown in Fig. 8.4 with conductor heights equal to 15 m above ground, separation of phases 11 m and the conductor of each phase consists of double-Moose (1.59 cm radius) at 45-cm bundle spacing. What are the self- and mutual capacitances between the 3 phases per kilometre length of line? Write them in matrix form.

Ans. The calculation of capacitance matrix of a multi-phase line is more complicated than that of its inductance matrix, or the capacitance of a single-phase line, or a bipolar d.c. line where there are only two conductors or poles to consider. The complication arises because the Maxwell's Potential Coefficients will occur in the denominator for the capacitance, and not in a straightforward manner in the numerator as for the inductance calculation as in Question 11.

We observe that: $V_{11} = (Q_1/2\pi\varepsilon_0)\cdot \ln(2H_{11}/R_{eq})$

$$V_{21} = (Q_2/2\pi\varepsilon_0)\cdot \ln(D_{12}/A_{12}) \text{ etc.}$$

Therefore, the potential of conductor 1 consists of 3 parts: Due to its own charge and self-capacitance, and due to charges of the other two phase-conductors and the mutual capacitance coefficients. Thus, for equal heights H and equivalent conductor radii $R_{eq} = \sqrt{(r\cdot B)}$, the three equations for the voltages and charges can be written down as follows:

$$\begin{aligned} V_1 &= V_{11} + V_{21} + V_{31} \\ &= (Q_1/2\pi\varepsilon_0)\cdot\ln(2H/R_{eq}) + (Q_2/2\pi\varepsilon_0)\cdot\ln(D_{12}/A_{12}) \\ &\quad + (Q_3/2\pi\varepsilon_0)\cdot\ln(D_{13}/A_{13}) \\ V_2 &= V_{12} + V_{22} + V_{32} \\ &= (Q_1/2\pi\varepsilon_0)\cdot\ln(D_{12}/A_{12}) + (Q_2/2\pi\varepsilon_0)\cdot\ln(2H/R_{eq}) \\ &\quad + (Q_3/2\pi\varepsilon_0)\cdot\ln(D_{23}/A_{23}) \\ V_3 &= V_{13} + V_{23} + V_{33} \\ &= (Q_1/2\pi\varepsilon_0)\cdot\ln(D_{13}/A_{13}) + (Q_2/2\pi\varepsilon_0)\cdot\ln(D_{23}/A_{23}) \\ &\quad + (Q_3/2\pi\varepsilon_0)\cdot\ln(2H/R_{eq}) \end{aligned}$$

In matrix form, we can write this as:

$$\begin{bmatrix} V_1 \\ V_2 \\ V_3 \end{bmatrix} = (1/2\pi\varepsilon_0)\begin{bmatrix} P_{11} & P_{12} & P_{13} \\ P_{21} & P_{22} & P_{23} \\ P_{31} & P_{32} & P_{33} \end{bmatrix}\begin{bmatrix} Q_1 \\ Q_2 \\ Q_3 \end{bmatrix};$$

or, $$[V] = [P]\cdot[Q]/2\pi\varepsilon_0$$

By symmetry, $P_{11} = P_{22} = P_{33}$; $P_{12} = P_{21}$; $P_{13} = P_{31}$; $P_{23} = P_{32}$

$\therefore$ The capacitance matrix per kilometre is, using $[Q] = [C][V]$,

$$[C] = 2\pi\varepsilon_0\,[P]^{-1} = 2\pi\varepsilon_0\begin{bmatrix} P_{11} & P_{12} & P_{13} \\ P_{12} & P_{22} & P_{23} \\ P_{13} & P_{23} & P_{33} \end{bmatrix}^{-1}$$

Now, $$P_{11} = P_{22} = P_{33} = \ln(30/0.0846) = 5.86$$

since $$R_{eq} = \sqrt{(0.0159\times 0.45)} = 0.0846 \text{ m}$$

$$D_{12} = D_{23} = 32 \text{ m}$$

$$D_{13} = 37.2 \text{ m}$$

$$A_{12} = A_{23} = 11 \text{ m}$$

$$A_{13} = 22 \text{ m}$$

$$\therefore \quad P_{12} = P_{21} = P_{23} = P_{32} = \ln(32/11) = 1.068$$

$$P_{13} = P_{31} = \ln(37.2/22) = 0.525$$

$$\therefore \quad [P] = \begin{bmatrix} 5.86 & 1.068 & 0.525 \\ 1.068 & 5.86 & 1.068 \\ 0.525 & 1.068 & 5.86 \end{bmatrix},$$

giving
$$[P]^{-1} = \begin{bmatrix} 0.176 & -.03 & -.01 \\ -0.3 & 0.18 & -.0 \\ -0.1 & -.03 & 0.176 \end{bmatrix}$$

$$\therefore \quad [C] = 2\pi\varepsilon_0\ [P]^{-1} = (10^{-9}/18)\ [P]^{-1} \text{ F/m}$$

$$= [P]^{-1}/18 \text{ nF/m}$$

$$= (1000/18)\ [P]^{-1} \text{ nF/km. } (1 \text{ nF} = 10^{-9} \text{ F}).$$

$$= \begin{bmatrix} 9.77 & -1.65 & -0.58 \\ -1.65 & 10.02 & -1.65 \\ -0.58 & -1.65 & 9.77 \end{bmatrix} \text{nF/km}$$

[The reader should check these calculations!].

We can observe the following properties from the capacitance matrix:

1. The self-capacitance of overhead high voltage lines is nearly 10 nF/km;
2. The middle conductor has a slightly higher self-capacitance than the two outer phase-conductors;
3. The mutual capacitances are all negative, while the self-capacitances are positive;
4. For a bundle, the self-capacitance is

$$C_{bc} = 2\pi\varepsilon_0/\ln(2H/R_{eq})$$

while for a single conductor it is

$$C_{sc} = 2\pi\varepsilon_0/\ln(2H/r).$$

Since $R_{eq} > r$, we have $\ln(2H/R_{eq}) < \ln(2H/r)$,

since $2H/r > 2H/R_{eq}$

But, since $\ln(2H/r)$ and $\ln(2H/R_{eq})$ occur in the denominators, $C_{bc} > C_{sc}$.

Consequently, bundling increases the capacitance of a phase-conductor.

However, it decreases the inductance since

$$L_{bc} = (\mu_0/2\pi)\cdot \ln(2H/R_{eq})$$

and
$$L_{sc} = (\mu_0/2\pi)\cdot \ln(2H/r)$$

$\therefore L_{bc} < L_{sc}$. As we will see in a later Chapter, decrease of series reactance (positive-sequence reactance $= L_s - L_m$) has some beneficial effects by increasing the power-transfer capacity according to the "tie-line" equation which we encountered before in Chapter 7, Question 11, namely, $P = E_s \cdot E_r \cdot \sin\delta / X_s$.

Q.19. What will be the self- and mutual-capacitances for a completely transposed line in Question 18?

Ans.

$$C_s = (9.77 + 10.02 + 9.77)/3 = 9.85 \text{ nF/km}$$

$$C_m = -(4 \times 1.65 + 2 \times 0.525)/6 = -1.275 \text{ nF/km}$$

The matrix of capacitances for complete transposition of the conductors is:

$$[C]_t = \begin{bmatrix} 9.85 & -1.275 & -1.275 \\ -1.275 & 9.85 & -1.275 \\ -1.275 & -1.275 & 9.85 \end{bmatrix} \text{ nF/km}$$

Q.20. What are the zero-, positive-, and negative-sequence capacitances of a fully-transposed line? What are their values in Question 18 ?

Ans. Following the argument of Question 13 for the sequence inductances, the three sequence capacitances are:

$$C_0 = C_s + 2C_m$$

and $$C_1 = C_2 = C_s - C_m$$

But we note that C_m is a negative quantity. Therefore, $|C_0| > |C_s|$, and $|C_1| = |C_2| < |C_s|$. This is the reverse of inductance relations where L_s and L_m are both positive quantities. For the line of Question 18 before:

$$C_0 = 9.85 - 2 \times 1.275 = 7.3 \text{ nF/km}$$

and $$C_1 = C_2 = 9.85 + 1.275 = 11.125 \text{ nF/km}$$

We also obtain the following relations:

$$C_s = (C_0 + 2C_1)/3$$

and $$C_m = (C_0 - C_1)/3$$

Similarly, $$L_s = (L_0 + 2L_1)/3$$

and $$L_m = (L_0 - L_1)/3$$

Q.21. What is meant by "Coupling Factor" between two overhead conductors? How is this used to determine the voltage induced in a low-voltage telephone line when it is running near a high voltage line?

Ans. Referring to Fig. 8.9, the self-capacitance of a conductor is

$$C_s = 2\pi\varepsilon_0/\ln(2H/R_{eq})$$

Consequently, if the voltage of the conductor with respect to ground is V, its charge is

$$Q = V \cdot C_s = V \cdot 2\pi\varepsilon_0/\ln(2H/R_{eq})$$

This will induce a charge on a second conductor located in the electrostatic field of this charge which, as has been seen from the capacitance matrix, is of opposite polarity giving a negative sign for the mutual capacitance. The voltage induced in the second conductor is thus:

$$v_I = Q/C_m = V \cdot C_s/C_m = V \cdot k$$

where k is called the "coupling factor".

$$k = \ln(D/A)/\ln(2H/R_{eq})$$

This is very important when considering the voltage induced in:

1. Low-voltage telephone lines running close to a high voltage line;
2. Voltage induced in unenergised circuit on a tower of a D/C line when the other circuit is already energised;
3. When lightning strikes one conductor on a tower, it induces a voltage in all other conductors on the same tower;
4. Voltage induced in an "overhead ground wire" or "shielding wire" (or wires) in the steady state by the phase voltages if the ground wire is insulated from the tower;
5. Many more induction problems.

Example: A conductor of radius 1 cm is strung 10 m above ground and energised by 132 kV 50 Hz. A telephone line is located 7 m away and 5 m above ground. What is the voltage induced in the telephone line?

Solution:

$$H = 10 \text{ m}$$

$$r = 1 \text{ cm} = 0.01 \text{ m}$$

$$D = \sqrt{(7^2 + 15^2)} = 16.55 \text{ m}$$

$$A = (5^2 + 7^2) = 8.6 \text{ m}$$

$$\ln(2H/r) = \ln(2000) = 7.6$$

$$\ln(D/A) = \ln(1.924) = 0.655$$

$$\therefore \quad v_I = 132 \times 10^3 \times 0.655/7.6$$

$$= 11.375 \times 10^3$$

$$= 11.375 \text{ kV}$$

This is a very high voltage for a low-voltage telephone line. This happens because the a.c. line is a single-phase line. In actual practice, 3-phase a.c. lines in the steady-state are energised by balanced set of 50-Hz voltages. There will then be very little voltage induced in the telephone circuits because of phasor addition of three voltages at 120° from each other. The slight voltage will be due to the distance of the telephone wires not being equal from the conductors of the high voltage line.

Where telephone lines cross high voltage power lines, it is normal practice to make them do so at right angles to their lengths and provide a mesh overhead on the telephone circuit. This mesh is grounded so as to shield the low-voltage line from induction of voltage from the power line. Crossing at right angles eliminates induction due to electro-magnetic effects by the large current carried by the power conductors.

VOLTAGE GRADIENTS ON CONDUCTORS

Q.22. A cylindrical conductor of radius r has a charge of q coulombs per metre length. What is the voltage gradient on the surface of the conductor? Assume the charge of the conductor to be concentrated at its centre, see Fig. 8.10.

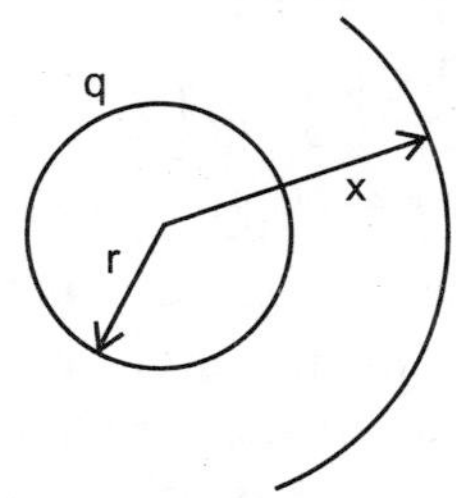

Fig. 8.10: Voltage gradient calculation on surface of round conductor.

Ans. Applying Gauss's Law to a cylinder of radius x from the centre of the conductor, as in Chapter 5 for cables, in Question 8, we have:

$$2\pi x\ \varepsilon_0 E_x = q$$

$$\Rightarrow \quad E_x = (q/2\pi\varepsilon_0)\cdot(1/x)\ \text{volt/metre}$$

Thus, on the conductor surface, the voltage gradient is

$$E_r = (q/2\pi\varepsilon_0)\cdot(1/r)\ \text{volt/metre}$$

Q.23. An overhead conductor of radius r metre is strung H metres above ground and energised by a voltage V to ground. What are the maximum and minimum surface-voltage gradients on the conductor? Gradients on the conductor?

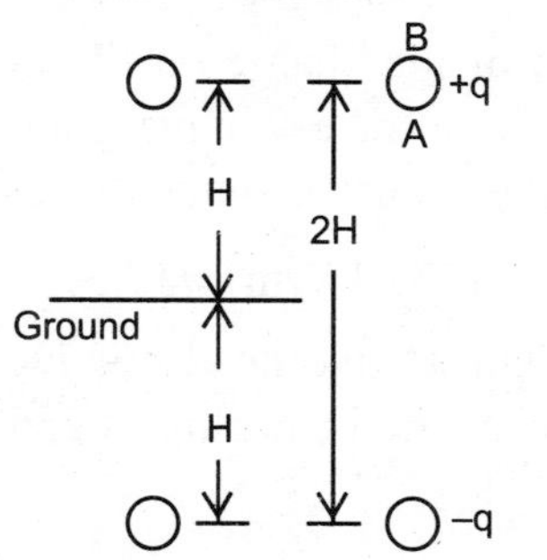

Fig. 8.11: Voltage gradient calculation on surface of single overhead line conductor above ground plane.

Ans. Referring to Fig. 8.11, the ground surface is replaced by an image conductor of charge $-q$.

$$\therefore \quad q = C \cdot V = 2\pi\varepsilon_0 \cdot V/\ln(2H/r)$$

By definition, the electric field strength at any point is the force exerted on a unit positive charge placed at that point. At point A, the bottom of the overhead conductor, the positive charge of the conductor, which we have assumed to be located at the conductor centre, repels the positive test charge, while the negative charge of the image conductor attracts the test charge. Consequently the two forces are in the same direction. It is here that maximum surface voltage gradient occurs. The gradient at point B on the top of the conductor is minimum since the two forces are in opposite directions.

$$\therefore \quad E_{max} = (q/2\pi\varepsilon_0) \cdot [(1/r) + 1/(2H - r)]$$

$$= [V/\ln(2H/r)] \cdot [(1/r) + 1/(2H - r)]$$

and

$$E_{min} = (q/2\pi\varepsilon_0) \cdot [(1/r) - 1/(2H - r)]$$

$$= [V/\ln(2H/r)] \cdot [(1/r) - 1/(2H - r)]$$

In a 3-phase line, there are other energised conductors such as 3 in a S/C line and 6 in a D/C line on the same tower. In d.c. transmission lines, there are 2 poles at $\pm V$ to ground. The individual contributions must be added vectorially in such cases. Also, if a phase-conductor or pole consists of a bundle with N sub-conductors, the charges of all the N conductors will contribute to the surface voltage gradient on any one of the sub-conductors. This will be considered in Question 25.

Example: A conductor of 1 cm radius is strung 10 m above ground and energised at 132 kV. Calculate the maximum and minimum surface voltage gradients on the conductor.

Solution:

$$\ln(2H/r) = \ln(2000) = 7.6$$

$$2H - r \approx 20$$

$$\therefore \quad E_{max} = (132/7.6)[(1/0.01) + (1/20)]$$

$$= 132(100 + 0.05)/7.6$$

$$= 1737 \text{ kV/m} + 0.05\%$$

$$E_{min} = 1737 \text{ kV/m} - 0.05\%$$

Therefore, the contribution of the charge on the image conductor below ground can be ignored when calculating the surface voltage gradient on the overhead conductor. Only its own charge need be taken into account in most cases of overhead lines.

Thus,

$$E = 1737 \text{ kV/m} = 17.37 \text{ kV/cm}$$

The surface voltage gradient of overhead conductors usually is of this value so that corona-inception does not occur in normal steady-state operation of the line in fair weather.

Q.24. In the previous Question, what is the voltage at which corona-inception will take place? Refer to Question 27 in Chapter 3 for Peek's Formula for corona-inception gradient. Take $\delta = 1$ here.

Ans. The formula that can be used for this case is:

$$E_0 = 30 \cdot \delta\,[1 + 0.301/\sqrt{(r\delta)}\,] \text{ kV/cm peak, (with } r \text{ in cm),}$$
$$= 30\,(1 + 0.301/1)$$
$$= 39.03 \text{ kV/cm}$$

But, $E = V/r{\cdot}\ln(2H/r)$

$\Rightarrow$ $V = r{\cdot}E{\cdot}\ln(2H/r)$

$\therefore$ $V = 1 \times 39.03 \times 7.6 = 296.63$

peak $= 296.3/\sqrt{2}$

$= 209.8$ kV, r.m.s.

The general formula for the corona-inception voltage of an overhead conductor is:

$$V = 30 \cdot \delta\,[1 + 0.301/\sqrt{(r\delta)}] \cdot r{\cdot}\ln\,(2H/r), \text{ kV peak,}$$

when all dimensions are in cm. When all dimensions are in metre units,

$$E_0 = 3000 \cdot \delta\,[1 + 0.0301/\sqrt{(r\delta)}], \text{ kV/m peak, and}$$
$$V = E_0 \cdot r \cdot \ln(2H/r)$$
$$= 3000 \cdot [1 + 0.0301/\sqrt{(r\delta)}] \cdot r{\cdot}\delta \cdot \ln(2H/r) \text{ kV peak.}$$

Recall that δ = air-density factor

$= [273/(273 + t)] \cdot (p_{mm}/760)$,

where t = temperature in °C,

and p_{mm} = pressure in mm Hg.

Q.25. What is the expression for surface voltage gradient on a sub-conductor of radius r in a bundled conductor with 2 sub-conductors spaced at a bundle spacing B, if the bundle height above ground is H (to the centre of the bundle)?

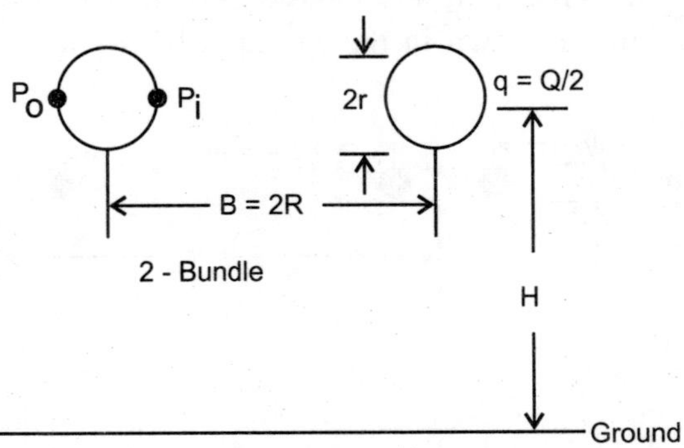

Fig. 8.12: Voltage gradient calculation on sub-conductor surface of 2-bundle conductor above ground.

Ans. Refer to Fig. 8.12. Take the total charge of the bundle = Q, and the charge on each sub-conductor as $q = Q/2$. Neglect the effect of the image conductors as shown in Question 22. Consider two points on a conductor surface on the line joining the two sub-conductors, shown as P_I and P_o, which can be easily seen to be the internal and external points. P_I is closer to the charge of the second conductor than P_o. Since the charges of both sub-conductors are of the same polarity, it is clear that the voltage gradient at the external point P_o will be higher than at the internal point P_I, since the force on a unit test charge is additive at P_o.

We now have to use a property of an electrostatic field which states the following: "When a metallic cylinder is placed inside a uniform electric field caused by an external charge or charges, the electric stress on the surface of the metallic conductor is twice the normal uniform electric stress when this conductor is absent." Using this property, we obtain the following result: consider the charge q on the right sub-conductor. The electric stress at a distance B from this charge is normally $E = (q/2\pi\varepsilon_0)/B$. But keeping a second metallic conductor at distance B will double the electric stress on the surface of this sub-conductor to the value:

$$E = (q/2\pi\varepsilon_0)\cdot(2/B)$$
$$= (q/2\pi\varepsilon_0)/R, \text{ since } B = 2R$$

where R = bundle radius. Therefore:

At P_o: $$E_{\max} = (q/2\pi\varepsilon_0)\,[1/r + 1/R]$$
$$= (Q/4\pi\varepsilon_0)\cdot(1 + r/R)/r \text{ kV/m}$$

and At P_I: $$E_{\min} = (q/2\pi\varepsilon_0)\,[1/r - 1/R]$$
$$= (Q/4\pi\varepsilon_0)\cdot(1 - r/R)/r \text{ kV/m}$$

The bundle charge is calculated from the voltage V, height H and the geometric mean radius $R_{eq} = \sqrt{(r\cdot B)}$ for the two-conductor bundle. Refer to question 8.9. The bundle charge is $Q = 2\pi\varepsilon_0\cdot V/\ln(2H/R_{eq})$ when it is alone. But if there are several phase-conductors in a S/C line, or more in a D/C line, or 2 poles in a bipolar d.c. line on the same tower, then the charge of any bundle must be calculated from Maxwell's Potential Coefficients, as will be shown in the example below for a bipolar ±400 kV d.c. line.

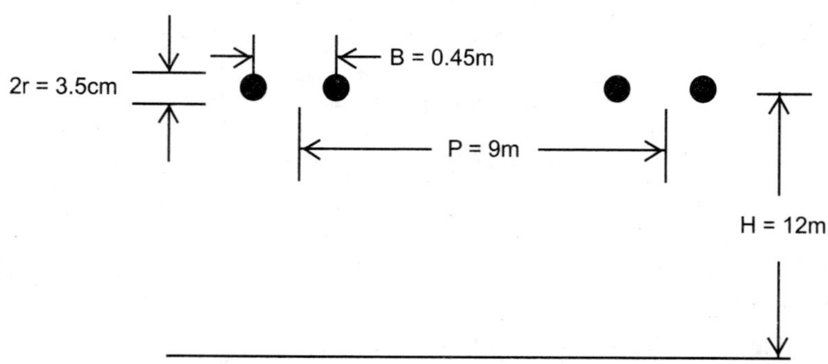

Fig. 8.13: D.C. bipolar line. Voltage-gradient calculation.

Example: A bipolar ±400 kV d.c. line has the dimensions shown in Fig. 8.13. The poles are at a height $H = 12$ m above ground, and each pole consists of a two-conductor bundle with 3.5 cm diameter of each sub-conductor, and the bundle spacing is $B = 45$ cm $= 0.45$ m. The pole spacing is 9 m. $r = d/2 = 3.5/2$ cm $= 1.75$ cm $= 0.0175$ m.

What are the maximum, minimum and average voltage gradients on the surface of the sub-conductors? Neglect the effect of charges on the image conductors, and omit the effect of charges of one pole on another. Consider only the charge on one bundle in calculating the voltage gradients on the sub-conductor surfaces.

Solution: We do this step by step.

Step 1: Find the self- and mutual Maxwell's Potential Coefficients. Number the poles 1 and 2.

$$R_{eq} = \sqrt{r \cdot B} = \sqrt{0.0175 \times 0.45} = 0.08874 \text{ m}$$

Then

$$P_{11} = P_{22} = \ln(2H/R_{eq}) = \ln(24/0.08874) = 5.6$$

$$P_{12} = P_{21} = \ln(D/A) = \ln(25.6/9) = 1.0465$$

since

$$D = \sqrt{(24^2 + 9^2)} = 25.6 \text{ m}$$

The pole-voltages are $V_1 = +\,400$ kV to ground, and $V_2 = -400$ kV.

Step 2: Set up the charge-voltage relationships and solve for Q_1 and Q_2. It should be evident that Q_1 and Q_2 are equal but opposite in polarity. Also, we need the quantity $(Q/2\pi\varepsilon_0)$ for calculating the voltage gradients. We can write

$$Q_1 = -Q_2 = Q \text{ for convenience}$$

$$V_1 = 400 \times 10^3 = (1/2\pi\varepsilon_0) \cdot (P_{11} \cdot Q_1 + P_{12} \cdot Q_2) = 5.6\, Q/(2\pi\varepsilon_0) + 1.0465\, (-Q/2\pi\varepsilon_0);$$

$$V_2 = -400 \times 10^3 = (1/2\pi\varepsilon_0) \cdot (P_{12} \cdot Q_1 + P_{22} \cdot Q_2) = 1.0465(Q/2\pi\varepsilon_0) + 5.6\,(-Q/2\pi\varepsilon_0)$$

These equations result in: $Q/2\pi\varepsilon_0 = 87.84 \times 10^3$ volt.

If need be, we have the actual charges of each bundle (equal and opposite for the two poles) to be:

$$Q = 2\pi\varepsilon_0 \times 87.84 \times 10^3 = (2\pi \times 10^{-9}/36\pi) \times 87.84 \times 10^3 \text{ farad/metre}$$

$$= (87.84/18) \times 10^{-6} \text{ F/m}$$
$$= 4.88 \ \mu\text{F/m}$$

Consequently, the charge of each sub-conductor in the two-conductor bundle will be

$$q = Q/2 = 2.44 \ \mu\text{F/m}$$

and the quantity $q/2\pi\varepsilon_0 = 43.92 \times 10^3$ V

Step 3: Then, the voltage gradients are:

$$E_{max} = (q/2\pi\varepsilon_0) \cdot (1/r) \cdot (1 + r/R)$$
$$= 43.92 \times 10^3 \times (1/0.0175) \cdot (1 + 0.0175/0.225)$$
$$= 2510 \times 10^3 \ (1 + 0.078) \text{ volt/metre}$$
$$= 2510(1 + 0.078) \text{ kV/m}$$
$$= 2705 \text{ kV/m}$$
$$= 27.05 \text{ kV/cm}$$
$$E_{min} = 2510(1 - 0.078) \text{ kV/m}$$
$$= 2314 \text{ kV/m}$$
$$= 23.14 \text{ kV/cm}(\pm 7.8\% \text{ from } 2510 \text{ kV/m})$$
$$E_{av} = (q/2\pi\varepsilon_0)/r$$
$$= (E_{max} + E_{min})/2$$
$$= 2510 \text{ kV/m} = 25.1 \text{ kV/cm}$$

The contribution of the charge of the second sub-conductor of the bundle on the surface voltage gradient on the first is 7.8% .

[**Note:** For a general formulation of surface voltage gradients on an *N*-conductor bundle, please refer to the author's book, *Extra High Voltage A.C. Transmission Engineering*, New Age International (P) Ltd., New Delhi, (2nd Ed., 1990), Chapter 4].

❑❑

CHAPTER 9

LIGHTNING PHENOMENON AND LIGHTNING PROTECTION

Q.1. What is the purpose of understanding lightning phenomenon?

Ans. Prior to the terrible damages humanity learnt to inflict by mass weapons of destruction, natural disasters such as earthquakes, volcanoes, floods, hurricanes, cyclones, lightning, etc., have claimed immense loss of property and life which they still do. The purpose of understanding lightning and other natural phenomena that cause devastation is to take adequate protection against these destructive forces generated by nature. It is entirely by understanding the scientific principles of lightning that mankind can protect property and life. It is the highest-voltage phenomenon in nature and since the 1920s has been studied systematically with all scientific tools available to the scientific society, and at much expenditure of money.

Q.2. What are some of the more important scientific tools utilised for the study of lightning which have enabled engineers to devise protective measures against its destructive effects?

Ans. These are so numerous that a complete list is too long to be given here. The scientific literature on lightning alone occupies a major portion of physics, called atmospheric physics, meteorology and electrical engineering. It is the most important branch of "Atmospheric Electricity" and the reader should consult the exciting literary and scientific output. But the most useful of the scientific methods have been the following, after the famous "Kite Experiments" of Benjamin Franklin in the 1750s at Philadelphia in USA.

1. Instrumentation placed aloft balloons to measure wind velocity, temperature, electric-charge intensity and voltage gradients in thunderclouds;
2. Observation by flyers in aircraft inside and near thunderclouds;
3. Systematic investigation of lightning phenomena from the top of very tall buildings and mountain tops. The more famous ones are the Empire State Building in New York City, Mt. Salvatore in Switzerland, etc.;

4. Instrumentation carried in ships at sea specially built with scientific exploration in view;
5. Measurements with high voltage power-transmission lines; and last but not least,
6. Investigations all over the world in high voltage laboratories equipped with "Impulse Generators" using the Marx voltage-multiplier circuits.

One can observe from the above list that the tools and methods used are very extensive and original, not to mention the cost.

Q.3. What are the meteorological and electrical conditions inside a thundercloud?

Ans. A thundercloud carries enormous amount of positive and negative electric charge in what are known as "charge centres". There may be up to ten bunches or concentration of charges, some positive and some negative. Figure 9.1 shows one of the most famous thundercloud pictures (schematic) based on the theories of G. C. Simpson, one of the pioneers in Atmospheric Electricity and thundercloud activity.

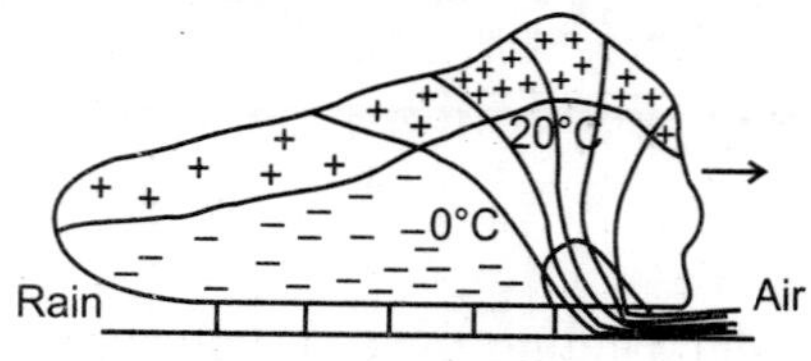

Fig. 9.1: G.C. Simpson's model of meteorological and electrical conditions inside a thundercloud.

The cloud's base can be 1 to 5 kilometres and its top 5 to 30 km above ground. Most of the cloud is negatively charged with its top portion carrying positive charge centres. Such a picture of the distribution of charge in the cloud is the result of theories based on observations of ground-level electric-field gradients and balloon experiments. When a cloud is approaching a point on ground from far away, as well as when it has receded for a long distance, the electric field is measured to be positive on ground. This is caused by positive charge centre at the top of the cloud. When the cloud is overhead, the field intensity is negative showing that the base of the cloud carries a negative charge up to nearly its middle or a little above it. The intensities of measured charges and electric stress at ground level are indications of the intensity and nature of electric charge up in the cloud, see Fig. 9.2.

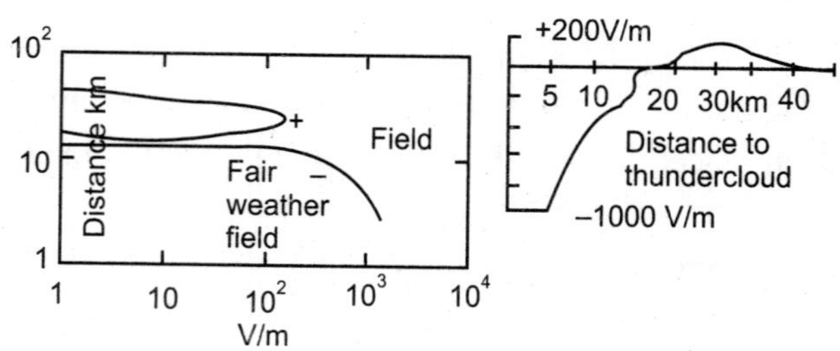

Fig. 9.2: Potential gradient at ground with distance of thundercloud.

Q.4. What are the factors which govern the formation and distribution of electric charge in a thundercloud?

Ans. The most important factors, ascertained from balloon-borne sensors and aircraft are the following:

(*a*) Water droplets in the atmosphere and inside the bottom portions of a cloud;

(*b*) Air currents, at the base and inside the cloud;

(*c*) Super-cooled water drops and vapour;

(*d*) Formation of ice crystals and snow flakes inside the cloud;

(*e*) Blizzard-like conditions with air currents at high speeds blowing the snow and ice particles causing violent friction;

(*f*) Separation of charges due to temperature effects, phase transition of water into ice and snow, releasing latent heat of cooling into the atmosphere in the cloud causing winds at high velocity and friction; and many other factors.

Because of many factors which are based on results of air-borne instruments, aircraft, ground-level observations, several theories for the mechanism of electrification of thunderclouds have been advanced from time to time. There is no single theory that encompasses the entire mechanism.

Q.5. What are the most important theories accepted until now for thundercloud-electrification?

Ans. These are:

1. Wilson's Theory based on availability of large number of charged ions in the atmosphere;
2. Simpson's Theory of break-up of large water drops into smaller ones which are charged during the process of breaking-up;
3. Charge separation caused by freezing of water into ice crystals and friction due to high-velocity wind.

Q.6. What are the details of Wilson's Theory of electrification of thunderclouds?

Ans. C. T. R. Wilson (Nobel Prize winner for his "Cloud Chamber" in atomic and nuclear physics), based the electrification process of water drops by friction with air currents. A large number of ions exist naturally in the atmosphere. These can be both positive or negative. These ions attach themselves to minute particles of dust and water drops. By electrical attraction, they form larger particles. The theory postulates that the bottom part of the particle acquires a positive charge and the top part a negative charge, which is caused by the electric field existing between cloud and ground, the ground usually being negative with respect to the cloud. This field is of the order of 10 kV/cm. By collision with the bottom positive charge of a large drop, a smaller drop acquires a negative charge. This is carried by air currents upward into the cloud which then acquires a negative charge. Inside the cloud, the introduction of these negative charges disturbs the neutrality of the atmosphere so that the top portion of the cloud acquires a positive charge.

Q.7. What are the principles of Simpson's Theory of charging of a thundercloud?

Ans. Simpson's Theory is based on the results of several types of experiments such as: (*a*) Laboratory experiments; (*b*) Meteorological observations by balloons; (*c*) Observations of electrical conditions during blizzards in the Antarctic.

The first process taking place is that a water drop changes its shape, becoming flat when its diameter is greater than 0.5 cm and becomes unstable when it falls by gravity through air currents which are usually rising upwards. For this to happen, the air current must be below 8 m/s (about 30 km/hour). If air has a higher velocity, no rain drops fall to ground. The breaking up of larger water drops makes them acquire positive charge while the air becomes negatively charged. The wind velocity inside the cloud is very high and the smaller drops are carried upwards so that the top of the cloud acquires a negative charge. The larger drops stay at the bottom of the cloud with their negative charge.

We note that Wilson's Theory offers a more convincing explanation for the cloud base and lower portions to acquire a negative charge while Simpson's Theory explains the presence of positive charge-centres in the upper portion of the cloud, not by air currents, but by another process.

Simpson has also advanced the theory that with ice crystals and swift-moving air, the opposite is true; namely, the air is positively charged while the ice is negatively charged. These differences in polarity of water drops and ice crystals lie in the manner the bonds between oxygen and hydrogen in these two phases are oriented or polarised. Thus, inside the cloud where water droplets freeze into ice crystals, the ice gives rise to a negative charge centre while the rising air carries positive charge upwards and creates positive charge-centres at the top of the cloud. This is a more convincing mechanism.

These processes may therefore be summarised in the words of one of the most honoured lightning investigators, Sir Basil Schonland. "It is at first sight evident that these great electrical machines, capable of developing potentials of thousands of millions of volts and of producing sparks many miles long, are made up by nothing but water and ice blown by the wind. The effectiveness of the thundercloud as an electric generator arises from the enormous number of water drops and ice particles in it, even though each of them is very feebly charged with electricity." ["The Flight of Thunderbolts" by B.F.J. Schonland. Oxford University Press, 1964].

Q.8. What are the orders of magnitude of charge, voltage to ground, current, and energy in a lightning stroke and heights of thunderclouds?

Ans. (*a*) **Heights.** The base of thunderclouds (cumulo nimbus) can range from 300 metres (1000 ft) to 1500 m (5000 ft) from which lightning strokes can be initiated to ground. Clouds higher than these are very unlikely to flash to ground, but generate discharges inside the cloud (intra-cloud) and between other clouds (inter-cloud).

(*b*) **Charge.** A charge-centre which can initiate a discharge to ground has a charge of the order of 20 coulombs and higher. (Charge of one electron = -1.602×10^{-19}C) (*c*) **Voltage to Ground.** This is estimated at 10 million to upwards of 100 million volts. To initiate a lightning stroke from a cloud whose charge centre is 300 m above ground with an average electric field gradient of 1 kV/cm or 100 kV/m, the average potential is nearly 30,000 kV = 30 MV;

(*d*) **Current.** From ground-based experiments, the maximum current measured by instruments is about 250,000 amps = 250 kA. The magnitude of current experienced at ground level is highly statistical in nature, but lightning arresters for 200 kA have been designed and manufactured. In high-current impulse-testing laboratories, generators can develop up to this current for testing lightning arresters.

(*e*) **Energy.** With the magnitudes of charge and voltage given before, it is clear that the energy in a lightning stroke is enormous, but it lasts for only a milli-second. With $Q = 20$ C and $V = 10^7$ volts, energy $E = Q \cdot V = 200$ mega joules. If this energy is released in 10 ms, the power is $P = 200 \times 10^6/10 \times 10^{-3} = 2 \times 10^{10}$ W = 20 GW. Unfortunately, there is no way of using this energy and it can only cause destruction. [It is of scientific interest of a general nature that billions of years ago, due to constant and incessant lightning hitting the oceans (the primordial soup) on the surface of the globe, this energy was the cause of the birth of amino acids and of the life process on earth! Thus, lightning stroke is the father and mother of all life forms on earth, including man].

Q.9. What are the different components and stages of a lightning stroke between cloud and ground?

Ans. About 80% of all lightning discharges take place as intra-cloud and inter-cloud flashovers. Because the thin water vapour acts like a veil or curtain which partially hides these flashovers when viewed from ground, they appear as sheaths and are thus named "sheath lightning strokes".

Of the remaining 20% which reach the ground as we witness them, 90% are due to lowering of negative charge from the lower portion of a thundercloud. We will describe only this type of stroke.

There are two broad types of lightning strokes: (1) Single stroke; and (2) Multiple strokes. The latter are the usual type, but let us first see what happens in a single stroke. There are basically 4 components to it:

1. Lowering of negative charge through a "pilot streamer" with a few Amps of current and velocity of 1/2000-th of light velocity. This appears as a continuously-moving discharge on photographs;
2. A stepped leader or forked leader which follows the pilot streamer, with velocity of 1/6-th light velocity and jumps in steps of about 50 m. The cloud-charge is lowered and collects at the tip of the stepped leader as a luminous point which causes sufficiently high electric field gradient to break down further path of the moist air;
3. A short flow of intense positive charge from the ground shooting up to meet the channel of the downward-coming stepped leader. This is about 10 to 50 metres in height and is usually initiated from a tall structure or a transmission-line tower, etc.; and finally

4. The return stroke which is the intense flow of positive charges from ground through the ionised channel of the stepped leader which eventually reaches the cloud, and neutralises a charge-centre from which the pilot streamer and stepped leader were initiated. The current in the return stroke can range from 1 to 200 kA, and the velocity is about 10% of light velocity.

Fig. 9.3: Four stages of progress of a single stroke of lightning. Pilot leader, Stepped leader, charge from ground meeting the stepped leader, and return stroke.

The sequence of events is depicted pictorially in Fig. 9.3. It takes about 10 ms = 10,000 μs for the pilot streamer and stepped leader to reach ground from cloud, but the return stroke takes only about 50–100 μs.

The mechanism in a multiple stroke contains all of the above. Further on, in the cloud, a second charge-centre can discharge through this intensely ionised channel of the first return stroke and lower further negative charge. Since the channel is already ionised, this charge shoots down like an arrow or dart and is thus called the "dart leader". Once again, positive charge flows rapidly after the dart leader reaches ground. If the ionisation of the channel is maintained, several back and forth strokes take place giving rise to multiple strokes, as shown in Fig. 9.4.

Therefore, what we on ground witness as the intensely-bright lightning stroke is caused by the extremely heavy current that constitutes the flow of positive charges from ground to cloud. A single stroke normally lasts for 1 milli-second while a multiple stroke can have a duration of up to 1 second with usual duration of 100 ms = 0.1 second.

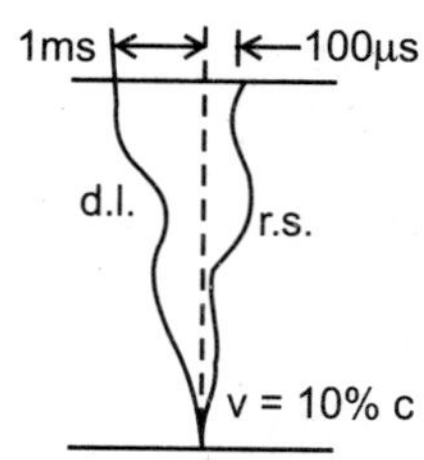

Fig. 9.4: Progress of multiple stroke after Fig. 9.3. Dart leader and second return stroke.

It is estimated that the energy of charged particles in the return stroke is equivalent to a temperature in the channel of at least 10,000°C. Since the air heats up to this temperature in the channel of about 0–20 cm in diameter, it expands extremely rapidly into the cooler air at tremendous rates or velocities giving rise to supersonic shock waves. In addition to the temperature which can destroy trees and buildings by heat, the supersonic expansion of air gives explosive conditions. The sound caused by the boom is what we hear as thunder, which is a succession of ear-splitting and terrifying supersonic booms in air.

It is clear that the return stroke originates at ground level through the point of contact of the stepped leader from the thundercloud. This forms the basis for principles of protection of structures against damage due to lightning. By erecting tall buildings with pointed rods on top, the stepped leader is attracted to the tallest point in the vicinity and the return stroke can only originate through this "lightning rod". It is normally estimated that all stepped leaders which are coming down in an area whose radius equals the height of this tallest structure are attracted to this lightning rod. This gives a zone of protection which is a right-circular cone whose height and radius are equal i.e., the semi-apex angle is 45°, as shown in Fig. 9.5. Structures with metallic roofs are safer than concrete roofs. Also, steel reinforcement rods help. All these conductors must be earthed very well for the current and are called "lightning grounds".

This zone of protection depends also on the magnitude of current in the return stroke so that in regions of intense lightning activity the angle is lower than 45°. An example is 40 m height for 10 kA current and 100 m for 50 kA current. These are also called "striking distances".

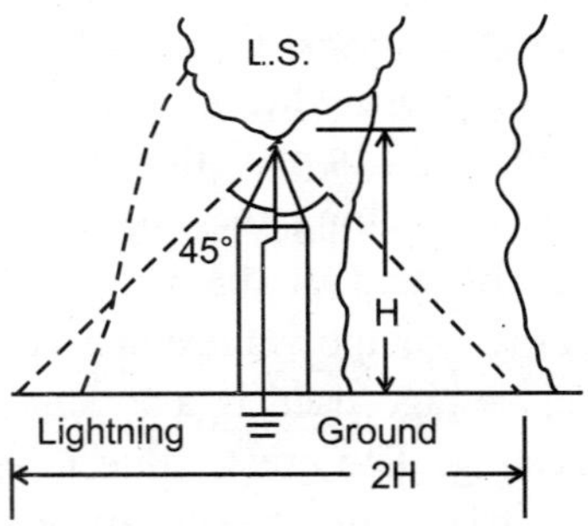

Fig. 9.5: Protective zone of lightning rod.

Q.10. What are the magnitudes of current and waveshapes of lightning strokes?

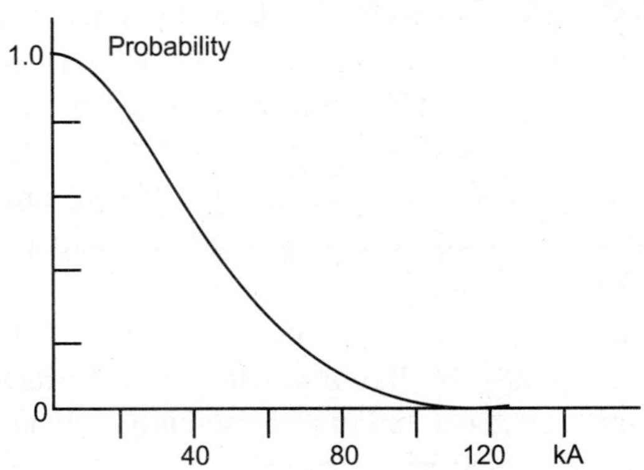

Fig. 9.6: Probability-distribution of lightning-current magnitudes.

Ans. These are extremely variable and therefore can only be described in statistical terms. A typical probability curve is shown in Fig. 9.6. As the current in the stroke increases, the probability of finding this magnitude decreases. An equation expressing this probability is, with I_c in kA:

$$p = 1.175 - 0.015 I_c, \; 25 \text{ kA} < I_c < 70 \text{ kA}$$

For example, a current of 25 kA crest value occurs with a probability of 80% or in 80% of the strokes the crest value of current can be expected to be 25 kA and lower. A crest value of 50 kA can be expected in 42.5% of the number of strokes contacting ground; etc.

Lightning arresters are designed normally for currents from 5 kA to 20 kA for transmission-line and equipment protection. Thus, testing laboratories must be equipped to test lightning arresters at these currents.

Waveshapes. There are two types of waveshapes:

(*a*) Lightning Voltage Waveshapes; and (*b*) Lightning Current Waveshapes. Once again, experiments have revealed that these show a large range of variation. But for testing purposes they have to be standardised.

(*a*) **Voltage Waveshape.** The standard waveshape of lightning voltage has been given in Question 2 in Chapter 1. According to I. E. C. and I. S., the voltage has a wavefront of $t_f = 1.67(t_{90} - t_{30})$, where t_{90} and t_{30} are the timings at 90% and 30% of crest value on the rising portion of the wavefront. As per the standards, this is 1.2 μs. The voltage, after reaching the crest, decreases to 50% crest value in the time t_t, measured from the zero of the wave. This time is standardised at 50 μs. Thus, a standard lightning-voltage waveshape is designated as 1.2/50 μs wave. The crest value of lightning voltage used is also standardised. This is considered in the Chapter on Impulse Testing. The crest value is usually controlled by the flashover voltage of an insulator-string on a tower, since when it flashes over the lightning stroke is grounded and the voltage collapses to zero. Always refer to Standard Specifications.

(*b*) **Current Waveshapes.** A lightning stroke is the intense movement of charge at high velocities. Thus, $I = Q \cdot v$. The higher the velocity v, the greater the current for a given amount of charge Q. This current is caused by the return stroke which flows in the ionised channel caused by the stepped leader. Estimates near tall buildings and transmission-line towers give a charge of 5 μC/cm travelling at a velocity of 10% of light = 3×10^9 cm/sec. This yields a current of $I = 5 \times 10^{-6} \times 3 \times 10^9$ = 15,000 A = 15 kA.

The rise time of this crest value occurs in about 2–10 μs, and falls to zero from crest in 10–30 μs. Thus, the lightning-current waveshape has been standardised at 8/20 μs which must be generated in impulse-current generators in a testing laboratory meant for testing lightning arresters.

Q.11. How are transmission lines protected from a lightning stroke by using overhead ground wire or wires? What is meant by the term "protective angle"?

Ans. It is not a good principle or practice to allow lightning strokes to strike the phase-conductor or pole on a transmission line, although it is not dangerous to the system, because lightning arresters are provided at both ends of the transmission line connected between the live conductor and ground. If a stroke contacts a live conductor, the voltage to ground of that phase or pole rises which may be sufficient to sparkover an insulator on a tower. This represents a line-to-ground fault or short-circuit into which the normal voltage of line (a.c. or d.c.) can drive a current. The normal protective system of the power system can easily take care of this by opening a circuit breaker. However, this causes interruption to the power supply. Until "single-pole reclosing" scheme of protection was introduced, at extra cost, power interruptions during stormy days (electrical storm) were frequent and lengthy. But with the installation of 3 single-pole circuit breakers, only the faulted phase or pole need be opened and since the lightning stroke lasts for a maximum length of only 1 milli-second (1/20-th of a cycle in 50-Hz systems) the circuit-breaker pole can be arranged to be opened and then re-closed automatically after say 2 cycles. This is a topic in system protection and system stability.

However, a useful and more certain remedy (and less expensive) is to intercept the stepped leader before contacting the live phase-conductors. This is carried out by overhead ground wires which are made from galvanised steel or could be ACSR conductors, strung above the live conductors and connected to earth at every tower. These serve the same purpose as lightning rods as described before in Question 9.

From vast experience accumulated since 1920s when C. L. Fortescue gave his theory of direct strokes and protection against them, the following rule is followed as good practice, see Fig. 9.7.

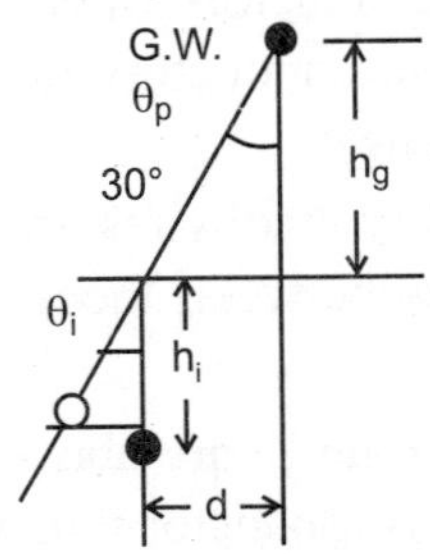

Fig. 9.7: Protective angle of an overhead ground wire.

An overhead ground wire is located at such a height above the phase conductors or poles so that at maximum swing of the insulator in wind, the live conductors or protected conductors fall within an angle of 30°.

Let h_I = height of insulator string below the cross arm,

h_g = height of ground wire above the cross arm,

θ_I = maximum allowed swing of insulator string,

and d = horizontal distance between ground wire and protected conductor.

Then, for 30° protective angle, at maximum swing of insulator, we have:

$$\tan 30° = 1/\sqrt{3} = (d + h_i \cdot \sin \theta_I)/(h_g + h_i \cdot \cos \theta_I).$$

This gives: $$h_g = \sqrt{3}d + h_I(\sqrt{3} \sin \theta_I - \cos \theta_I).$$

As an example, if the insulator swing is $\theta_I = 30°$, then for 30° protection angle, $h_g = \sqrt{3}d$.

A single ground wire on top of the tower along its centre line is adequate for 132 kV, S/C or D/C. Normally, for a 220 kV D/C line, two ground wires are required so that all the six phase-conductors will fall within 30° of the vertical from the ground wire as shown in Fig. 9.8.

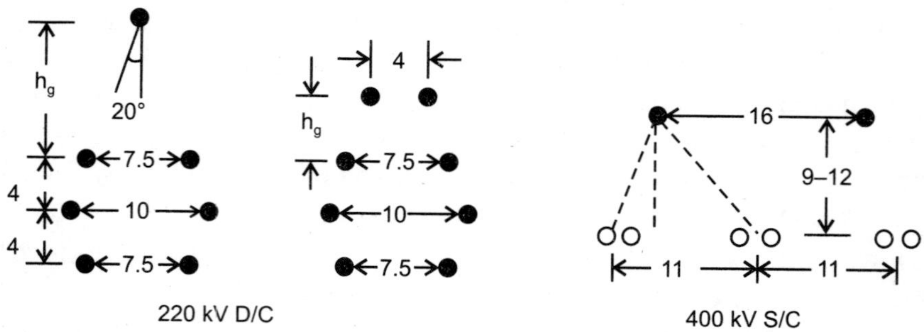

Fig. 9.8: Examples of protective angle of single- and double overhead ground wires on double-circuit 220 kV line. Double ground wire protection of S/C 400 kV e.h.v. line.

Example: A 220-kV 3-phase D/C line has the phase conductors disposed as in Fig. 8.5. Calculate

(*a*) The height above the top conductors where a single ground wire should be located such that none of the phase conductors fall outside an angle of 20° in the steady state (without insulator swing);

(*b*) Repeat the same when two ground wires are located above the top phase-conductors with a spacing of 4 metres between them.

Solution:

(*a*) Let h_g = height of single ground wire above the top conductors. Let us first try the value for h_g such that the top conductor falls within 20° of the ground-wire protection zone. Then:

$$3.75/h_g = \tan 20° \Rightarrow h_g = 3.75/\tan 20° = 10.3 \text{ m}$$

Now let us check if the middle-phase conductor will fall within 20° from the vertical. Vertical height from ground wire = 10.3 + 4 = 14.3 m

$$\therefore \quad d = 14.3 \tan 20° = 5.2 \text{ m} > 5 \text{ m}$$

Thus, a single ground wire located 10.3 m above the top-phase conductors along the centre-line of the tower will offer the safe specified protective angle.

(*b*) With two ground wires at 4-metre spacing, the middle conductor will then be 5 – 2 = 3 m from the ground wire horizontally and $(h_g + 4)$ metres vertically.

$$\therefore \quad \tan 20° = 3/(h_g + 4),$$

giving
$$h_g = (3/\tan 20°) - 4 = 4.24 \text{ m}$$

However, the top-phase conductor is now outside this 20° protective angle. The angle is $\tan^{-1}[(3.75 - 2)/4.24] = 22.4°$.

Therefore, the ground-wire height is still determined by the requirement of the top conductors falling inside the 20° angle specified.

$$\therefore \quad \tan 20° = (3.75 - 2)/h_g$$

$$\Rightarrow \quad h_g = 1.75/\tan 20°$$

$$= 4.8 \text{ metres above the top conductors.}$$

Thus, if a single overhead ground wire is used, for 20° protective angle, it should be located 10.3 m above the top phase conductors on the centre-line of the tower. If two ground wires at 4 m spacing are used, they should be located 4.8 m above the top-phase conductors to offer the same 20° protective or shielding angle. The tower height is reduced by 10.3 – 4.8 = 5.5 metres in the latter case at the expense of stringing one more shielding wire.

For e.h.v. lines, with S/C horizontal configuration, 2 overhead shielding wires are always used to keep the tower height low. The same applies to bipolar d.c. lines.

The reader should work out the protective angles offered by ground wires for the following case, see Fig. 9.8.

Horizontal 400-kV S/C line with following details:

Phase conductors: Height at tower = 12 m
Height at midspan = 9 m
Phase spacing = 11 m.

Ground wires: Height at tower = height at midspan = 21 m (no sag);
Horizontal spacing = 16 m between the ground wires.

Neglect insulator swing. Find protective angles at tower and midspan.

$$[\theta_T = 18.4°,$$
$$\theta_{MS} = 14° \text{ for outers;}$$
$$\theta_T = 41.6°,$$
$$\theta_{MS} = 33.7° \text{ for middle].}$$

Q.12. What are the most important methods of offering good lightning protection to structures and transmission lines other than overhead ground wires and lightning arresters?

Ans. For protection of structures, a lightning rod is in most cases sufficient. The cone of protection is between 30° and 45°. Where a structure is very wide, several masts erected on the roof will be necessary in order to obtain this protective angle everywhere.

An alternative, which is preferred, is to string a set of wires over the roof and connect the mesh to ground. Such methods are used for protecting receiving stations and switchyards where a very vast area is to be shielded from direct strokes. Very tall masts are erected inside and outside along the boundary of the area to be protected and connected at their tops by a mesh of wires, usually of the same size as the line conductors coming in or leaving the station. But this is not so always. Steel wires are also used.

Materials used for these conductors can be steel, copper or aluminium. Their cross-sections must be adequate so that they do not melt at the high lightning currents. These are all governed by Standard Specifications. It is suggested that the cross-sectional area be at least 100 mm^2.

The lightning rod or conductor must be capable of carrying the current to ground so that it is most essential that the ground termination have a very low (practically zero) resistance. This applies even to transmission lines at towers to which the overhead ground wires are connected solidly.

Q.13. What are the types of grounds used for offering low resistance? What are the equations used for determining their sizes?

Ans. Refer to Fig. 9.9.

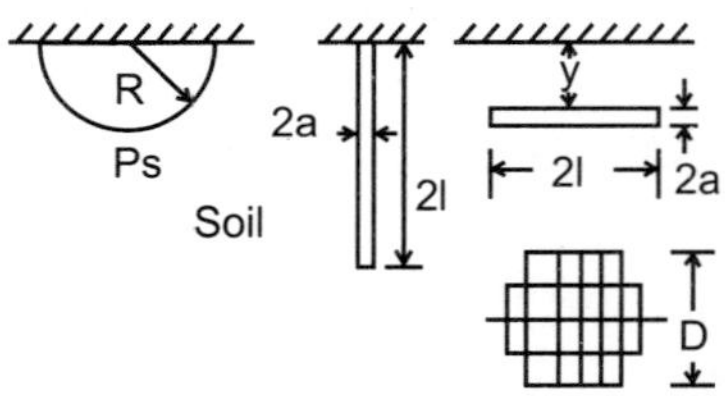

Fig. 9.9: Ground electrodes used in practice.

The shapes of ground electrodes used are: (*a*) Hemispheres; (*b*) Long vertical steel rods driven into ground called "driven rods; (*c*) Wires buried in the ground horizontally at a depth of about 0.5 metre and extending for about 15 m, called "counterpoises". These are used in hard rocky areas where it is difficult or impossible to drive rods vertically. The radius of driven rods and counterpoises is greater than 1 cm; (*d*) Buried grid of wires usually in substations and receiving stations.

The important point is to first measure and ascertain the soil resistance in ohm-metres. The following average values are found:

Sea Water	Moist Soil	Loose Soil and Clay	Rock
$r_s = 1$	10	100	>100 ohm-metre

(*a*) **Hemisphere.** With radius R, $R_g = r_s/2\pi R$ ohm.

(*b*) **Vertical driven rod.** Radius a, length $2l$.

Rudenberg's Formula: $R_g = (r_s/2\pi l) \cdot \ln(2l/a)$;

Dwight's Formula: $R_g = (r_s/2\pi l) \cdot [\ln(4l/a) - 1]$ ohm.

(*c*) **Horizontal buried wire or counterpoise.** Radius a, length $2l$, depth y.

$$R_g = (r_s/2\pi l)\cdot \ln(2l^2/a\cdot y) \text{ ohms}$$

(*d*) **Buried grounding grid.** Major dimension D, length of wire L:

Saraoja's Formula: $R_g = r_s\ (1/2D + 1/L)$ ohms.

In all cases, the principle involved is that the current must be provided as much area as possible to penetrate the soil. (Note that it is the return stroke from the ground that carries the heavy current, while the aim of the overhead lightning conductor is to attract the stepped leader coming down from the cloud to a desired point).

Example: Let it be required to obtain a ground-resistance of 10 ohms in a soil of resistivity of 50 ohm-metre (moist to dry soil). Calculate the required dimensions of: (*a*) Hemisphere; (*b*) Vertical driven rod with radius $a = 1$ cm; (*c*) Length of counterpoise with $a = 1$ cm and $y = 0.5$ m depth.

Solution: (*a*) Hemisphere: $10 = 50/2\pi R \Rightarrow R = 50/20\pi = 0.8$ m

(*e*) **Vertical driven rod.** $a = 1$ cm $= 0.01$ m.

(*i*) *Rudenberg's Formula*: $10 = (50/2\pi l)\cdot \ln(2\cdot l/0.01)$

or, $0.4\pi \cdot l = \ln(200\cdot l)$

Since this is a transcendental equation, the value of length l must be obtained by trial and error. A few trials gives $l = 5.6$ m.

(*ii*) *Dwight's Formula*: $10 = (50/2\pi\cdot l)\cdot \ln(4\cdot l/a - 1)$,

or, $0.4\pi \cdot l = \ln(4\cdot l/a - 1)$.

Trial and error gives $l = 5.3$ m

This length is quite reasonable and managable for driving it into loose or moist soil.

(*f*) **Horizontal buried wire or counterpoise.**

$$10 = (50/2\pi\cdot l)\cdot \ln(2l^2/a\cdot y)$$

$$= (50/2\pi\cdot l)\cdot \ln(2\cdot l^2/0.01 \times 0.5),$$

or, $$0.4\ \pi\cdot l = \ln(400\cdot\ l^2)$$

$$= \ln 400 + 2\ln(l) = 6 + 2\ln(l),$$

or, $$1\cdot 2\pi\cdot\ l - 3 = \ln(l)$$

Trial and error for length l gives 8.1 m .

One buried wire of 8 m in length would offer 10-ohm ground-resistance in a soil of 50 ohm-metre resistivity. It is essential to see that the metal be not eaten away by soil corrosion with lapse of time.

Q.14. What is known as a "step potential" and how is this important for danger to animals and humans?

Ans. When a human being or an animal is caught in an open field during a thunderstorm, the first impulse or reaction is to obtain shelter from the rain and lightning under the nearest tree. This may have serious consequences and pose inherent danger. If lightning strikes the tree directly, there is an intense current flow through the ground and the current density is very high at the base or root of the tree. The potential gradient = current density × specific resistance of soil, or, $E = \delta \times r_s$.

Therefore, the potential across the legs of an animal or human is $V = E \cdot s$, where s = step width. This may be high enough to give rise to primary shock currents and consequent loss of life. There are other reasons for loss of life due to high potential difference which are as follows:

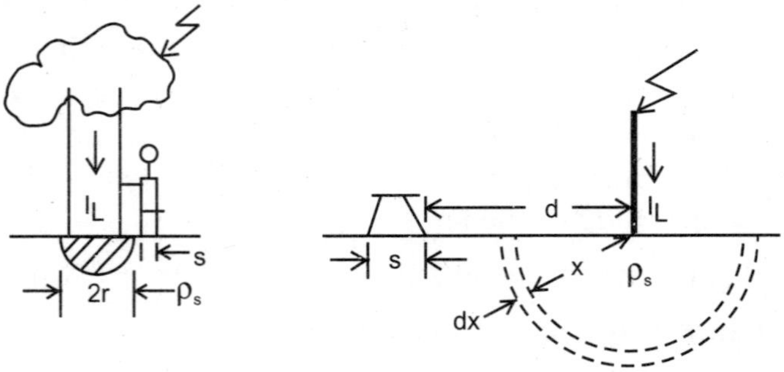

Fig. 9.10: Step potential during lightning stroke.

(*a*) Even if the man stays with feet together on the ground but is touching the tree trunk at a height of say 4 feet from ground (1.2 m), the flow of current through the resistance of tree can give rise to a high potential difference in the 4-foot height. This may cause primary shock currents to flow through the chest and heart between the hand and foot.

(*b*) Let the base of tree or a pole be represented as a hemisphere of radius r through which the lightning current flows in the ground, see Fig. 9.10.

If I_L = lightning current

and r_s = soil resitivity,

the current density is $\rho_s = I_L / 2\pi r^2$,

where $2\pi r^2$ = surface area of hemisphere.

∴ Potential gradient is: $E = r_S \cdot I_L / 2\pi r^2$ volt/m.

If s = step width of the man or animal, voltage appearing across the feet is:

$$V_s = r_s \cdot I_L \cdot s / 2\pi r^2 \text{ volt}$$

For r_s = 50 ohm-metre

I_L = 20 kA

s = 0.8 m,

and r = 0.5 m,

the step voltage is $V_s = 50 \times 20 \times 0.8/(2\pi \times 0.5^2)$

$= 509$ kV

(c) Let a man or animal be near a tall object (such as a pole through which lightning current flows from ground). Let d = distance of person or beast along ground from the pole, and s = step width. At a distance x from the base of the pole, the current density is

$$\rho_x = I_L/2\pi x^2 \text{ A/m}^2$$

Consider a width dx at x. The p.d. across dx is:

$$dV_x = r_s \cdot \rho_x \cdot dx = (r_s \, I_L/2\pi) \cdot dx/x^2 \text{ volts}$$

Therefore, the p.d. between the legs of the person or animal is:

$$\begin{aligned} V &= \int_d^{d+s} (r_s \cdot I_L/2\pi) \cdot (1/x^2) \cdot dx = (r_s \cdot I_L/2\pi) \cdot [1/x]_{d+s}^{d} \\ &= (r_s \cdot I_L/2\pi)[1/d - 1/(d+s)] \\ &= (r_s \cdot I_L/2\pi) \cdot s/[d \cdot (d+s)] \text{ volts} \end{aligned}$$

Evidently, larger the step width, the higher will be the voltage across the step. Also, farther the person or animal is from the base of the pole or tall object, the less the voltage experienced.

Example: Let

$$\begin{aligned} r_s &= 100 \text{ ohm-metres}, \\ s &= 0.8 \text{ m}, \\ I_L &= 20 \text{ kA}, \\ d &= 10 \text{ m from the base of the pole} \end{aligned}$$

What is the step voltage experienced?

Solution:

$$\begin{aligned} V &= (100 \times 20 \times 10^3/2\pi) \cdot [0.8/(10 \times 10.8)] \\ &= 2.36 \times 10^3 \text{ volt} \\ &= 2.36 \text{ kV} \end{aligned}$$

This will certainly be fatal, even if a direct hit is not experienced.

On account of the danger presented by the step potential, one should keep far away from places where there exists the possibility of lightning-current entering the ground. It is for this reason that at generating- and receiving station switchyards at the ends of a high voltage transmission line where lightning arresters are connected to draw the lightning current to ground, a fence is erected so that personnel cannot come close to the grounding point. This is necessary not only because of the danger of contact with high voltage terminals but also for reducing the step potential to safe values.

LIGHTNING ARRESTERS (OR SURGE ABSORBERS)

Q.15. What is the purpose of surge absorber or lightning arrester in the scheme of protecting a high voltage transmission system?

Ans. There are many factors that have to be considered:

(*a*) These are connected from line to ground at both ends of a transmission line at the generating station and receiving station (and at any intermediate sub-station where power is tapped off). Thus, each phase or pole has its own lightning arrester.

(*b*) When lightning strikes anywhere along a line, if an insulator on a tower does not flashover thereby ground the stroke at that point, the voltage and accompanying current travel on the line and reach both ends, after some attenuation.

(*c*) The lightning arrester has the lowest insulation strength among all high voltage equipment it has to protect, mainly the very expensive transformers, carrier communication and protection equipment. Therefore, it is set to drain the lightning current to ground without suffering damage to itself.

(*d*) Once the arrester gap (in gap-type arresters) breaks down, the system has a ground into which the lightning stroke is diverted, see Fig. 9.11. This causes a voltage drop in the arrester-resistance material;

(*e*) This voltage drop, called the *IR* drop, must be such that it does not exceed the "protective voltage" of the equipment insulation. No insulation in any part of the system should break down due to lightning when protected by the arrester;

(*f*) When the series gap inside the arrester has sparked over due to the over-voltage, the power-frequency voltage of the line also drives a current to ground through the arrester material. This is called the "power-follow current" and it is at power frequency (50 Hz or 60 Hz);

(*g*) During the flow of lightning current (which lasts for about 1 ms) and the subsequent power-follow current (whose half-cycle time is 10 ms or 8.33 ms for 50 or 60 Hz systems, respectively) there is an arc between the electrodes of the series gap. However, at a zero of the power-follow current in a.c. systems, the arc is extinguished and the arrester regains its normal open-circuit state. (We can appreciate that this is not possible in d.c. systems).

For a gap-type arrester, therefore, the sequence of operation is as follows:

(*i*) The high lightning voltage superimposed on the power-frequency voltage breaks down the series gap inside the arrester housing;

(*ii*) This draws the lightning current to ground but is accompanied by the power-follow current;

(*iii*) Due to these two currents, a voltage appears across the arrester material whose value is designed to be lower than the insulation breakdown value of the connected equipment to be protected;

(*iv*) At a current zero, the arc is extinguished and the arrester regains its original healthy condition.

These functions are general, but there are some variations depending upon the type of arrester used.

Q.16. What are the different types of arresters or surge absorbers used in high voltage power transmission systems?

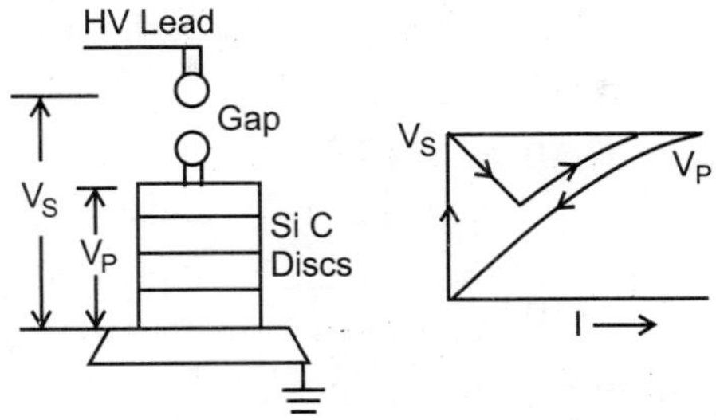

Fig. 9.11: Gap-type normal SiC lightning arrester or surge absorber, and its I–V characteristic during operation. V_S = sparkover voltage of series gap, V_P = voltage across resistor element during discharge.

Ans. Refer to Fig. 9.11.

(*a*) The components are: (*i*) A gap (usually air or compressed nitrogen for consistent value of sparking voltage) between two electrodes, the top one connected to the high voltage terminal of the line conductor, and the bottom electrode connected to the top of the arrester material; (*ii*) The arrester resistance material grounded at the bottom end.

(*b*) **Construction and Characteristics**

(*i*) **The electrodes** of the air gap can be small spheres or hemispheres. There should be a single large gap or a series of smaller gaps. They are all open under normal condition when only the power-frequency voltage appears across them. But under the impact of the lightning voltage which may be twice the peak value of power-frequency voltage (2 p.u.) the gap breaks down creating a conducting path to ground.

(*ii*) **Resistance Material.** This is the most important part of the arrester. It is made from sintered (like glass) silicon carbide (SiC), vitrified out of contact with air under extremely stringent atmospheric and temperature conditions. The atmosphere in which the material is sintered could be the noble gases but never oxygen.

Its most important electrical property, on which the entire scheme of protection against lightning depends, is its non-linear resistance characteristic. Its resistance decreases with increase in current through it. Ideally, for offering good protection, the voltage across the material must remain constant at all current values.

Thus, $V_{id} = I{\cdot}R_{id} = K$, a constant.

$\therefore$ $R_{id} = K{\cdot}I^{-1}$,

or, the resistance should vary inversely with current flow. Instead of showing the variation of resistance with *I*, the lightning arrester is characterised by its *VI* relation. This is shown in Fig. 9.11.

In practice, this ideal is not achieved for the SiC resistance material. Normal arresters have the VI characteristic given by $V = K_2 \cdot I^{1/6} = K_2 \cdot I^{0.167}$, or, $I = K_3 \cdot V^6$. Let us observe the change in voltage with current taking the variation of I over a few decades, or powers of 10.

For $I = 10$ A

$$V_{10} = K_2 \cdot 10^{0.167}$$

For $I = 100$ A

$$V_{100} = K_2 \cdot 10^{2\times 0.167}$$

$$V_{100}/V_{10} = 10^{0.167} = 1.4678$$

For $I = 1000$

$$= 10^3 \text{ A}$$

$$V_{1000}/V_{10} = 1.4678^2$$

$$= 2.1544$$

For $I = 10{,}000 \text{ A} = 10^4 \text{ A} = 10 \text{ kA}$

$$V_{10,000}/V_{10} = 1.4678^3$$

$$= 3.1623$$

Therefore, even though the current through the material has increased from 10 A to 10 kA, or a thousand times, the voltage across the material has increased by only 3.1623 times.

Lightning arresters are specified by Standards for different values of maximum current-discharge property which is different for different voltage levels. According to Indian Standards IS-3070 (Part I)-1974, the recommended practice is as follows:

System Voltage, (kV):	Up to 230	345 – 400	500	750 (735 – 765)
Crest Current, (amps): (8/20 μs waveshape)	5,000	10,000	15,000	20,000

When an arrester is selected for its current-discharge capability for a given system voltage level, it means that the voltage across the resistance material should not exceed a recommended value, which is higher than the system power-frequency voltage by a factor. This voltage is called the IR-voltage of the arrester and we will give values of these voltages under "Insulation Coordination" next. The insulation of all major equipment, most importantly the expensive transformer, must be above this IR-voltage by a safety factor called the "Protective Ratio". International Standards (I.E.C. and also I.S.) recommend that the protective ratio be not less than 1.2, that is, the withstand value of insulation of high voltage windings of a transformer protected by the arrester must be 20% higher than the IR-voltage across the arrester, when discharging the maximum recommended 8/20 μs lightning current through it. This IR-voltage is in its turn 2 to 3 times (p.u. value) the system maximum power-frequency voltage. This will be discussed under "Insulation Coordination". Note that in gap-type arresters, when we speak about the *IR*-voltage or arrester voltage, we must keep in mind that the sparkover of the series gap is also involved. If this is higher than the *IR*-voltage across the resistance material, then this is taken as the protective voltage of the arrester. By proper voltage-grading circuits inside the arrester housing, these two voltages are made equal. The voltage-grading circuit is not discussed here but it takes the form of series-connected capacitors connected across the line and ground but built integrally with the unit inside the housing of the arrester assembly.

Q.17. What are the components, constructional details and characteristics of a current-limiting type of arrester?

Ans. In the normal gap-type arrester, the current has to go through zero before the gap regains its insulation characteristics and the arc between the electrodes is extinguished. In many applications, such a delay is quite unnecessary and so efforts are made to extinguish the arc prior to attaining current-zero. This is achieved by having the current to flow between the poles of a magnet, whose field is created by the lightning current itself. This will blow the arc out even when it has energy.

An arrester of this type can extinguish higher currents and absorb higher energy which is the case with switching operations on lines. Therefore, current-limiting gap-type arrester works very well under lightning as well as under higher-energy longer-duration switching surges ($\approx$ 5 ms duration). The material is still SiC and the *I–V* characteristic on the rising portion is the same as for the normal gap-type arrester. But the extinction of arc takes place at a finite value of current, as shown in Fig. 9.12.

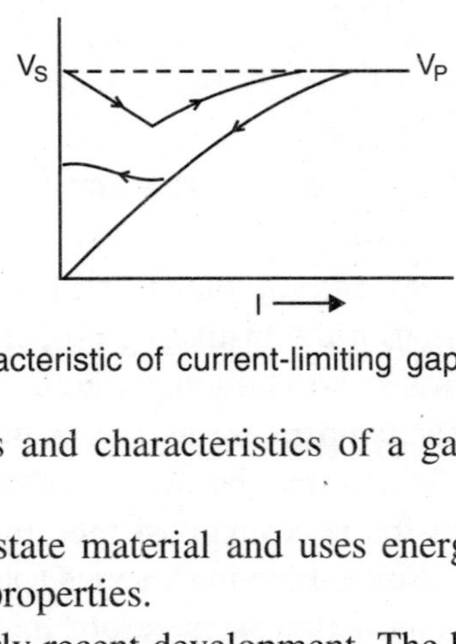

Fig. 9.12: I–V characteristic of current-limiting gap-type surge absorber.

Q.18. What are the components and characteristics of a gapless Metal Oxide surge absorber or arrester?

Ans. This is made from solid-state material and uses energy-level differences to offer either insulation or conduction properties.

This arrester is of fairly recent development. The basic material is *n*-type ZnO grains which are sintered in a flux of various insulating oxides such as Si_2O_3. Other materials such as CoO, MnO, Cr_2O_3 are added to control energy levels and change them under the input of the voltage across it and determine the insulating and conducting properties. With $I = K \cdot V^{35-50}$, the voltage is nearly constant even for a variation of current over 5 decades. Its characteristics are shown in Fig. 9.13.

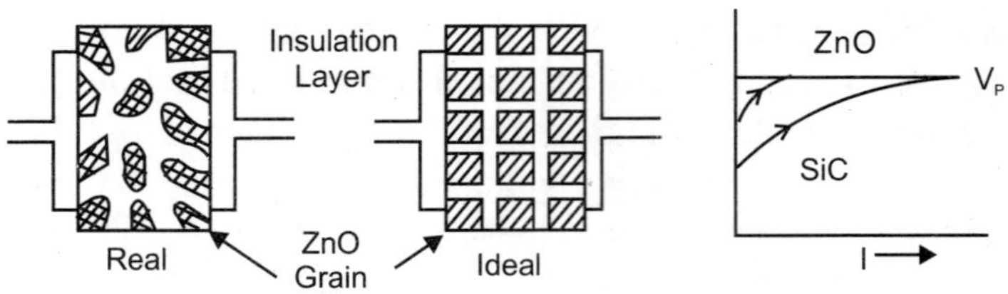

Fig. 9.13: Construction of ZnO gapless arrester material and its I–V characteristic. Comparison with SiC resistor material.

As an example, let the *I–V* characteristic be represented as $I = K{\cdot}V^{40}$.

Then, $V = K'\cdot(I)^{1/40} = K'\cdot I^{0.025}$

where $K' = K^{1/40}$ = constant

At I = 10 A

$V_{10} = K'{\cdot}10^{0.025}$

and at I = 100 A

$V_{100} = K'{\cdot}10^{0.05}$.

∴ For an increase of 10 times in *I* or by one decade, the increase in voltage is

$$V_{100}/V_{10} = 10^{0.025}$$
$$= 1.059 \text{ times}$$

For I = 10,000 A = 10 kA

$$V_{10,000}/V_{10} = 10^{4\times0.025} = 10^{0.1}$$
$$= 1.259$$

The *I–V* characteristic is now much superior and the voltage rise is only 25.9% more for current increase by 4 decades.

The arrester has no series gap to contend with so that it is continuously energised by power-frequency voltage. But the insulating properties of ZnO and its additives keep the current to less than 1 mA at normal temperature and power-frequency voltage level. As in all solid-state materials, temperature-increases produces run-away conditions. Under the impact of high voltage which may be due to lightning or switching surges, which is superimposed on the power-frequency voltage, the material becomes a good conductor by change of energy levels. Since there is no series spark gap, the time-lag of breakdown of the gap is eliminated. The action is therefore instantaneous. At the termination of lightning or switching transient, the current is reduced at power frequency and there is no heavy power-follow current as in the gap-type arrester.

Q.19. What is meant by "Insulation Coordination" based on lightning? What are Basic Impulse Insulation Levels (BIIL or simply BIL)?

Ans. A generating station, transmission line and receiving station consist of many components at high voltages. These equipments are normally:

(*a*) Transformers at both ends;

(*b*) Insulators on the towers and elsewhere such as switchyards;

(*c*) Potential and Current transformers;

(*d*) Circuit Breakers; and

(*e*) Last but most importantly, the Lightning Arrester or Surge Absorber.

Every one of these components is insulated to withstand the high voltage caused by (*i*) power-frequency energisation; (*ii*) lightning impulse; and (*iii*) switching surge voltage. It is evident that the lightning arrester has a resealable insulation and should be so designed that it conducts to ground or breaks down at a lower voltage than any other high voltage equipment. The transformer is the most expensive item and should be insulated for a higher withstand voltage level than the arrester-gap breakdown voltage or the subsequent *IR*-voltage.

The selection of insulation-withstand voltage levels of all these components is known as "Insulation Coordination". These levels are all governed by Standard Specifications, and we will give important examples.

Consider the following details of practice used in various countries.

(1)	(2)	(3)	(4)	(5)	(6)	(7)
Country	Nominal System kV, V_n (l–l)	Max. Sys. kV, V_m	Arrester Rating kV, V_a (ph-gr)	Earthing Coeff. $C_e = V_a/V_m$	Impulse Protective Level V_p, kV, Peak value	Station Equipment Impulse Withstand Level, V_S. kV, Peak
U.S.A.	345	360	258	0.72	750	1050
India	400	420	336	0.8	1100	1550
Sweden	400	400	360	0.9	1100	1425
U.S.A.	500	550	396	0.72	1130	1425 1550
Canada	735	735	636–686	0.87–0.93	1700	2050 2200

The following explanations about the above values must be studied:

Col. (3). For every system-voltage level, called the "nominal" system voltage in kV, Standard Specifications of each country prescribes a maximum voltage (r.m.s. line-to-line value) which should not be exceeded in any part of the system. Insulation levels of all equipment for power-frequency voltage are based on this maximum allowable system voltage. This is V_m.

Col. (4). When the system nominal voltage (r.m.s. line-to-line) and maximum voltage are fixed, the lightning arrester rating must be selected. The method of selecting this voltage rating is quite complicated and is not given here. (Please see the author's book: *Extra High Voltage A.C. Transmission Engineering*, 2nd Edition, Chapter 9, Wiley Eastern Ltd., 1990). This voltage, in principle, is selected by the voltage rise of the healthy phases when a line-to-ground fault occurs on one phase. This is V_a. Since this is a power-frequency voltage, V_a is still specified as r.m.s. value but is the line-to-ground voltage. It is also called the dynamic voltage rise.

Col. (5). The Earthing Coefficient $C_e = V_a/V_m$, where both V_a and V_m are r.m.s. values of power-frequency voltage, but V_a is line-to-ground voltage while V_m is line-to-line voltage. The Earthing Coefficient lies between 0.72 and nearly 1.0.

Col. (6). Having selected the power-frequency voltage rating of the arrester, its *IR*-voltage drop (or gap sparkover voltage also in gap-type arresters) under lightning-impulse current is specified. Since this is a lightning-impulse voltage, it is described as the peak or crest value, V_P. This is the protective level of the arrester. For the given system voltage, we have already given the maximum current to be used which the material has to discharge in Question 17. The arrester must maintain this voltage as ceiling or maximum. This is also the lowest voltage of any equipment insulation level. All other equipment must be designed such that their insulation must withstand a higher impulse voltage.

The two voltages, V_P and V_a, determine the "protective level ratio": $N_P = V_P/V_a$. Note that both are line-to-ground voltages, but V_P is the peak value of lightning-impulse type of voltage while V_a is the r.m.s. value of power-frequency voltage. The value of N_P is nearly 3. (The reader should work this out from the values given in the Table).

Col. (7). Now, the station equipment insulation must withstand a lot higher impulse voltage V_S of the lightning-type than the arrester impulse-protection level V_P. This is prescribed by Standard Specifications and are given in this column.

Once the station equipment impulse withstand voltage level V_S is known and the arrester protective level V_P is also known, they determine the "Protective Ratio": $C_P = V_S/V_P$. Both voltages are line-to-ground peak values of impulse voltage. In practice, V_S is determined by the transformer insulation, and the waveshape used in testing laboratories and design offices is 1.2/50 μs. The voltage V_P, however, is determined by the maximum prescribed discharge current through the arrester resistance material of 8/20 μs waveshape.

The protective ratio is between 1.2 and 1.4. The recommended minimum value is 1.2 by the I.E.C. The reader should also work out the ratio $C_I = V_S/V_m$ = equipment impulse withstand voltage/system maximum voltage. This is called the "Impulse Withstand Ratio" and lies between 2.8 and 4 in practice.

An engineer involved in design or selection of high voltage equipment insulation levels must be very familiar with Standard Specifications and understand their implications. In addition to the Basic Impulse Levels (BIL), for system voltages above 220 kV, switching-surge requirements become more important. Insulation levels based on switching-surge requirements are called "Switching Impulse Levels" or SIL. This aspect is not discussed here.

Q.20. What is meant by several levels of rated lightning-impulse withstand voltage of transformers? What is meant by 2-levels down or 2-step reduction when selecting transformer insulation levels?

Ans. In the previous Question we observed in Column (7) that a voltage V_S is specified as the peak value of line-to-ground lightning-impulse withstand voltage of station equipment. This refers particularly to transformer insulation although all other equipment is also included.

However, since a transformer is the backbone of the system, so to speak, and is very expensive and complicated in design of its insulation-structure, a system engineer must be doubly sure that a transformer insulation never breaks down or suffers damage. Since we are dealing with a powerful and variable source such as lightning, it is evident that its intensity is not the same everywhere in the country or the world. Lightning intensity is more severe in certain areas than in others. To determine the intensity, the engineer must have wide experience.

Having decided whether the *isokeraunic* level or lightning-activity level is abnormally severe or normal, the engineer is provided with choice of transformers with several impulse-insulation levels, V_S. For example, for 400-kV system, there are 5 levels of rated lightning-impulse withstand voltage along with 2 levels of switching-impulse withstand voltage. Examples from some standards are as follows:

Nominal system kV, r.m.s.: $V_n = 400$.

Max. Equipment Voltage, kV: 420.

Switching-Impulse Withstand Voltage, kV, crest:	950	1050
Lightning-Impulse Withstand Voltage, kV, crest:	1050	1175
	1175	1300

Several other values according to IS 2026, Part III (1981) are shown in the Table below. It is evident that there are several values of protective ratio $N_P = V_S/V_P$ at the choice of an engineer. According to the judgement of the engineer regarding the severity of lightning activity in a given region where the transformer will be located, the most suitable transformer BIL can be selected.

Table: BIL and SIL values as per IS 2026, Part III (1981)

Nominal kV V_n	Max. kV V_m	Switching Impulse Withstand Voltage, kV, crest V_{SS}	Lightning Impulse Withstand Voltage, kV, crest V_S	Lightning Arrester Protective Level, kV, crest V_P
275	300	750	850 and 950	550
		850	950 and 1050	625
345	362	850	950 and 1050	625
		950	1050 and 1175	750

Contd...

400	420	950 1050	1050 and 1175 1175, 1300 and 1425	750 850
500	525	1050 1175	1175, 1300 and 1425 1425 and 1550	900 1100
750	765	1425 1550	1550 and 1800 1800 and 1950	1300 1465

It is evident that the lower the L.A. protective level i.e., lower the *IR*-voltage (or gap sparkover voltage, whichever is the higher value), the lower the equipment insulation need be. With progress of time, industry has come up with improvements in lightning-arrester resistance material which has brought about lowering of the arrester voltage for the same discharge current. Thus, suppose for a 275 kV system, the transformer BIL was 1050 kV crest by using an arrester of 10,000 ampere discharge current and arrester protective level of 625 kV. But an improved arrester for the same discharge current is manufactured in time with 500 kV protective level. Then the transformer BIL need not be 1050 kV but 850 kV giving a protective level ratio of 850/500 = 1.7, the same as 1050/625 = 1.68. This is called a "step reduction of BIL". Since there is also an intermediate 950 kV level prescribed, lowering from 1050 to 850 kV is a "two-step" reduction in BIL. Thus, the reduction in equipment BIL goes hand in hand with improvement in lightning-arrester performance.

Note from the Table given in Question 19, for the 400 kV system in India, the values of V_P = 1100 kV and V_S = 1550 kV are higher than those prescribed by IS 2026 given above, where the highest level is V_P = 850 kV and V_S = 1425 kV. This is because the U.P.S.E.B. and N.T.P.C. lines from Singrauli to Delhi are very long (850 km) and the systems are new to India. Therefore, safety margins are kept quite high and the design is conservative. In the U.S.A. for a 500 kV system, the values used are V_P = 1130 kV and V_S = 1425 and 1550 kV, almost equal to the 400 kV system in India. Such differences always exist between countries and they are all based solely upon experience.

Q.21. How are spark gaps or spill gaps used for coordination of insulation?

Ans. In order to protect expensive porcelain bushings and internal insulation of equipment such as transformers and circuit breakers, spark gaps or spill gaps are provided as shown in Fig. 9.14.

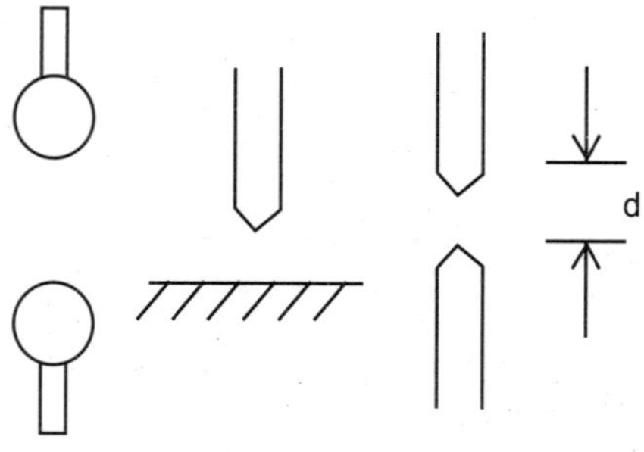

Fig. 9.14: Spill-gap geometries: Sphere-to-sphere, rod-to-plane and rod-to-rod.

They can be in the form of sphere-to-sphere, sphere-to-plane, rod-to-plane or rod-to-rod. The flashover voltage of these gaps is adjusted to be less than the equipment BIL or SIL. In case of any malfunction of a lightning arrester, these gaps sparkover and protect the equipment internal insulation. These gaps must not flashover under normal power-frequency voltage which is applied continuously and their impulse breakdown voltage must be higher than that of the lightning arrester so that they only act as safety valves. The arrester maintains the minimum voltage of any equipment in the system when it operates under impulses.

The following formulas apply for power-frequency and impulse flashover voltages for spill-gap geometries, with gap spacing in metres.

Electrode Geometry	Power Frequency kV, peak	Lightning Impulse kV, peak or crest	Switching Impulse kV, peak, positive
Rod-Plane	$V_{SO} = 650 \cdot d^{0.576}$	$500 \cdot d$	$500 \cdot d^{0.6}$
Rod-Rod	$V_{SO} = 850 \cdot d^{0.576}$	$650 \cdot d$	$650 \cdot d^{0.6}$

All these V_{SO} values are the results of extensive laboratory experiments. Note that the rod-to-rod gap flashover voltages are higher than those of rod-to-plane flashover voltages by a factor of 1.3. These gaps have lower flashover voltages under positive switching impulses with waveshape of 250/2500 μs as compared to lightning-impulse waveshape of 1.2/50 μs for gap spacings $d > 1$ m. For example, for $d = 2$ m, V_{SO} for lightning/V_{SO} for switching = $2/2^{0.6} = 2^{0.4} = 1.322$, which is 32.2% higher for lightning impulse than for switching surge.

Q.22. What are the principal breakdown and withstand voltage characteristics of porcelain insulator discs or strings? What is meant by time to breakdown?

Ans. A standard porcelain insulator disc is designated as 5.75 × 10 inch (14.6 × 25.4 cm). The thickness of porcelain between the cap and pin is 5.75" (14.6 cm) and the diameter of the disc is 10" (25.4 cm). The creepage distance over the porcelain surface is about 12.5" (32 cm). Its flashover strength is described by a time-to-breakdown characteristic which means that a voltage magnitude which can break down the porcelain depends upon how long it takes for the voltage to rise to flashover value. Thus, there is no such voltage as a universal flashover value. An example is provided here for one disc for lightning-type voltage impulse in fair weather:

Time to breakdown, μs	0.5	1.0	2.0	3.0	4.0	6.0	8.0
kV/disc, crest	188	150	125	110	105	97.5	92.5

This difference arises because the breakdown mechanisms take time to act and settle down. Thus, if a voltage is maintained for only 1 μs, it requires 150 kV crest value for the insulator to flash over. On the other hand, if a voltage is maintained for 8 μs (with lower rate of rise) then flashover takes place at a lower voltage of 92.5 kV. The time-to-breakdown and crest voltage magnitude characteristic is important for every type of insulation on account of the following reasons.

When a lightning stroke hits a transmission line, the voltage at the tower struck or near it becomes the full lightning voltage. This will travel on the line conductors along both sides of the point hit. At the adjacent towers, the voltage encounters a low-resistance ground and reflects back to the original point of incidence of the stroke with a negative sign. At this instant and subsequently, the voltage across the insulator decreases. This takes 2 μs for a span between towers of 1 μs at light velocity (= 300 m). Until the negative reflection arrives, the insulator string must withstand high voltage and should not flash over.

The flashover voltage shows a highly linear characteristic with number of discs. Thus, for 20 discs in the string used for 220 kV line, the sparkover voltage is 20 times the values given before for a single disc.

For a full wave of 1.2/50 μs impulse, the flashover voltage per disc is 37.5 kV crest in fair weather and 60 kV crest in rain.

Power-Frequency Voltage: In fair weather the flashover voltage is 75 kV/disc crest value = 53 kV, r.m.s. value. Since the leakage distance over the porcelain between the cap and pin is 32 cm, this yields 53/32 = 1.67 kV/cm, r.m.s. In rain, this is reduced to about 60% of the fair-weather value or 1 kV/cm, r.m.s.

Withstand Voltages: The withstand voltage of insulator discs is taken as 20% lower than the flashover values given above. All such values are based on probability considerations and this topic is not discussed here. In the field of Statistics, there is an important property called the Standard Deviation from the Mean Value. Thus, extensive experimental investigations are necessary before deciding on flashover and withstand voltage values.

Q.23. What are the instruments used for measuring and obtaining the different characteristics of natural lightning, out of which our knowledge of this devastating natural phenomenon has been based?

Ans. These are very numerous and ingenuous, commencing from Benjamin Franklin's very daring and bold experiments with kites, piano wire, key, Leyden Jar and his assistant's shocking reactions, to the modern Digital Storage Oscilloscopes with Pre-Transient triggering facility and Storage memory. Only a few modern instruments are described here briefly. For laboratory measurements and testing of high voltage equipment by generating artificial lightning, the reader is referred to Chapter 17 on High Voltage Impulse Testing and Measurement, later on.

(*a*) **Digital Storage Oscilloscope:** Until the advent of this modern version of the normal Oscilloscope, it was essential that a camera had to be kept in readiness by a human operator with its shutter open. As soon as a lightning stroke or flash registered itself on the screen, which was fed as a voltage across a resistance at the ground end with a suitable delay line between the horizontal and vertical deflection plates, the operator had to wind the film for the next shot. But now-a-days, this is unnecessary. However, because of the very high cost it is clear that oscilloscopes in general and Storage Oscilloscopes in particular can be used at only one or a few locations in a scientific investigation. It is thus not possible to cover a vast area in order to determine the statistical properties of the severity of lightning activity or what is called the

"Isokeraunik Level". The oscilloscope, however, measures nearly all of the vital properties, such as waveshape, current magnitude, etc.

(*b*) **Magnetic Links:** When it is necessary to measure only the crest value of current over a vast area to obtain probability values, magnetic links are very inexpensive and versatile. These are soft-iron pieces attached at towers in the ground lead, but removed sufficiently far from the tower steel so as not to disturb the magnetic properties of the link. When a lightning stroke contacts a tower and is led to ground, its current passes through the soft-iron piece and magnetises it. The strength of the magnetic field retained by the iron (in the hysteresis loop) is proportional to the crest value of current. This is measured in a laboratory.

(*c*) **The Klydonograph:** This was invented by J.F. Peters in the 1920s at the Westinghouse Laboratories. In principle, it uses a rod electrode rounded at the tip and a flat base which serves as the ground electrode. A photographic film or plate is inserted in between the electrodes with the emulsion side on top on which rests the rounded rod electrode.

It utilises the principle that when a lightning voltage is applied between the electrodes, a flower-like pattern called a Lichtenberg Figure appears on the emulsion. The size and pattern of the figure depend upon the voltage magnitude and duration as well as on the polarity of the lightning voltage. These are of course compared and calibrated with laboratory pictures of Lichtenberg Figures obtained under controlled conditions.

(*d*) **Other Instruments:** Many instruments were developed in the 1920s, and subsequently, which have been as follows:

(*i*) *Fulchronograph.* This consisted of a rotating wheel on the rim of which small iron pieces were attached. When the current flows in between the electrodes through which these pieces are made to pass, they are all magnetised and the strength of magnetisation depends upon the current magnitude at the instant a piece has passed between the electrodes. Thus, by measuring the field strengths of every piece in proper sequence, the waveshape with magnitude and timing can be ascertained. The oscilloscope has replaced this instrument.

(*ii*) *Magnetic Surge Crest Ammeter.* Again, this uses iron pieces which are magnetised by the lightning current and it is evident that this is indicative of the crest value of the lightning impulse current. It cannot indicate the complete waveshape.

(*iii*) *Boys Camera.* Historically, the use of Boys Camera invented by Sir Charles Boys in England and its improved version used extensively by Sir Basil Schonland in South Africa led to the photograph of the lightning stroke, its components and characteristics. It has a rotating mirror which reflects the luminous lightning on to a curved photographic film. It clearly shows the progress of the stepped leader and the more intense return stroke step-by-step. It has also been improved by J. D. Malan and R.D. Linton in South Africa and others elsewhere until we now have a full knowledge of the different components of a complete lightning stroke, both single and multiple.

Q.24. What are meant by "Ball Lightning" and "Bead Lightning"?

Ans. These are also known as "extreme forms of atmospheric electricity". Both are the results of the remnants of primary lightning strokes which leave behind intense ionised particles and channels in their wake. Very few observers have witnessed these forms of lightning, like U.F.Os!

(*a*) **Ball Lightning:** This is usually a single globular type of highly luminous entity, and is mobile. Its shape has been observed to be spherical, oval, pear-shaped or even rod-shaped. Their usual size is 10–40 cm and they are red to yellow in colour. They give a hissing sound, and have been observed inside buildings as well as in the open. They have a sulphurous odour or like ozone. Therefore, they are caused by intense ionisation of oxygen. They last for only a couple of seconds and move at about 2 m/s. Certain observers have seen ball lightning rolling on high voltage lines. Its energy density has been estimated at about 2000–4000 joule/cm^3 and temperature about 5000°C which gives yellow light at a wavelength of 5800 Å . It has sufficient energy and temperature to burn wood and start fires in farms and elsewhere.

(*b*) **Bead Lightning:** This is also known as "pearl lightning" or "chain lightning" from its colour and appearance. It is common between cloud and cloud, forming luminous beads strung like a necklace with dark gaps. The normal lightning stroke does not decay uniformly so that it may leave bunches of luminous remnants which look like beads. They last for 1 or 2 seconds.

❑❑

CHAPTER 10

PARTIAL DISCHARGE (CORONA) AND ITS EFFECTS

Q.1. What is meant by "corona"? How and where does it occur in high voltage equipment and transmission lines?

Ans. Corona, meaning a crown, is also known as "partial discharge". It occurs whenever the electric field stress reaches a high enough value at an electrode or electrodes so as to cause a local breakdown of the insulation surrounding the electrodes. For example, if the electric field gradient exceeds 30 kV/cm around a sphere in air at N.T.P., then the air breaks down due to generation of an electron avalanche as described in Chapter 3. It gives a bluish discharge which is caused by ionisation of oxygen molecules in the air surrounding the sphere.

Corona occurs at the following points in h.v. equipment and lines:

(*a*) In solid and liquid insulations where gases are trapped giving rise to voids and bubbles;

(*b*) In gases used as insulation, particularly oxygen. Corona discharge does not occur in pure nitrogen;

(*c*) Around transmission-line conductors in air.

Q.2. Differentiate between glow-type corona and pulse-type corona discharges. What are Trichel Pulses?

Ans. There are in general two modes or types of corona discharges:

(*i*) Glow or Pulseless Corona;

(*ii*) Pulse Type or Streamer Corona.

Both are caused by energy fed into the discharge from the high voltage power source. Glow corona occurs without pulses on very thin wires and in moist conditions. Pulse-type corona occurs around larger-size electrodes.

On thin wires, there exists a pulse-type corona but only when the wire is at negative potential. These pulses are known as Trichel Pulses after the initial discoverer of pulses, caused by excessive field gradients. On large-sized electrodes, such as transmission-line conductors, in voids in solid and liquid insulations, both polarities yield pulse-type corona. Usually, the negative-polarity pulses are of much smaller amplitudes than positive-polarity pulses. Their charge content and duration also differ, the positive pulse being much longer and containing larger charge than the negative pulses. Corona pulses are current pulses because they are caused by movement of electric charges in the discharge. However, they can only be detected and measured by the voltage generated across a resistance which is fed to the plates of an oscilloscope or into a noise meter whose input must be a voltage. The usual order of timings and current amplitudes are:

Negative: Front time–1 to 5 ns (1 ns = 10^{-9} s); 50% value on tail: 10–50 ns.

Positive: Front time–5 to 50 ns; Tail–50 to 200 ns.

Conductors with larger radius of curvature yield pulses with larger amplitudes and longer timings. This is because the electric field falls off more gradually around a larger electrode than around an electrode with smaller radius of curvature. In fact, as the electrode becomes smaller and smaller, there may not develop any pulses at all but the discharge mode becomes the pulseless glow corona.

Q.3. What are the equations for corona-inception voltage gradients for the following electrode types?

(*a*) Sphere; (*b*) Concentric Cylinders; (*c*) Cylinder above Plane; and (*d*) Parallel Cylinders.

Ans. The corona-inception gradient is also known as "Corona Threshold Gradient". Refer also to Question 27 in Chapter 3. In all cases, the air-density factor is important. It is given by:

$$\delta = [(273 + t_r)/(273 + t)] \cdot (p_{mm}/p_r),$$

where t_r = reference temperature could be 0°C or 20°C,

t = temperature of atmosphere under consideration in °C,

p_r = reference pressure in mm Hg usually 760 mm, and

p_{mm} = pressure of atmosphere in mm Hg under consideration.

Normally, the reference temperature and pressure are taken as t_r = 0°C and p_r = 760 mm Hg. Some other sources may use t_r = 20°C = 68°F, this being the usual environmental temperatures in testing laboratories. The reader must always be aware of these differences whenever using formulas. A student cannot be punished for not knowing these, since there is no uniformity in the world of high voltage whenever atmospheric conditions are concerned.

All voltage-gradient values are peak or crest values. They apply to all types of usually-encountered excitations, namely: (*i*) D.C.; (*ii*) Sinusoidal A.C. with symmetry in both half-cycles; (*iii*) Lightning Impulses; and (*iv*) Switching Surges, see Fig. 10.1.

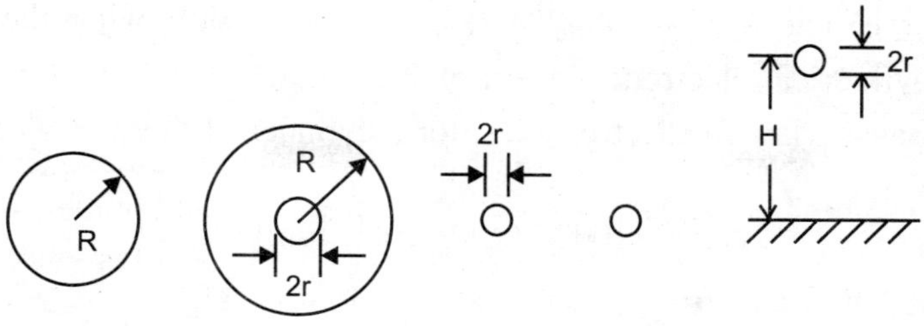

Fig. 10.1: Electrodes encountered in high voltage equipment and lines. Sphere, Coaxial cylinders, Parallel cylinders, Conductor above plane.

(*a*) **Sphere:** $E_0 = 30 \cdot \delta$ kV/cm $= 3000 \cdot \delta$ kV/m $= 3 \cdot \delta$ MV/m.

(*b*) **Concentric Cylinders:**

Inner Cond. radius in cm: $E_0 = 31 \cdot \delta[1 + 0.308/\sqrt{(r_{cm} \cdot \delta)}]$ kV/cm

Inner Cond. radius in m : $E_0 = 3100 \cdot \delta[1 + 0.0308/\sqrt{(r_m \cdot \delta)}]$ kV/m

(*c*) **Parallel Cylinders and Cylinder-to-Plane:**

Cylinder rad. in cm: $E_0 = 30 \cdot \delta\ [1 + 0.301/\sqrt{(r_{cm} \cdot \delta)}]$ kV/cm

Cylinder rad. in m: $E_0 = 3000 \cdot \delta\ [1 + 0.0301/\sqrt{(r_m \cdot \delta)}]$ kV/m

The corresponding corona threshold or inception voltage magnitudes depend on the electrode geometries and dimensions of the insulation structures. These are considered in Question 4. Also refer to Chapter 8 for Voltage Gradient calculations for Overhead Transmission-Lines and Chapter 5 for Concentric Cylindrical Cables.

Q.4. How are corona-inception voltages calculated for the following practical cases of co-axial cylinder and overhead transmission-line conductors above ground? Consider specific examples.

Example 1: For conducting experiments with corona discharges in labs and research projects on overhead transmission-lines, small-scale models using a coaxial cylindrical arrangement or mock-up are utilised. The conductor is of the same size as the overhead line but the outer cylinder can be of any diameter. We only discuss a single conductor and not a bundle.

Consider the following numerical case: The conductor used for a 500-kV 3-ph or d.c. line has outside diameter of 6.35 cm = 0.0635 m. It is strung concentrically inside an outer cylinder of diameter 4 m. Calculate the corona-inception voltage for this set-up. Take $\delta = 1$.

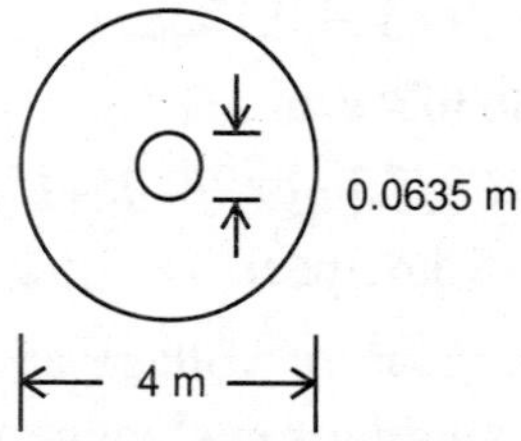

Fig. 10.2: Calculation of corona-inception voltage on conductor inside a cylinder. Conductor radius r, cylinder radius R.

Solution: Let us first derive a general expression for the corona threshold voltage for this concentric-cylindrical electrode system of Fig. 10.2.

Using Gauss's law in Electrostatics, for a cylinder of radius x with charge Q coulomb metre, there is,

$$[\varepsilon_0 = (1000/36\pi) \times 10^{-12} \text{ F/m}]$$

$$2\pi x \cdot \varepsilon_0 E_x = Q$$

or,

$$E_x = (Q/2\pi\varepsilon_0)/x$$

$\therefore$ Across a differential distance dx, the voltage difference or p.d. is:

$$dV_x = E_x \cdot dx = (Q/2\pi\varepsilon_0) \cdot (1/x) \cdot dx$$

If the voltage between the inner conductor and outer cylinder is V, then

$$V = (Q/2\pi\varepsilon_0) \int_r^R dx/x$$

$$= (Q/2\pi\varepsilon_0) \cdot \ln(R/r) \text{ volts}$$

$$\therefore \quad (q/2\pi\varepsilon_0) = V/\ln(R/r)$$

Then, the voltage gradient in terms of the voltage is

$$E_x = [V/\ln(R/r)] \cdot (1/x)$$

On the surface of the inner conductor, i.e., $x = r$

$$E_r = [V/\ln(R/r)] \cdot (1/r) \text{ volt/m}$$

This expression yields the voltage in terms of the voltage gradient

$$E_r : V = E_r \cdot r \cdot \ln(R/r)$$

If now, $E_r = E_0$, the corona-inception gradient, then, the corona-inception voltage or threshold voltage or disruptive voltage will be:

$$V_0 = E_0 \cdot r \cdot \ln(R/r)$$

$$= 3100 \cdot \delta\ [1 + 0.0308/\sqrt{(r \cdot \delta)}] \cdot r \cdot \ln(R/r) \text{ kV}$$

For the given dimensions $r = 0.03175$ m,

$$R = 2 \text{ m}$$

and

$$\delta = 1$$

Peeks's Formula for E_0 becomes:

$$E_0 = 3100 \cdot 1\ [1 + 0.0308/\sqrt{0.03175}]$$

$$= 3100(1 + 0.17285)$$

$$= 3636 \text{ kV/m}$$

$$\therefore \quad V_{0s} = 0.03175 \cdot \ln(2/0.03175) \cdot 3636$$

$$= 478 \text{ kV, peak}$$

For d.c. excitation, this is the value of voltage at which corona discharge commences. For a.c. excitation, the corresponding r.m.s. value of voltage will be $478/\sqrt{2} = 338$ kV.

We may remark here at this stage that the outer radius or diameter of the inner conductor has been used in the calculation as if it has a smooth surface. Hence the voltage is shown as V_{os}. In actual practice, an ACSR conductor when used for overhead transmission lines has a rough surface which is usually called a "rosette" pattern. On account of this non-smooth surface, corona-inception takes place at a lower voltage and the inception gradient is also lower. This gives the concept of a "roughness factor". For normal transmission-line work, the roughness factor is about 1.3. Thus, corona-inception on the above conductor would occur at the voltage: $V_{or} = V_{os}$/(roughness factor) = 478/1.3 = 368 kV, peak. For a.c., the corresponding r.m.s. or effective value is $368/\sqrt{2} = 260$ kV.

The voltage V_{os} is known as the nominal smooth-conductor inception voltage for corona. Therefore, the procedure for evaluating the threshold gradient and voltage would be to calculate them for a conductor with the same outer radius as the given rough conductor, and then divide these values by the roughness factor. This factor should not be introduced into Peek's Formula to start with. It will give completely erroneous results. It is not as if the radius of the conductor is reduced to start with.

This conductor is the most famous in high voltage power transmission. It has been used in the U.S.A. for a 500 kV a.c. line by the Bonneville Power Administration (B. P. A.). It is the largest diameter ACSR conductor manufactured in the world, and is called "Expanded ACSR Conductor".

We next consider the case where the conductor is strung above a ground plane, as happens in an actual transmission line.

Example 1: The average height of conductor above ground is 15 m for the conductor considered above with $r = 0.03175$ m. For $\delta = 1$, calculate the corona-inception voltage.

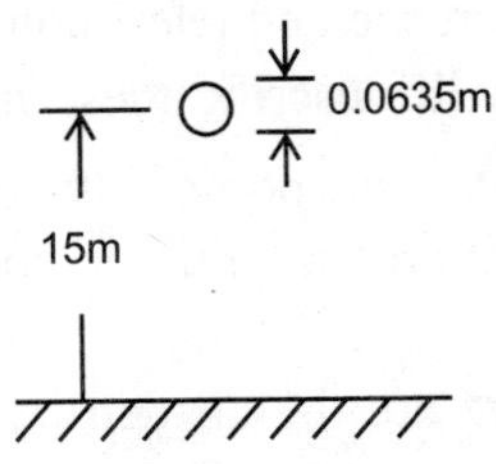

Fig. 10.3: Calculation of corona-inception voltage on conductor above ground plane. Conductor radius r, height above ground H.

Solution: See Fig. 10.3. Consider the nominal smooth-conductor value. The Maxwell's Potential Coefficient is $\ln(2H/r)$. See also Questions 15 and 22 in Chapter 8.

Then, $$V = (Q/2\pi\varepsilon_0) \cdot \ln(2H/r)$$

Also, $$E = (Q/2\pi\varepsilon_0) \cdot (1/r)$$

$\therefore$ $$V_{os} = E_{os} \cdot r \cdot \ln(2H/r)$$

For the cylinder-to-plane geometry, Peek's Formula for corona-inception gradient is:

$$E_0 = 3000 \cdot \delta[1 + 0.0301/\sqrt{(r \cdot \delta)}] \text{ kV/m, peak}$$

With $\delta = 1$

$r = 0.03175$ m

$H = 15$ m

we obtain: $E_{os} = 3000\,(1 + 0.0301/\sqrt{0.03175})$

$= 3000 \times 1.169$

$= 3506.8$ kV/m, peak

$\therefore$ $V_{os} = 0.03175 \times \ln(2 \times 15/0.03175) \times 3506.8$

$= 762.8$ kV, peak

For an a.c. system, the r.m.s. value of threshold voltage is 539.5 kV.

The line-to-ground voltage for the 500 kV system is $500/\sqrt{3} = 288.7$ kV, r.m.s. (with maximum of $550/\sqrt{3} = 317.54$ kV). Therefore, on a smooth conductor, the corona-inception voltage is much higher than normal power-frequency voltage. Therefore, in fair weather, no corona can be present.

But if we consider roughness on the conductor surface using a roughness factor of 1.3, corona-inception will occur at a line-to-ground voltage of $539.5/\sqrt{3} = 415$ kV, r.m.s., which is still higher than 288.7 kV for the 500 kV system. Even with a 10% excess voltage of 550 kV line-to-line = 316.54 kV line-to-ground, corona will not be present.

However, in foul weather, corona occurs at about 60% of the fair-weather corona threshold voltage. This will be $0.6 \times 415 = 249$ kV. We now observe that this is lower than 288.7 kV and we may therefore expect corona to be present. This will yield power loss, audible noise, radio interference and television interference. Engineers must be aware of all these calculations. We will describe these interference problems in Question 6.

Q.5. What are the corona-threshold or inception voltages for the two examples in Question 4 before at the following temperature and altitude conditions? (*a*) $H = 2000$ m, $t = -10$°C; and (*b*) Sea level and 30°C.

Ans. The only quantities that change are the air-density factors. We have to calculate δ. The equations are:

(*i*) Pressure at height *H*: $p_{mm} = 760 - 70 \times 10^{-3}$ H;

(*ii*) Air-density factor: $\delta = [(273 + t_r)/(273 + t)] \cdot (p_{mm}/p_r)$

We will take the reference values as: $t_r = 0$°C, and $p_r = 760$ mm Hg

Then $\delta = [273/(273 + t)] \cdot (p_{mm}/760)$

$= 0.3592 \cdot p_{mm}/(273 + t)$

(*a*) $H = 2000$, which gives $p_{mm} = 760 - 70 \times 10^{-3} \times 2000 = 620$ mm Hg

$\delta = 0.3592 \times 620/(273 - 10) = 0.8468$

(*b*) $H = 0,$ $p_{mm} = 760$ mm Hg

$\therefore$ $\delta = 0.3592 \times 760/(273 + 30) = 0.9$

The corona-inception gradients and corresponding voltages become:

(*a*) With

$$r = 0.03175 \text{ m}$$
$$R = 2 \text{ m}$$
$$\delta = 0.8468$$
$$E_{0s} = 3100 \times 0.8468\ [1 + 0.0308/\sqrt{(0.03175 \times 0.8468)}]$$
$$= 3118 \text{ kV/m, peak}$$
$$V_{0s} = r \cdot \ln(R/r) \cdot E_{0s}$$
$$= 0.03175 \times \ln(2/0.03175) \times 3118 = 410 \text{ kV, peak}$$
$$= 290 \text{ kV, r.m.s.}$$

At $\delta = 1$, the values were:

$$V_{0s} = 478 \text{ kV, peak} = 338 \text{ kV, r.m.s.;}$$

and $$V_{0r} = 369 \text{ kV, peak} = 260 \text{ kV, r.m.s.}$$

(*b*)

$$r = 0.03175 \text{ m}$$
$$R = 2 \text{ m}$$
$$\delta = 0.9$$

$\therefore$ $$E_{0s} = 3298.35 \text{ kV, peak}$$

This gives $$V_{0s} = 410 \times 3298.35/3118$$
$$= 433.7 \text{ kV peak}$$
$$= 306.73 \text{ kV, r.m.s.}$$

Also, $$V_{0r} = 433.7/1.3$$
$$= 333.6 \text{ kV peak}$$
$$= 235.95 \text{ kV, r.m.s.}$$

Example 2:

(*a*) With

$$r = 0.03175 \text{ m}$$
$$H = 15 \text{ m}$$

$\delta = 0.8468$, the corona-inception gradient will be:

$$E_{0s} = 3000 \times 0.8468[1 + 0.0301/\sqrt{(0.03175 \times 0.8468)}]$$
$$= 3006.74 \text{ kV/m peak}$$
$$= 2126.4 \text{ kV/m, r.m.s.}$$

The corresponding voltage is

$$V_{0s} = 0.03175 \times \ln(30/0.03175) \times 3006.74$$
$$= 654 \text{ kV peak}$$
$$= 462.5 \text{ kV, r.m.s.}$$

Then $V_{0r} = V_{0s}/1.3$

$= 503$ kV peak

$= 355.8$ kV, r.m.s.

(*b*) With $r = 0.03175$

$H = 15$ m

$\delta = 0.9$ at sea level and 0°C

The corona-inception gradient is:

$$E_{0s} = 3000 \times 0.9\ [1 + 0.03175/\sqrt{(0.03175 \times 0.9)}]$$
$$= 3193.8 \text{ kV, peak.}$$

$$\therefore \quad V_{0s} = 0.03175 \times \ln(30/0.03175) \times 3193.8$$
$$= 694.7 \text{ kV, peak}$$
$$= 491.3 \text{ kV, r.m.s.}$$

Then, $$V_{0r} = V_{0s}/1.3$$
$$= 491.3/1.3$$
$$= 378 \text{ kV, r.m.s.}$$

Therefore, comparing the two situations, we have:

(*a*) $H = 2000$ m, $t = -10$°C, $\delta = 0.8468$, $V_{0r} = 355.8$ kV, r.m.s;

(*b*) Sea level, $t = 30$°C, $\delta = 0.9$, $V_{0r} = 378$ kV, r.m.s.

Thus, in cold mountainous areas, corona commences at lower voltages than at lower altitudes and higher temperatures. Also, as the air-density factor increases, the inception voltage also increases, as can be compared with Question 4, where it is 539.5/1.3 = 415 kV at $\delta = 1$ for $r = 0.03175$ m and $H = 15$ m.

Such data are very important, because testing laboratories for high voltage equipment and line materials are located at one place while transmission lines can run at different elevations and have different temperatures from the laboratory location. Thus, conversion factors must be available to apply to the true situations. Even a single long transmission line can run along different climatic conditions, such as the Himalayan Mountains where generating stations are located to Delhi which is extremely hot and is at an elevation of only 300 metres above sea level.

Q.6. What are the effects of corona in high voltage equipment and transmission lines?

Ans. Nearly all the effects of corona are harmful to insulation. There are other effects also which are important from several non-electrical aspects of life but depend on the comforts and leisure brought about by the use of electricity such as radio, TV, etc. We only give the more important engineering effects. These are primarily:

(*a*) Deterioration of solid and liquid insulations due to partial discharges in gaseous voids;

(*b*) Electrical energy or power loss from transmission-line conductors. This is called Corona Power Loss.

(*c*) Excessive audible noise emanating from overhead h.v. lines. This is more serious from e.h.v. lines. The noise can disturb residents living closeby. Since the noise is present even during night time when ambient noises are very low or non-existent, residents tend to lose sleep, giving rise to tendency for insanity. This problem or condition is known as the Psycho Acoustics Problem. It is most pronounced in moist weather and rain. The frequency range of acoustic noise is up to 20 kHz.

(*d*) **Carrier Communication Interference:** Telephone and system protection signals utilise the conductors of overhead lines between generating station and receiving station. Therefore, transmitters and receivers of these signals called Carrier Signals suffer interference from corona discharges of the pulse type from the conductors. Powers of the transmitters and receivers must be high enough to combat (override) the interference level of corona-generated noise. The frequencies used for carrier signals is up to 250 kHz. Signals of higher frequency suffer large attenuation due to high-frequency resistance of the ACSR or other types of conductors caused by skin effect.

(*e*) **Radio Interference:** Amplitude-modulated radio signals (MW) are in the range of 0.15 MHz to 1.6 MHz (150 kHz to 1600 kHz). Pulse-type corona with a.c. and on the positive half-cycle of a.c. give rise to copious interference to these signals when they enter the antenna input end of radio receivers, located close to transmission lines. The only way to reduce the noise level other than using conductor sizes on the lines (to prevent the generation of noise-generating pulse-type corona) is to locate receivers far enough away from the lines in the lateral direction. This means that the power company must locate the line away from residential areas and thus acquire a larger width for the right-of-way for the transmission line. This is quite expensive. Interference levels are prescribed or designated by the minimum signal to noise ratio (S/N ratio) by government regulations. The limits are different in different countries of the world and the reader should be aware of these Standard Specifications, where they exist, before designing lines.

(*f*) **Television Interference:** Frequency Modulation (FM) and Television signals are in the 50–100 MHz range, or even higher. Several signals from corona-type discharges interfere with good or satisfactory TV reception. These discharges are most pronounced from corona emanating from water drops sticking to an overhead conductor and therefore is primarily a phenomenon in rainy conditions or foul weather and snow. Wet snow gives rise to TV interference. The type of interference can take the form of dark streaks drifting across the picture tube or white spots called "snow".

Normally, Radio Interference is caused by positive corona pulses when the highly-stressed electrode is going through the positive half-cycle of a.c. excitation or is the positive pole of a d.c. line. However, for equal potential gradients (peak value) an a.c. line yields a higher level of interference than a d.c. bipolar or monopolar line. Furthermore, under foul-weather conditions, the noise level of an a.c. line increases from its fair-weather value, while it is opposite in the d.c. case; the noise level of a d.c. line actually goes down in rain from the fair-weather value. This is because of the difference in the physics of their effect on electron avalanche mechanisms.

Audible Noise is also lower from a d.c. line than from an a.c. line for the same values of voltage gradients on the conductors. Positive corona pulses do not generate interference signals beyond 3 MHz of any concern so that they do not cause TVI.

However, it is otherwise with negative corona pulses. These can generate interference up to 200 MHz. But their level is much lower than positive corona streamer pulses up to 3 MHz. Therefore, TVI is mainly due to negative corona pulses. An additional source of TVI is what is known as "Micro discharges" from cracked porcelain insulators (or glass). This is a nuisance near metropolitan areas where TV receivers cannot be located too far away from the transmission-line corridor. It is not a corona-type discharge but a spark discharge which is a complete breakdown of a gap and not local. This type of breakdown has been discussed in detail in Chapter 3. Another frequent cause of interference is what is known as "loose guy wire" holding transmission-line poles which may come too close to the line conductors and cause sparking especially in rainy weather. This can occur even from low-voltage distribution lines in urban areas to be of concern to TV receivers.

Q.7. What are the equations available for calculating energy or power loss due to corona on high voltage conductors?

Ans. Energy and power loss were first given in equation form by F.W. Peek, Jr., based on extensive experiments at the laboratories of the General Electric Company in the U.S.A. Many others have followed suit. All such formulas are empirical although certain attempts have been made to derive them theoretically. We shall only give some empirical formulae here.

The available formulas can be classified broadly as follows, where E_0 and V_0 are corona-threshold or inception electric gradient and voltage, respectively, and P_c is the corona loss in proper units as indicated:

I. Those based on voltages: V is actual voltage and ks are constants.

(*a*) Skilling's Formula: $P_c = k_S (V - V_0)$;

(*b*) Peek's Formula: $P_c = k_P (V - V_0)^2$;

(*c*) Ryan and Henline Formula: $P_c = k_{RH} V (V - V_0)$;

(*d*) Peterson's Formula: $P_c = k_{PS} V^2 \cdot f(V/V_0)$, where f is a function;

(*e*) Foust and Menger Formula: $P_c = k_{FM} \cdot V^3$;

(*f*) Prinz's Formula: $P_c = k_{PR} V^2 (V - V_0)$.

In all cases, they apply only when the actual voltage V is higher than the corona-inception voltage V_0.

II. Those based on voltage gradients: $E > E_0$.

(*a*) Ontario Hydro or Nigol-Cassan Formula:

$$P_c = k_{NC} \cdot E^2 \cdot \ln(E/E_0);$$

(*b*) Project EHV or Anderson-Baretsky-McCarthy Formula:

$$P_c = k_{ABM} \cdot E^5$$

The actual loss in watts must include some or all the factors given below:

(*a*) Corona-inception voltage, V_0;

(*b*) The actual voltage of the conductor, V, at which the loss is calculated;

(*c*) The excess voltage $(V - V_0)$;

(*d*) Conductor surface voltage gradient E at voltage V;

(*e*) Corona-inception voltage gradient E_0 at voltage V_0;

(*f*) The excess voltage gradient $(E - E_0)$;

(*g*) Frequency, f;

(*h*) Conductor size or diameter, d;

(*i*) For a bundle conductor, the number of conductors, N, bundle spacing, B;

(*j*) Atmospheric condition: fair-weather, rate of rainfall, snow, etc.;

(*k*) Conductor surface condition: new or weathered. This is described by the "roughness factor".

We will only give two formulas applicable to single conductors. The reader is referred to specialised literature on this very extensive topic. Some examples of calculation of the power loss are also provided.

I. Peek's Formula: This is the classic formula relating the corona power loss in watts per unit length of conductor with nearly all the variables affecting the loss. It is applicable to a.c. excitation only for conductor of radius r strung at a height H above ground. Voltages and gradients are r.m.s. values.

Let V = voltage of conductor with respect to ground,

V_0 = corona-inception voltage = $r \cdot \ln(2H/r) \cdot E_0$,

E_0 = $(30/\sqrt{2}) \cdot [1 + 0.0301/\ \sqrt{(r \cdot \delta)}]$, 1 kV/cm, r.m.s., with r in metre unit.

Then Peek's expression for the loss in kW/km is, with V and V_0 in kV:

$$P_c = 5.16 \times 10^{-3} \cdot f \cdot \sqrt{(r/2H)} \cdot V^2\ (1 - V_0/V)^2 \text{ kW/km}$$

Example: Let

r = 1 cm radius

= 0.01 m

H = 5 m

f = 50 Hz

δ = 1

E = 1.1 E_0

V = 1.1 V_0

(10% excess voltage above corona-inception voltage). Calculate the corona loss in kW/km by Peek's Formula.

We first calculate E_0 and V_0, the corona-inception values.

$$E_0 = 21.4\,(1 + 0.0301/\sqrt{0.01})$$
$$= 27.84 \text{ kV/cm, r.m.s. (Note: } 30/\sqrt{2} = 21.4)$$
$$V_0 = r \cdot \ln(2H/r) \cdot E_0$$
$$= 0.01 \times \ln(10/0.01) \cdot 27.84 \times 10^2$$
$$= 192.3 \text{ kV}$$

Then, $V = 1.1\, V_0 = 211.5$ kV

$\therefore$ $P_c = 5.16 \times 10^{-3} \times 50 \times \sqrt{(0.01/10)} \times 211.5^2\,(1 - 1/1.1)^2$

$$= 3.016 \text{ kW/km} = 3.016 \text{ watt/metre length.}$$

We can also calculate the corona-loss current at voltage V:

$$I_c = P_c/V$$
$$= 3.016/211.5$$
$$= 0.014 \text{ A/km} = 14 \text{ mA/km}$$
$$= 14\ \mu\text{A/m}$$

II. Ryan and Henline Formula: This formula takes into account the capacitance of the conductor to ground:

$$C = 2\pi\varepsilon_0 / \ln(2H/r) \text{ farad/m}$$
$$C_0 = (1000/36\pi) \times 10^{-12} \text{ F/m}$$
$$= 8.84194 \times 10^{-12} \text{ F/m}$$

Then, the power loss in kW/km is given by Ryan and Henline as:

$$P_c = 4 \cdot f \cdot C \cdot V\,(V - V_0) \times 10^6 \text{ kW/km}$$

For the previous conductor,

$$C = 2\pi \times (1000/36\pi) \times 10^{-12}/\ln(10/0.01)$$
$$= 8.04 \times 10^{-12} \text{ F/m}$$

$\therefore$ $P_c = 4 \times 50 \times 8.04 \times 10^{-12} \times 211.5^2\,(1 - 1/1.1) \times 10^6$ kW/km

$$= 6.54 \text{ kW/km} = 6.54 \text{ W/m}$$

The corresponding corona current is $I_c = 6.54/211.5$ A/km $= 30.9\ \mu$A/m.

Since all such formulae are empirical, they are bound to differ by certain factors, since they are based on different premises, but the order of magnitudes is the same.

Q.8. How is Partial Discharge measured from solid and liquid insulations inside h.v. equipment during testing?

Ans. When the thickness of insulation increases, the probability of voids and gas bubbles formed in the insulation increases. This is harmful to the very high voltage equipment. Standard Specifications prescribe partial-discharge measurement, such as a transformer undergoing a routine test at prescribed voltage for 30 minutes.

These voids give rise to corona-type pulses over a wide frequency range up to 5 MHz. These pulses are propagated through the capacitance network of the transformer winding and along the connectors. Partial discharge measurement is recommended by I.E.C. Standard Publication No. 270, 1980, and I.S. 2026: Power Transformers. Among many circuits available, one is shown in Fig. 10.4.

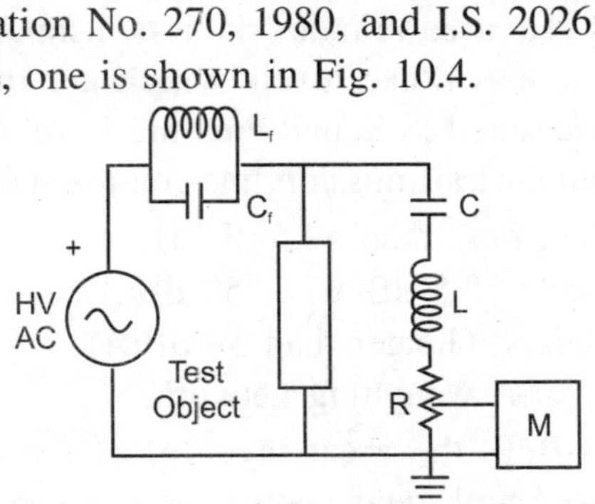

Fig. 10.4: Circuit for measurement of partial discharges.

The measuring device M consists of an oscilloscope or a radio noise meter. On an oscilloscope, the charge-voltage (Q-V) diagram is displayed as a Lissajous Figure. In using a noise meter, the reading appears on a scale as micro-volts, μV. The pulses are directed through the voltage-divider circuit shown as C-L-R. A series filter consisting of parallel L_f-C_f circuit makes sure that no pulses get back into the source transformer, but all pulses reach the voltage divider and measuring circuit.

When a partial discharge occurs, it amounts to release of charge from the voids and a Lissajous Figure appears on the screen of the oscilloscope, whose horizontal plates are fed by the source voltage (through the voltage divider) and the vertical plates by the charge. When using a Noise Meter, a reading develops on its scale. Any figure on the 'scope screen or reading on the noise meter is an indication of the existence of a partial discharge.

Q.9. How is audible noise (AN) measured from a transmission line?

Ans. Audible noise occurs from an e.h.v. line in foul weather. Figure 10.5 shows the block diagram of an AN meter. The input end has a standard microphone which is directed towards the line conductors and its housing is protected from the weather and environment.

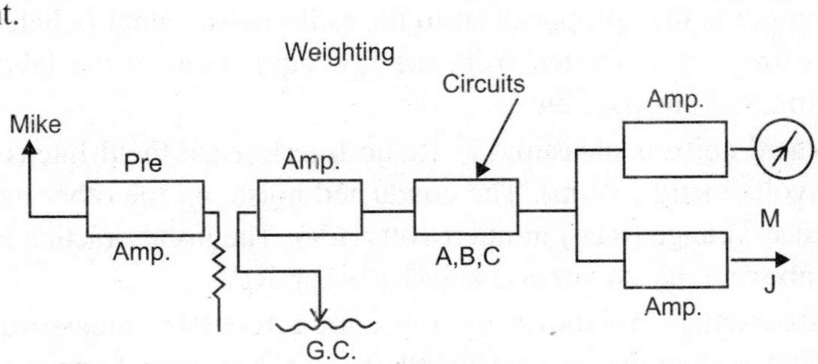

Fig. 10.5: Circuit for measurement of audible noise.

The output of the microphone is amplified and fed to a "weighting circuit" which is a filter with prescribed amplitude-frequency characteristics. There are essentially three basic networks, designated as *A*, *B*, *C* networks. The *A*-network has the same response as the human ear and is also used in industrial establishments such as machine shops and generating stations to assess noise levels which are important to Labour Relations Departments. Noise is measured as Sound Pressure Level (SPL). The following limits are usually followed for outdoor transmission lines on the *A*-weighting network:

No complaints: Less than 52.5 dB(*A*);
Few complaints: 52.5 dB(A) to 59 dB(*A*);
Many complaints: Greater than 59 dB(*A*).

The symbol (*A*) stands for *A*-weighting network.

Alexander Graham Bell, the great inventor of the telephone and a scientist and humanitarian who worked with deaf people, gave the basic unit of SPL as 20×10^{-6} newton/m^2 or 20 μpascals or 20 μbars. This relates to the pressure of air which is carrying the vibrations. The dB value at any pressure level is: SPL (dB) $= 10 \cdot \log_{10} (\text{SPL}/20 \times 10^{-6})$.

It is also known as the "Acoustic Power Level" (PWL) or simply as Audible Noise Level (AN).

Example: A meter indicates 50 dB. What is the SPL in pascals?

Solution: $50 = 10 \cdot \log(\text{SPL}/20 \times 10^{-6})$, giving SPL = 2 pascals or newtons/m^2.

Q.10. How is radio noise measured from h.v. equipment in the laboratory, and from outdoor transmission lines?

Ans. A radio noise meter (RN Meter) is a very sensitive super-heterodyne radio receiver whose input is through an antenna, or else the signal is fed through a cable into the input jack of the meter. The antenna can be either (*a*) a whip or rod antenna, usually 0.5 or 1 metre long, or, (*b*) a loop antenna of 0.25 or 0.5 metre on side. The rod picks up the vertical component of electric field but the loop antenna picks up the magnetic field of the radio wave.

The IF is 455 kHz after superheterodyning the input signal at MHz frequencies by a local oscillator. This IF is amplified and fed to a filter whose output is measured on a meter. The output is thus proportional to the radio noise signal radiated from the outdoor transmission line, or conducted from the h.v. equipment in the laboratory, such as an insulator string, transformer, etc.

The radiated noise is measured as Radio Interference Field Intensity (RIFI or simply RI) in microvolts/metre (μV/m). The conducted noise, on the other hand, is measured as Radio Influence Voltage (RIV) in microvolts (μV). The usual practice is to designate them as decibels above 1 μV/m for RI, and 1 μV for RIV.

A circuit used in a laboratory for conducted RIV measurement is shown in Fig. 10.6. The test object is energised from a h.v. transformer through a parallel L_f-C_f filter which blocks all corona pulses emanating from the test piece from entering the source transformer, but channels them into the noise-measuring circuit. A voltage divider C_c-L applies a low voltage to the noise-measuring circuit for safety so that no more than 10 volts at power frequency appear there even if the source transformer voltage may be 1 million volts (1 MV).

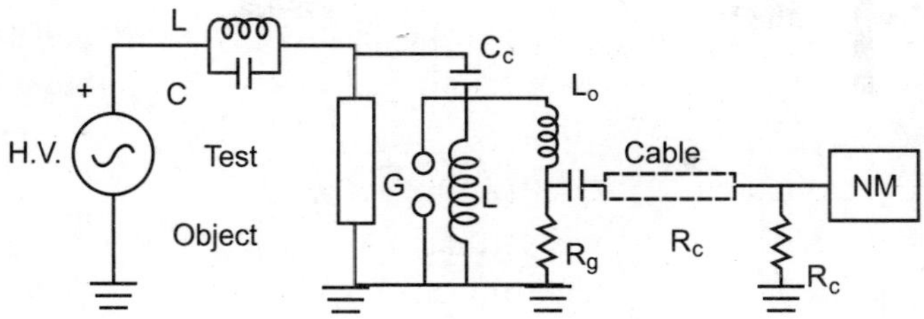

Fig. 10.6: Circuit for measurement of conducted radio noise in laboratory.

The inductance L acts as an open circuit at high frequency in the MHz range, but the coupling capacitor C_c is so designed that it acts like a short circuit at these high frequencies. Therefore, all interfering signals get into the meter circuit. Since radio noise is a high-frequency phenomenon (the usual measuring frequency is 1 MHz), the meter input impedance (usually 50 ohms or 75 ohms) and the measuring cable surge impedance (50 or 75 ohms) must be matched with the voltage divider. This is carried out by the divider shown as L_o-R_g. The noise meter (NM) reads directly in microvolts in such a circuit or the meter scale can be calibrated in dB above 1 μV.

When using rod or loop antenna at the input end in the outdoors or inside laboratories to measure radiated interference values, the meter reads μV/m, or it can be calibrated to indicate dB above 1 μV/m.

The conversion from μV to dB above 1μV is as follows:

Let RIV be in μV. Then its dB value above 1 μV is:

RIV (dB) = $20 \cdot \text{Log}_{10}$(RIV in μV).

Note the difference between conversion to dB values between Audible Noise and Radio Noise. AN is the power level whereas RN is in volts.

Example: The radio noise level in an experiment is measured as 100 μV on a RN meter. Convert it to dB above 1 μV.

Solution: dB value = $20 \cdot \log_{10}(100) = 40$ dB above 1 μV.

The same conversion relation holds for RI also in μV/m and dB value above 1 μV/m.

❑❑

CHAPTER 11

ELECTROSTATIC FIELD OF HIGH VOLTAGE TRANSMISSION LINES

Q.1. What are the factors that are important to consider which are caused by the electrostatic (e.s.) field near high voltage lines?

Ans. These are very many and only a few very important items can be given here. The space near a high voltage line is charged by the voltage which induces voltage in all objects coming close to it. Some indication of the "static" has been given in Chapter 1 about a shock resulting from touching an object such as a vehicle parked on insulated tyres near a h.v. transmission line. Also, the voltage induced in a telephone line running close to a h.v. line was considered in Chapter 8, Question 20.

Other important effects of the e.s. induction caused by h.v. lines near its vicinity are listed below:

(*a*) Electric field at ground level and vicinity: This affects human, animal and plant life near the line. This will be detailed in the next Question;

(*b*) On a D/C tower, when one circuit is energised, it induces a voltage in the second circuit. This is important when carrying out repairs on the tower to the unenergised circuit, or when stringing a second circuit when there is already an energised circuit on the tower. If both circuits are strung simultaneously during erection of a new transmission line, this problem does not arise;

(*c*) When a D/C tower carries two circuits at different voltages, there will be induced voltages in the circuits which are harmful to transformers and lightning arresters connected to the lower-voltage circuit;

(*d*) One very beneficial effect where the electrostatically-induced voltage from line conductors is utilised is as follows:

Where overhead shield wires or "ground wires" are used for offering protection against lightning, and these wires are insulated from the tower by making them run on porcelain insulators on top of the tower, a voltage is induced in these shielding wires. This voltage can be tapped at suitable points along the long transmission line and utilised for feeding small village loads, farms or other "rural loads". This scheme is very useful and worthwhile for India where rural electrification has assumed great importance from socio-economic-humanitarian points of view. It is estimated, by calculation which is not considered here, that about 24 kV is induced from a 400-kV line in India.

We will consider some of these points in more detail now and give some numerical examples encountered in practice to give an idea of the dangers arising from e.s. induction from h.v. lines. The reader is referred to the geometry of a 400-kV line with shielding wires in Fig. 9.8 and following the examples to be given in this Chapter, should calculate the voltage induced in the shield wires.

(*e*) While we will consider only the e.s. effects here, it is wise to remember that there are certain problems caused by electro-magnetic (e.m.) induction due to the heavy currents carried by the line conductors. These are caused by the mutual inductance effect and the magnetic field surrounding the line. It is also being realised lately that these magnetic fields affect the human body, especially babies, when homes are built too close to the line and the exposure is continuous over long periods of time. This aspect is still in experimental stages but is important, since the action of nerves in the body depends upon transmission of electrical signals in the neurons which are generated by the chemical ions and which in turn are affected by the external magnetic fields.

Q.2. What are the effects of high e.s. field on humans, animals and plant life in the vicinity of a very-high voltage overhead transmission line? What are their limits so that clearances may be suitably designed?

Ans. We have already discussed primary and secondary shock currents in Chapter 1 which apply to human beings. They will not be repeated here. But some more points may be of importance.

(*a*) When a farmer is ploughing a field under and near a h.v. line with an umbrella shading him from the sun, the pointed tip of the umbrella fixed to his tractor gives rise to partial discharges due to high electric field present under the line. The vehicle may also be charged and when the farmer tries to climb on the vehicle, he may experience a shock if the tractor tyres are made of insulating material. This "umbrella effect" came into prominence in Japan since farm workers have an umbrella fixed permanently on their heads while working in the fields, unlike in India where rural population wear cloth turbans wound on their heads as protection against the very hot sun.

It has been experimentally determined that the limit for the e.s. field at ground level should be 15 kV/m, r.m.s., for safe non-shocking working of human beings near a line. This is controlled by the minimum clearance of the line conductors above the head of a man. As an example, in the U.S.A., a power company (the B.P.A.) utilising 1200 kV kept the limit for the ground-level gradient as 9 kV/m and so they had to use a minimum clearance of 23 metres at midspan for the 3 phase-conductors. From insulation point of view against lightning and switching-surge voltages, the clearance could have been selected as 17 m. The extra 6 m was dictated entirely by e.s. field point of view at ground level where human beings and vehicular traffic could experience high induction levels

(*b*) **Animals:** Where poultry farms exist in rural areas over which a transmission line can run, the limit for adequate healthy growth of chickens is experimentally determined to be 30 kV/m. At these levels, hens cannot pick up feed grains because of vibration of their beaks. Therefore, they are undernourished and do not grow to big enough size for them to be commercially useful. Pigs and cows pick up electric charge because their hides and skins and hoofs are insulating materials; and when they touch grounded objects such as drinking-water troughs, a spark jumps from their noses and mouths to the metallic trough.

(*c*) **Plant Life:** Plants such as wheat, rice, sugar cane, etc., underneath a line experience the following type of damage due to high e.s. field: Their grain-bearing parts or the upper portion of a sugar cane plant are very pointed and dry. At high e.s. field of 20 kV/m (r.m.s.), they give rise to corona and the grain or cane gets damaged due to local heating. At 30 kV/m, the by-products of corona being ozone and N_2O, the plant suffers further chemical damage.

From the above description, we can see that if the ground-level field is kept below 15 kV/m as determined by danger to human beings, animals and plants also do not experience damage. Thus, all power lines base the clearance from ground on such limits. National Electrical Codes prescribe such limits.

Q.3. What is the minimum clearance prescribed by the National Electrical Code or Codes for high voltage transmission lines from safety considerations?

Ans. These vary from country to country, and the reader must refer to the codes of individual countries. Indian practice has been based upon British practice historically, and the rule is as follows:

Minimum clearance should be 17 feet (5.2 m) for voltage up to 33 kV. Beyond 33 kV, the clearance should be increased by 1 foot for each additional 33 kV or part thereof.

According to this rule, the reader can work out the minimum clearance for any given transmission-line voltage level. Examples are given below for important voltage levels:

Transmission Voltage, (kV)	Extra Clearance (kV – 33)/33, (ft)	Rounded to Nearest Higher Integer, (feet)	Total Clearance (feet and metres)
33 kV			17′ = 5.2 m
66 kV	1		18′ = 5.5 m
132 kV	(132 – 33)/33 = 3	3	20′ = 6.1 m
220 kV	(220 – 33)/33 = 5.67	6	23′ = 7.0 m
345 kV	(345 – 33)/33 = 9.45	10	27′ = 8.23 m
400 kV	(400 – 33)/33 = 11.12	12	29′ = 8.84 m
525 kV	(525 – 33)/33 = 14.9	15	32′ = 9.76 m
750 kV	(750 – 33)/33 = 21.72	22	39′ = 11.9 m
1200 kV	(1200 – 33)/33 = 35.36	36	53′ = 16.16 m

Q.4. A conductor of radius r is at a height H above ground. At a point P on ground at a distance x from under the conductor, what is the expression for the electric field? In which direction is it?

Ans. Refer to Fig. 11.1, please.

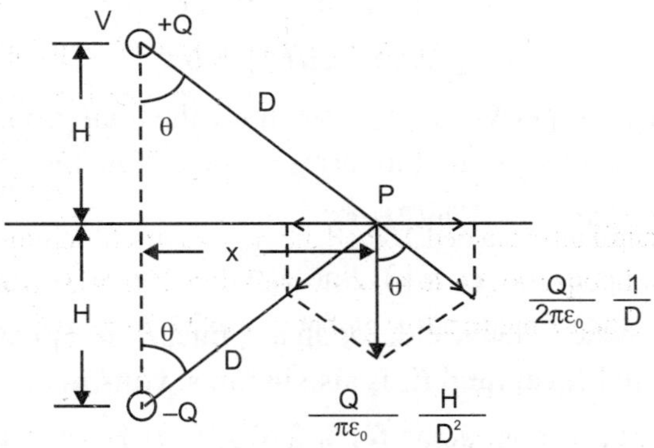

Fig. 11.1: Conductor above ground plane. Calculation of vertical component of electric field at ground level.

As discussed in Chapter 8, Question 14, when the ground plane is considered an equipotential surface, its effect can be replaced by Maxwell's Method of Images by placing an image conductor at depth H below ground with its charge being equal but of opposite polarity. One property of an equipotential surface is that the electric-field component along its surface is zero. The electric field is always perpendicular to the equipotential surface. We will prove this property from the method of images also. Thus, in this problem, only the vertical component of field is present; the horizontal component of electric field along the ground surface is zero. However, above the ground in air, both components are present.

We observed in Chapter 8 that the charge on the conductor at a voltage V is given through Maxwell's Potential Coefficient thus:

$$V = (Q/2\pi\varepsilon_0) \cdot \ln(2H/r)$$

Now, the distance between the conductor and point P is $D = \sqrt{(H^2 + x^2)}$. This is also the distance of P from the image conductor. Also, $\cos\theta = H/D$.

There are two electric-field vectors to add at P:

(*i*) Electric field vector at x due to conductor charge $+ Q$ is $E_c = (Q/2\pi\varepsilon_0)/D$. Since by definition, the electric field strength is the force exerted on a unit positive test charge, the force is along PA.

(*ii*) Similarly, due to the image charge of $-Q$, the electric field at P is $E_i = (Q/2\pi\varepsilon_0)/D$, but is directed along the line joining P to the charge and is an attraction towards $-Q$.

It is evident from the diagram that the horizontal components of electric field vectors due to $+Q$ and $-Q$ are equal in magnitude but opposite in direction. They cancel each other.

Therefore, along the ground surface, $E_h = [(Q/2\pi\varepsilon_0)/D] \cdot (\sin\theta - \sin\theta) = 0$. On the other hand, the vertical components of electric field due to both $+Q$ and $-Q$ are directed in the same sense. Therefore, the total vertical component of ground-level field at P is:

$$E_v = 2 \times [(Q/2\pi\varepsilon_0)/D] \cdot \cos\theta$$
$$= (Q/\pi\varepsilon_0) \cdot (H/D^2) \text{ V/m}$$

We will use this property and expression for the total ground-level field strength from now on. We also observe that in terms of the conductor voltage and line dimensions,

$$Q/2\pi\varepsilon_0 = V/\ln(2H/r)$$

$$\therefore \quad E_v = E_g = [2V/\ln(2H/r)] \cdot (H/D^2) \text{ volt/m}$$

Since this is a steady-state field, for an a.c. line, E_g is a phasor and if V is in r.m.s. volts or kV, the ground-level field E_g is also in r.m.s. volts/m or kV/m, respectively.

Example: An overhead conductor for a 220-kV 50 Hz a.c. line is at a height of 10 m above ground (minimum clearance). Its radius is 0.554" = 1.407 cm = 0.01407 m. (This is a single Drake ACSR conductor 1.108" outside diameter). Calculate the electric field at ground level underneath the line conductor if its voltage is $245/\sqrt{3}$ kV, r.m.s.

Solution: Note that as per Indian Standards, the maximum operating voltage for 220 kV level is 245 kV.

$$\therefore \quad V = 245/\sqrt{3} = 141.45 \text{ kV, r.m.s.}$$

which will be used for calculation of the ground-level field due to one phase-conductor only. We will consider the combined effect of all three phases next.

$$\therefore \quad E_g = [2 \times 141.45/\ln(2 \times 10/0.01407)] \cdot (1/10)$$
$$= 3.9 \text{ kV/m, r.m.s.}$$

Q.5. A S/C horizontal configuration a.c. line has all phase-conductors at equal height H above ground. The phase spacing between adjacent phases is S.

Their voltages are: $V_a = V\angle 0°$, $V_b = V\angle -120°$,

and $V_c = V\angle 120°$.

Calculate the ground-level field strength at: (*a*) Under the line centre; and (*b*) Under any of the outer phases.

Ans. See Fig. 11.2.

(*i*) The first step is to find the relation between voltages and conductor charges through Maxwell's Potential Coefficients.

$$\begin{bmatrix} V_a \\ V_b \\ V_c \end{bmatrix} = (1/2\pi\varepsilon_0)\begin{bmatrix} \ln(2H/r) & \ln(D_{12}/S) & \ln(D_{13}/2S) \\ \ln(D_{21}/S) & \ln(2H/r) & \ln(D_{23}/S) \\ \ln(D_{31}/2S) & \ln(D_{32}/S) & \ln(2H/r) \end{bmatrix}\begin{bmatrix} Q_a \\ Q_b \\ Q_c \end{bmatrix}$$

Here, $D_{12} = D_{21} = D_{23} = D_{32} = \sqrt{(4H^2 + S^2)}$;

$D_{13} = D_{31} = \sqrt{(4H^2 + 4S^2)}$.

In Matrix notation: $[V]_{ph} = (1/2\pi\varepsilon_0)\ [Q]_{ph}$.

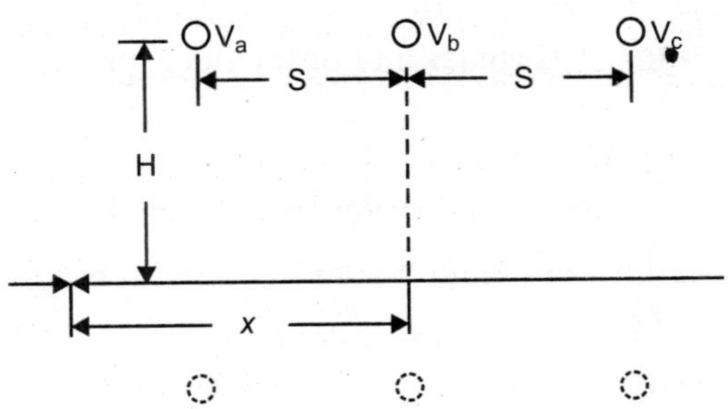

Fig. 11.2: 3-phase a.c. line. Height H, phase spacing S, conductor radius r. Calculation of ground-level field.

(*ii*) By inverting the Maxwell Potential Coefficient Matrix, we will obtain:

$$(1/2\pi\varepsilon_0)\cdot[Q]_{ph} = [P]^{-1}\cdot[V]_{ph}$$

We must note that both the voltages and resulting charges are phasors with magnitude and angle. In other words, they are complex quantities. If the reader is unfamiliar with matrix operations, then she/he can write the three simultaneous equations between voltages and charges, and then solve for the unknown phasor charges, but keep the quantity ($Q/2\pi\varepsilon_0$) intact. This procedure is shown at the end later on.

(*iii*) The total electric field along ground at distance x from the line centre will consequently be the sum of the three phasors:

$$E_g = E_a + E_b + E_c$$
$$= (Q_a/\pi\varepsilon_0)\cdot[H/\{H^2 + (x - S)^2\}] + (Q_b/\pi\varepsilon_0)[H/(H^2 + x^2)] + (Q_c/\pi\varepsilon_0)\cdot[H/\{H^2 + (x + S)^2\}]$$

We consider an example of a numerical calculation next. For a bundle conductor with N sub-conductors, it is clear that the Maxwell Self-Potential coefficient is obtained by replacing conductor radius r with the geometric mean radius $R_{eq} = R_{gm} = (N\cdot r\cdot R^{N-1})^{1/N}$, where the bundle radius $R = B/\{2\cdot\sin(\pi/N)\}$ with B = bundle spacing or spacing between adjacent sub-conductors. For a 2-bundle, $R_{eq} = \sqrt{(r\cdot B)}$.

(*a*) Under the line centre, $x = 0$.

$$\therefore \quad E_{gc} = (Q_a/\pi\varepsilon_0)\cdot[H/(H^2 + S^2)]\angle 0^\circ + (Q_b/\pi\varepsilon_0)\cdot(1/H)\angle -120^\circ + (Q_c/\pi\varepsilon_0)\cdot[H/(H^2 + S^2)]\angle 120^\circ.$$

(*b*) Under any of the outer phases, $x = S$.

$$\therefore \quad E_{go} = (Q_a/\pi\varepsilon_0)\cdot(1/H)\angle 0^\circ + (Q_b/\pi\varepsilon_0)\cdot[H/(H^2 + S^2)]\angle -120^\circ + (Q_c/\pi\varepsilon_0)\cdot[H/(H^2 + 4S^2)]\angle 120^\circ.$$

Q.6. A S/C horizontal configuration 400-kV line in India has the following details as shown in Fig. 11.3. H = 13 m, S = 12 m, conductor 2 × 3.18 cm diameter at bundle spacing of B = 45.72 cm (18"). $R_{eg} = \sqrt{0.0159\times 0.4572} = 0.08526$ m. The maximum operating voltage is 420 kV, r.m.s., line-to-line, = $420/\sqrt{3}$ kV, to ground. Calculate the ground-level field: (*a*) Under the line centre; (*b*) Under any of the outer phases. Comment on the result.

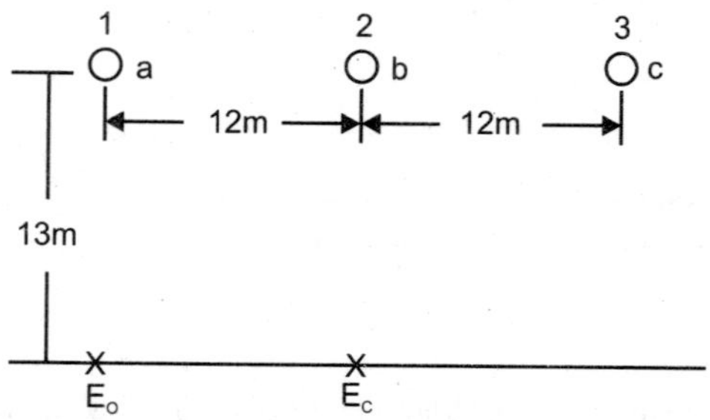

Fig. 11.3: 400-kV S/C line with horizontal configuration for Question 6.

Ans. Step 1: The Maxwell Potential Coefficient Matrix is first calculated. Conductors are numbered 1, 2, 3 for the phases *a*, *b*, *c*, respectively, as shown in Fig. 11.3.

Then,
$$P_{11} = P_{22} = P_{33} = \ln(2H/R_{eq}) = \ln(26/0.08526) = 5.72$$
$$P_{12} = P_{21} = P_{23} = P_{32} = \ln[\sqrt{(26^2 + 12^2)}/12] = 0.87$$
$$P_{13} = P_{31} = \ln[\sqrt{(26^2 + 24^2)}/24] = 0.39.$$

$$\therefore \quad [P] = \begin{bmatrix} 5.72 & 0.87 & 0.39 \\ 0.87 & 5.72 & 0.87 \\ 0.39 & 0.87 & 5.72 \end{bmatrix}$$

Step 2: Invert $[P]$.

$$[P]^{-1} = \begin{bmatrix} 0.1793 & -0.026 & -0.008 \\ -0.026 & 0.1827 & -0.026 \\ -0.008 & -0.026 & 0.1793 \end{bmatrix}$$

Step 3: Find $[Q/\pi\varepsilon_0] \cdot [V]_{ph} = [P] \cdot [Q/2\pi\varepsilon_0]_{ph} \cdot V_{ph} = 420/\sqrt{3}$ kV

$$\therefore \quad (1/2\pi\varepsilon_0) \begin{bmatrix} Q_a \\ Q_b \\ Q_c \end{bmatrix} = [P]^{-1}\,[V]_{ph}$$

$$= \begin{bmatrix} 0.1793 & -0.026 & -0.008 \\ -0.026 & 0.1827 & -0.026 \\ -0.008 & -0.026 & 0.1793 \end{bmatrix} \begin{bmatrix} 1 & +j\,0 \\ -0.5 & -j\,0.866 \\ -0.5 & +j\,0.866 \end{bmatrix} 420\sqrt{3}$$

$$= \begin{bmatrix} 47.782\,\underline{|4.46^\circ} \\ 50.62\,\underline{|-120^\circ} \\ 47.782\,\underline{|115.2^\circ} \end{bmatrix}.$$

$$\therefore \quad (1/\pi\varepsilon_0) \begin{bmatrix} Q_a \\ Q_b \\ Q_c \end{bmatrix} = \begin{bmatrix} 95.564\,\underline{|4.46^\circ} \\ 101.24\,\underline{|-120^\circ} \\ 95.564\,\underline{|115.2^\circ} \end{bmatrix}$$

Step 4: Electric field under the line centre: The general expression for E_g was found to be:

$$E_g = (Q_a/\pi\varepsilon_0) \cdot H/\{H^2 + (x - S)^2\} + (Q_b/\pi\varepsilon_0) \cdot H/(H^2 + S^2) + (Q_c/\pi\varepsilon_0) \cdot H/\{H^2 + (x + S)^2\}$$

With $H = 13$, $S = 12$ and $x = 0$, and the value of $[Q/\pi\varepsilon_0]$ found above:

$$E_{gc} = \{13/(13^2 + 12^2)\} \times 95.564\,\underline{|4.5^\circ} + (1/13) \times 101.24\,\underline{|-120^\circ} + \{13/(13^2 + 12^2)\} \times 95.564\,\underline{|115.2^\circ}$$

$$= -1.6269 - j\,2.8445 = 3.28\,\underline{|-120^\circ}$$

The magnitude of ground-level field under the line centre is thus

$$|E_{go}| = 3.28 \text{ kV/m}.$$

(*b*) Under any of the outer phases: $H = 13$ m, $S = 12$ m and $x = 12$ m

$$\therefore \quad E_{go} = (1/13) \times 95.564\,\underline{|4.5^\circ} + \{13/(13^2 + 12^2)\} \times 101.24\,\underline{|-120^\circ} + \{13/(13^2 + 24^2)\} \times 95.564\,\underline{|115.2^\circ}$$

$$= 4.515 - j\,1.56$$

$$= 4.78\,\underline{|-19^\circ}$$

The magnitude of ground-level field under any of the outer phases is thus

$$|E_{go}| = 4.78 \text{ kV/m}.$$

Comment: We observe that the field strength is higher under any of the outer phases at ground level than under the line centre.

The reader should be able to calculate and plot the magnitude of ground-level field strength at any distance x along ground varying from 0 to $3H$ and and observe the interesting results. The field is of course symmetrical about the line centre. The result obtained by this author by using a Digital Computer is shown in Fig. 11.4. We observe the following properties from this figure:

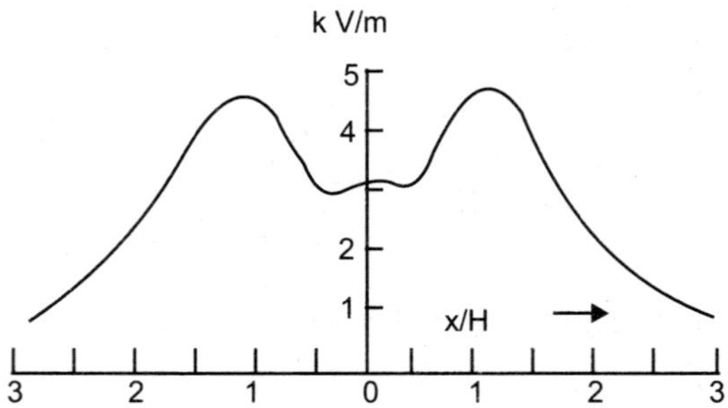

Fig. 11.4: Calculated values of ground-level field of 400-kV line for x = 0 to 3H.

(*a*) The field pattern shows a saddle under the line, being least on either side of the line centre;

(*b*) The field reaches maximum value a little beyond the outer phase-conductors, but not right under them;

(*c*) The field decreases monotonically as one moves away from the outer phase-conductors along ground.

Certain other properties may also be noted by an interested reader, particularly if Digital Computer facility is available.

(*d*) The field strength increases with increase in voltage for the same line configuration.

(*e*) The field strength at ground level decreases if line height H is increased. This type of calculation will yield the minimum clearance required for any value of limit set for the problem under consideration and forms the basis for design of line height from e.s. field point of view.

Note: If the reader is not familiar or conversant with solving simultaneous equations by one-step matrix-inversion procedure, the following notes will prove useful. The voltages and charges are complex quantities and the algebra is of necessity lengthy.

We have the following three equations between phase voltages and the quantity (charge $Q/2\pi\varepsilon_0$). Denote $(Q/2\pi\varepsilon_0) = e$ with suitable suffixes a, b, c.

$$5.72e_a + 0.87e_b + 0.39e_c = (1 + j\,0)\ 420/\sqrt{3} \qquad ...(1)$$

$$0.87e_a + 5.72e_b + 0.87e_c = (-0.5 - j\,0.866)\,420/\sqrt{3} \qquad ...(2)$$
$$0.39e_a + 0.87e_b + 5.72e_c = (-0.5 + j\,0.866)\,420/\sqrt{3} \qquad ...(3)$$

Omit the common multiplier $420/\sqrt{3}$ for the moment.
Multiply (3) by 0.87, and (2) by 5.72, and subtract one from another:
Then we obtain: $4.6371e_a + 31.9615e_b = -2.43 - j\,5.7069$ ...(4)

Next, multiply (1) by 0.87 and (2) by 0.39, and subtract one from another.
Then: $4.6371e_a - 1.4739e_b = 1.065 + j\,0.3377$ ... (5)
Subtract (5) from (4). Then,

$$33.4354\,e_b = -3.495 - j\,6.0446$$
$$\therefore \quad e_b = -0.010453 - j\,0.18078$$
$$= 0.20883\angle 240°$$
$$= 0.20883\angle -120°$$

Using this in (4), there results: $e_a = 0.196445 + j\,0.01533 = 0.197\angle 4.46°$

Using these values of e_a and e_b in equation (1), we finally obtain e_c:

$$e_c = (1 - 5.72e_a - 0.87e_b)/0.39$$
$$= -0.08385 + j\,0.178436$$
$$= 0.197\angle 115.17°$$

In order to obtain the quantities ($Q/2\pi\varepsilon_0$), we multiply the above values of e's by $420/\sqrt{3}$. The results are:

$$Q_a/2\pi\varepsilon_0 = 47.77\angle 4.46°\,;$$
$$Q_b/2\pi\varepsilon_0 = 50.64\angle -120°\,;$$
and
$$Q_c/2\pi\varepsilon_0 = 47.77\angle 115.17°\,.$$

As a comparison, the reader can calculate the ground-level field profile when the line is assumed to be perfectly transposed. The author's calculation yields: $|E_{gc}| = 3.43$ kV/m, and $|E_{go}| = 5.03$ kV/m.

Q.7. In the numerical example shown in the previous Question, consider a telephone line running parallel to the 400-kV line whose height is 5 metres above ground. Calculate the voltage induced in it if the distance of the telephone line is: (*a*) 20 metres from the outer phase; and (*b*) 100 metres from the outer phase.

Ans. See Fig. 11.5. The telephone wire is numbered 4.

From previous calculations, the quantities ($Q/2\pi\varepsilon_0$) were:

$$(Q_a/2\pi\varepsilon_0) = 47.78\angle 4.5°$$
$$(Q_b/2\pi\varepsilon_0) = 50.62\angle -120°$$
and
$$(Q_c/2\pi\varepsilon_0) = 47.78\angle 115.2°$$

The Maxwell's Potential Coefficients between the telephone wire and the phase conductors of the 400-kV line are calculated as follows:

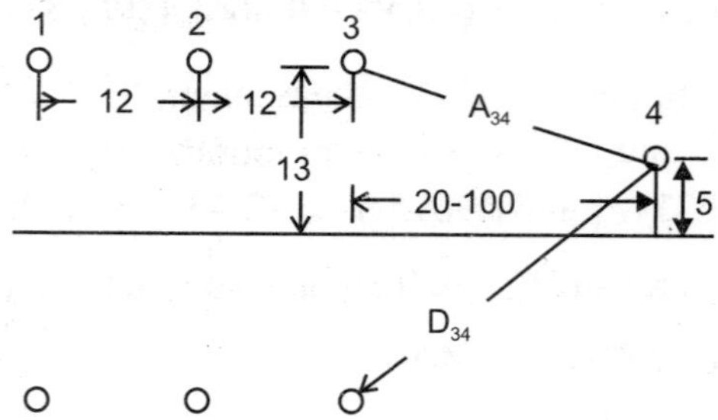

Fig. 11.5: 400-kV line and telephone line for calculation of induced voltage in the latter for Question 7.

(*a*) **20-metre separation:**

$$P_{14} = \ln(D_{14}/A_{14}) = \ln[\sqrt{(18^2 + 44^2)}/\sqrt{(8^2 + 44^2)}] = \ln(47.54/44.72) = 0.061$$

$$P_{24} = \ln(D_{24}/A_{24}) = \ln[\sqrt{(18^2 + 32^2)}/\sqrt{(8^2 + 32^2)}] = \ln(36.71/32.99) = 0.107$$

$$P_{34} = \ln(D_{34}/A_{34}) = \ln[\sqrt{(18^2 + 20^2)}/\sqrt{(8^2 + 20^2)}] = \ln(26.9/21.54) = 0.222$$

The voltage induced will be $V_I = \Sigma P_i \cdot Q_i/2\pi\varepsilon_0$.

Thus: $V_I = (Q_a/2\pi\varepsilon_0) \cdot P_{14} + (Q_b/2\pi\varepsilon_0) \cdot P_{24} + (Q_c/2\pi\varepsilon_0) \cdot P_{34}$

$= 2.915(0.997 + j\,0.0078) + 5.416(-0.5 - j\,0.866) + 10.6(-0.426 + j\,0.905)$

$= -4.316 + j\,4.915 = 6.54\underline{|131.3^\circ}$ kV.

(*b*) **For 100-metre separation:**

$P_{14} = \ln(125.3/124.26) = 0.0083,$

$P_{24} = \ln(113.44/112.29) = 0.01$, and

$P_{34} = \ln(101.61/100.32) = 0.01337.$

$\therefore \quad V_I = 0.4(0.997 + j\,0.0078) + 0.506(-0.5 - j\,0.866) + 0.639(-0.426 + 0.905)$

$= -0.1263 + j\,0.143 = 0.191\underline{|131.5^\circ}$ kV

Thus, if the telephone line is only 20 metres from the outer phase, a voltage of 6.54 kV is induced in it, while if it is moved to 100 metres from the outer phase of the 400-kV line, the induced voltage falls to 191 volts. These voltages are quite considerable, when we think about it.

Also see Question 21 for voltage induced from a single high voltage conductor into a telephone line.

❑❑

CHAPTER 12

High Voltage Transformers and Circuit Breakers

Q.1. What are the types of transformers used in high voltage power transmission and laboratory testing practice?

Ans. (*a*) Potential transformers; (*b*) Current transformers; (*c*) Auto transformers; (*d*) Two-winding transformers; (*e*) Three-winding transformers; (*f*) Tap-changing transformers; (*g*) Booster transformers; (*h*) Motor-starting transformers; (*i*) 3-phase to 6-phase transformation transformers for Thyristor Bridges for D.C. transmission; (*j*) High voltage testing transformers in single units; and (*k*) H.V. cascade-connected testing transformer units.

Q.2. What are the types of circuit breakers used in high voltage power transmission and distribution?

Ans. (*a*) Simple air circuit breakers; (*b*) Dead-tank oil circuit breakers; (*c*) Live-tank circuit breakers; (*d*) Oil-minimum circuit breakers; (*e*) Air-blast circuit breakers; (*f*) SF_6 breakers; and (*g*) Vacuum circuit breakers.

TRANSFORMERS

Q.3. What is the basic principle on which transformer action takes place? Derive the e.m.f. equation.

Ans. We consider only sinusoidal input with the instantaneous value: $v = \sqrt{2}\, V \sin 2\pi f t$, with frequency f hertz, and V = r.m.s. value in volt. A transformer has a "core" on which two separate coils are shown with N_1 and N_2. (The coils are shown as separately wound, but this need not be necessarily the case), see Fig. 12.1.

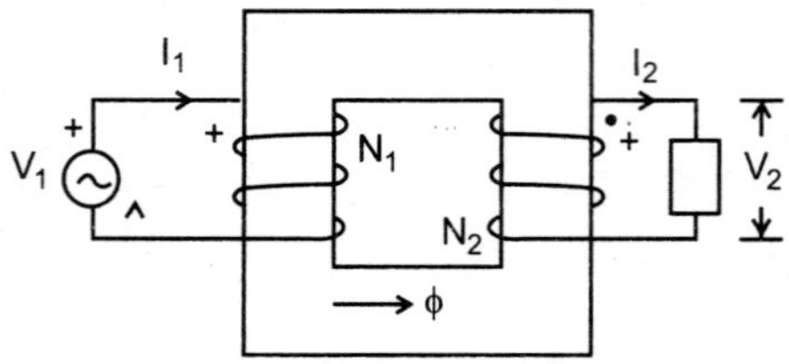

Fig. 12.1: Ideal two-winding transformer.

One coil is connected to a source of voltage V_1, current I_1 and power W. The other is connected to load drawing a current I_2 at voltage V_2. The power drawn by the load is also W. As such, we omit power losses in the transformer. Such a transformer is then known as an "ideal two-winding transformer".

The applied voltage to the "primary" winding sets up a flux ϕ in the core which also varies sinusoidally in time. It is given by the Faraday-Lenz Law for the induced voltage e in coil 1: $e_1 = -N_1 \cdot d\phi/dt$.

Let us consider the flux to be given by $\phi = \phi_m \cdot \sin 2\pi ft$.

Then, $e_1 = -2\pi f N_1 \cdot \cos 2\pi ft = 2\pi f N_1 \phi_m \cdot \sin(2\pi ft - 90°)$.

This shows that the induced voltage in any coil lags behind the flux in time by 90°. The source voltage v_1 has to oppose this induced voltage so that its equation is

$$v_1 = 2\pi f N_1 \phi_m \cos 2\pi ft$$
$$= \sqrt{2} V_1 \cos 2\pi ft$$

$$\therefore \quad V_1 = \sqrt{2}\pi f N_1 \phi_m$$
$$= 4.44 f N_1 \phi_m \text{ volt, r.m.s.}$$

with ϕ_m = peak or crest value of the sinusoidal flux in weber. This is the basic e.m.f. equation of a coil in a transformer. The voltage in a second coil linking the same flux is

$$e_2 = -N_2 \, d\phi/dt = -2\pi f N_2 \phi_m \cos 2\pi ft$$

It is this voltage which drives the current into a connected load. Thus, the voltage across the load is (r.m.s. value):

$$V_2 = 4.44 f N_2 \cdot \phi_m$$

It is clear that this voltage is 180° out of phase with the source applied voltage V_1.

(*a*) $\therefore V_1/V_2 = N_1/N_2$, which is the turns ratio of the transformer;

(*b*) We also have $V_1/N_1 = V_2/N_2 = 4.44 f \phi_m$ = equality of volts/turn

Therefore, the voltage per turn in both windings is equal;

(*c*) The power supplied and consumed are equal in the windings.

$$\therefore \quad V_1 \cdot I_1 = V_2 \cdot I_2$$

We must remember that all quantities are sinusoidally varying in time and they are phasor quantities. The capital letters denote r.m.s. values of voltage and current.

$$\therefore \quad I_1/I_2 = V_2/V_1$$

$$= N_2/N_1 = \text{inverse of voltage or turns ratio}$$

Also, $$N_1 \cdot I_1 = N_2 \cdot I_2$$

which shows that the ampere-turns of the two coils are equal.

(*d*) The impedance seen from the primary terminals is: $Z_1 = V_1/I_1$, while the load impedance connected to the secondary is:

$$Z_2 = V_2/I_2 = [(N_2/N_1) \cdot V_1]/[(N_1/N_2) \cdot I_1]$$

$$= (N_2/N_1)^2 \cdot (V_1/I_1)$$

$$= (N_2/N_1)^2 \cdot Z_1$$

$$\therefore \quad Z_1/Z_2 = (N_1/N_2)^2 = \text{square of the turns ratio.}$$

The winding with larger number of turns looks into a higher impedance, has higher voltage but lower current.

We, therefore, observe that a transformer can be used for the following purposes:

(*i*) Voltage transformation;

(*ii*) Current transformation;

(*iii*) Impedance transformation;

(*iv*) Isolation of one circuit from another electrically.

Q.4. Where are potential transformers used in high voltage work?

Ans. Potential transformers are known as PTs. They are normally used wherever a high voltage has to be stepped down to feed instruments and protective relays. They are not used for high powers. The standard winding has a rated voltage of 110 V or 220 V, no matter what the h.v. winding voltage is.

For example if the h.v. winding has to be connected to a 750 kV system ($750/\sqrt{3}$ kV = 433 kV to neutral or ground) and the l.v. winding or secondary has to feed 110 V to instruments, the turns ratio has to be ($433 \times 10^3/110$) = 3936/1.

Normally, the turn-to-turn insulation can take only 6 volts per turn, so that the number of turns in the two windings may be 72,000/18 or of this order. The current in the h.v. primary will not exceed a few milliamperes so that it is made from very thin insulated copper wire. It must be insulated from the secondary for the full voltage while the l.v. secondary is at nearly ground potential. PTs do not carry large amounts of power so that they are small in size although they can be very tall with a tall bushing mounted on the steel tank or fibre cylinder containing the windings in cooling oil.

Q.5. Where are current transformers used in h.v. work?

Ans. Current transformers are known as CTs. They are also used for feeding instruments and relays with low current, which is proportional to the current flowing in the h.v. equipment or transmission-line conductors. Normally, the standard secondary-side current is 5 amperes.

In almost all cases, the primary which carries the high value of power-current is at a very high voltage and is in the form of a thick round conductor. The secondary is a toroid with a hole in the centre through which the h.v. high-current lead is threaded. Therefore, the primary is considered as half a turn only as shown in Fig. 12.2. The CT poses grave danger when the secondary terminals are open. Thus, the secondary must ALWAYS be closed.

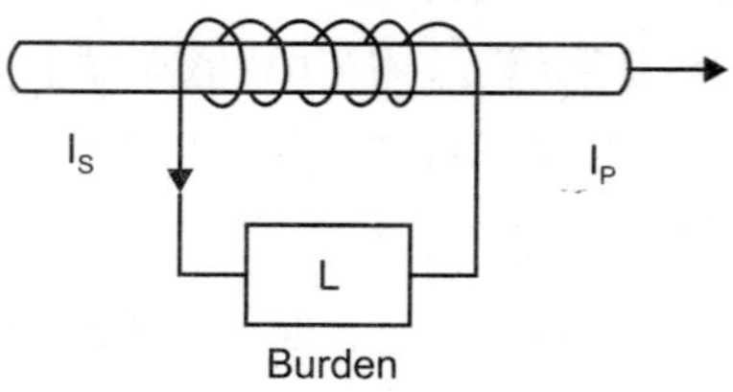

Fig. 12.2: Current transformer (CT) with ½-turn primary feed through.

The reason is as follows:

Let a CT be used for stepping down a current of 1000 amps to 5 amps. The turns ratio is inversely proportional to the current ratio. If the primary has half a turn, the secondary must have 100 turns. This is the result of ampere-turn balance. Therefore, the secondary voltage is 100 times the voltage across the primary conductor (which of course, is not the high system voltage). A shock may result when the secondary terminals are open and someone comes into contact with them. A CT is always provided with a short-circuiting switch which must be closed when instruments, meters and relays have to be pulled out without shutting off the power in the h.v. circuit.

Since instruments on switchboards are at nearly ground potential and the primary is at high voltage, the CT secondary must be insulated from the primary for the high voltage. This is known as the primary insulation that separates the two windings.

Q.6. What are the different losses taking place in a practical transformer? How are they represented in an equivalent electric circuit?

Ans. There are three types of major losses: (*a*) Power losses in the resistances of the windings; (*b*) Voltage losses due to difference in fluxes linked by the two windings called "leakage fluxes"; and (*c*) Current losses due to magnetisation of the core called "excitation current" which consists of both in-phase current giving watts loss and quadrature current (with respect to the applied primary voltage) which sets up the flux in the core. These are now detailed below:

(*a*) **Power Losses in Resistances:** The primary and secondary windings have resistances r_1 and r_2, respectively, and carry currents I_1 and I_2, respectively. The total heating losses are $(I_1^2 r_1 + I_2^2 r_2)$. The resistance values to be used are their power-frequency (50 or 60 Hz) a.c. values which are about 40% higher than any measured d.c. values. These losses reduce the efficiency of the transformer and contribute to temperature rise of the windings, their insulation and the cooling medium in the transformer.

(*b*) **Magnetic Core Losses:** Because the flux is varying in time cyclically at frequency *f*, a magnetic material such as silicon-steel exhibits: (1) Hysteresis loss, and (2) Eddy-current loss, which heat the core and further raise the temperature of the transformer. These losses are evaluated by the following formulae:

1. **Hysteresis loss.** $P_h = f \cdot V_c \cdot \sigma_h \cdot B_m^n$, watt

Here, f = frequency in Hz,

V_c = volume of core material in m^3,

B_m = peak of sinusoidal flux density in core = ϕ_m/A in tesla,

A = area of core perpendicular to flux direction in m^2,

and σ_h = hysteresis constant for the core material known as Steinmetz Coefficient.

The exponent n has the value 1.6 to 2. The addition of 4% silicon to magnetic steel reduces the hysteresis loss i.e., makes both σ_h and n low. In normal power transformers, the peak value of flux density B_m is nearly 1 to 1.2 tesla and usually the core works in the non-linear portion of its magnetisation characteristic.

2. **Eddy-current Loss.** Since the alternating flux flows in the core material, the core behaves as if it were one turn of a large coil, and can be considered as a "third coil". Therefore, a voltage is induced in the core material and a current flows depending on the specific resistance r_c of the material. The formula for the eddy-current loss is:

$$P_e = (\pi^2/6r_c) \cdot V_c \cdot f^2 \cdot B_m^2 \cdot t^2 \text{ watt},$$

where r_c = specific resistance of core material in ohm-m,

= 25×10^{-8} for silicon steel,

and t = thickness of stamping or sheet in metre.

We note that thinner the sheet thickness t, the lower the loss. The entire core is built up from thin sheets of 14 mils (0.35 mm) thickness and the sheets are insulated from each other, either by coating both sides with insulating varnish or else a coating of oxide is provided during rolling of the sheet by blowing steam on the hot sheet. This usually reduces the overall cross-sectional area of the built-up core by 5% and only 95% of the cross-section of the core carries the flux. This is known as the "stacking factor".

Example: A 50-Hz transformer core has overall dimensions of 15.24 cm × 7.62 cm × 1.83 m length for the flow of magnetic flux.

The core is made from 4% Si steel sheets with $r_c = 25 \times 10^{-8}$ Ω-m resistivity. The stacking factor is 0.95. If the hysteresis loss is equal to the eddy-current loss, determine: (*a*) The eddy-current loss and total core losses; and (*b*) the value of Steinmetz's hysteresis coefficient. Take the exponent n = 1.6. The volts per turn in the transformer is 5 volts.

Solution: First determine the peak value of flux density B_m on which the losses depend. From the e.m.f. equation:

$$E/N = 5$$

$$= 4.44 \, f \, \phi_m$$

$$= 4.44 \, f \, A_c \, B_m$$

Now, $A_c = 0.1524 \times 0.0762 \times 0.95 \text{ m}^2$

$= 0.011 \text{ m}^2$

$\therefore$ $B_m = 5/(4.44 \times 50 \times 0.011)$

$= 2.0415$ tesla

The volume of core is

$$V_c = 0.011 \times 1.83 \text{ m}^3 = 0.02 \text{ m}^3$$

With $t = 0.35 \times 10^{-3}$ m

the eddy-current loss is:

$$P_e = (\pi^2/6 \times 25 \times 10^{-8}) \times 0.02 \times 50^2 \times 0.35^2 \times 10^{-6} \times 2.0415^2$$
$$= 168 \text{ watt}$$

If $P_h = P_e = 168$ W,

then, $f\, V_c\, \sigma_h\, B_m^{\,n} = 50 \times 0.02 \times 2.0415^{1.6}\, \sigma_h$

$= 168$

This yields $\sigma_h = 53.63$

It is left as an exercise for the reader to provide the units for the Steinmetz Hysteresis Coefficient.

The total core losses amount to 336 watts. Since the core losses are caused by the full applied voltage responsible for setting up the flux, the resulting loss can be considered as a current in phase with the voltage flowing in an equivalent core-loss resistance connected across the applied voltage in an electrical equivalent circuit of a practical transformer.

(*c*) **Leakage Fluxes and Leakage Inductances:** The voltage applied to the primary winding sets up the flux ϕ in the magnetic core whose peak value is given by the basic e.m.f. equation. This flux links the secondary coil and induces a voltage which is fed to a load. However, in doing so, not all the flux generated by the primary winding links the secondary coil. Some flux "leaks" into other paths, such as the insulation between the coils since the two coils do not occupy the same physical space, and the cooling medium. These are non-magnetic materials and so the flux flowing in them is very low compared to the main flux in the core. Since the magnitudes of these fluxes depend on the currents carried by the coils, the resulting inductance-voltage drop has to be placed in series with the coil resistance as a voltage drop in each coil. These are denoted as jI_1x_1 and jI_2x_2 for the two coils, respectively. Thus, the total voltage drops in the two windings are: $I_1(r_1 + jx_1)$ and $I_2(r_2 + jx_2)$.

Now, we saw earlier that an impedance in any winding can be transformed to any other winding by the square of the turns ratio, $(N_1/N_2)^2$. We write $(r_2' + jx_2')$ = transformed impedance of the secondary impedance to the primary side $= (r_2 + jx_2) \cdot (N_1/N_2)^2$.

$\therefore$ The total series voltage drop will be:

$$I_1Z_1' = I_1(r_1 + jx_1) + I_1(r_2' + jx_2')$$
$$= I_l[\{r_1 + (N_1^2/N_2^2)r_2\} + j\{x_1 + (N_1^2/N_2^2)x_2\}].$$

The quantity in the square brackets is known as the equivalent impedance of the transformer viewed from the primary terminals.

Equally so, the equivalent series impedance of the transformer when viewed from the secondary terminals is

$$Z_2' = \{r_1(N_2^2/N_1^2) + r_2\} + j\{x_1(N_2^2/N_1^2) + x_2\}$$

When using per-unit values, both Z_1' and Z_2' have the same values.

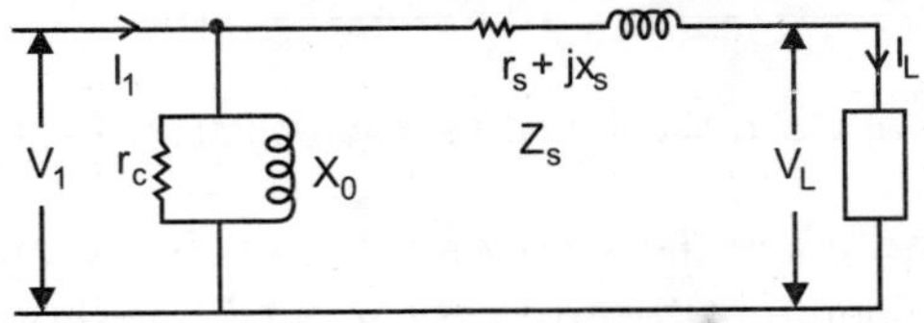

Fig. 12.3: Equivalent circuit of a practical transformer referred to winding #1.

The resulting equivalent circuit of the transformer, first given by C.P. Steinmetz over 120 years ago is still going strong and is shown in Fig. 12.3. A reactance X_0 is shown across the applied voltage to denote the pure magnetising component of the "no-load" current which flows in order to set up the core flux. The resistance r_c denotes the equivalent resistance of the core losses, namely, the hysteresis and eddy-current losses.

Refer to Chapter 7, Question 4–8 and Figs. 7.4–7.6 for application of the equivalent circuit for calculation of efficiency and regulation. In most problems connected with high voltage power transmission, the core loss and magnetising branches can be omitted. The total current drawn by these two in good transformers for voltages above 11 kV and power of 100 kVA and above is about 2% of full rated current. For low-voltage low-power transformers, it may go up to 5%.

Q.7. How are the components of the equivalent circuit of a two-winding transformer determined experimentally?

Ans. In Fig. 12.3, we have shown two loops: (1) The parallel branches r_c and X_0; (2) The series branch consisting of $Z_1 = r_1 + jx_1$, the load impedance Z_2 across a voltage V_L and a current I_L flowing through them. In order to determine the magnitude and phase angle of any impedance or its resistance and reactance connected in series, or parallel, we must measure 3 quantities, namely, the voltage, the current and the real power P. Then,

$$Z = V/I;$$

$$R = P/I^2$$

$$\cos\varphi = P/V{\cdot}I$$

$$X = \sqrt{(Z^2 - R^2)}$$

If R_0 and X_0 are connected in parallel, then:

$$I_c = P/V$$

$$I_m = \sqrt{(I^2 - I_c^2)}$$

$$X_0 = V/I_m$$

and $$R_0 = P/I_c^2$$

The components of the equivalent circuit of Fig. 12.3 are determined as follows, from the results of two important tests:

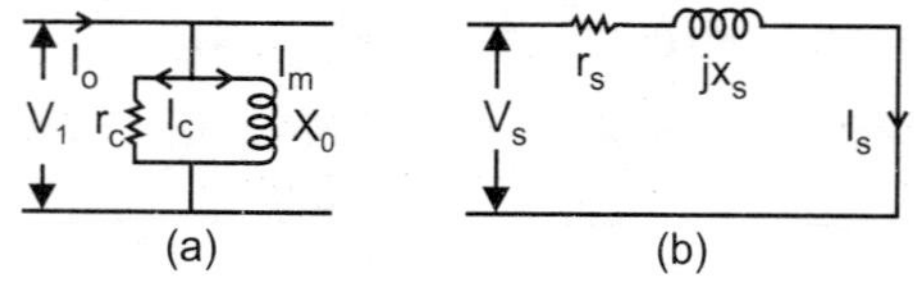

Fig. 12.4: (*a*) Open-circuit and (*b*) short-circuit test conditions.

1. **Open-circuit Test:** When the secondary terminals are open, the only current drawn from the primary source is the no-load current I_o, as shown in Fig. 12.4(*a*). This test is carried out with full rated voltage applied to the primary with voltmeter, ammeter and wattmeter connected. In h.v. work, a PT has to be used for feeding the voltmeter and the potential coil of the wattmeter. A CT feeds the ammeter and current coil of the wattmeter. Let the measured values on the meters be V_1, I_0 and P_0.

   Then, $$I_c = P_0/V$$
   $$I_m = \sqrt{(I_0^2 - I_c^2)}$$
   $$X_0 = V/I_m$$
   $$r_c = V/I_c$$
   $$= P_0/I_c^2 = V^2/P_0$$

2. **Short-circuit Test:** Referring to Fig. 12.4(*b*), if the load terminals are short-circuited, a very heavy current will flow which will be $I_{sc} = V_1/Z_1$. Now, Z_1 in general is not more than 0.1 per unit in good transformers so that if this test is carried out with full applied voltage of 1.0 p.u., then the current flowing in both windings will be 10 p.u. Therefore, the test is carried out so that the current does not exceed 1.0 p.u. This will require that the applied voltage be reduced to nearly 10% of rated voltage.

   Now, at this reduced voltage, the flux set up in the core is also reduced and the core losses being nearly proportional to B_m^2, they become negligible. Therefore, the only impedance controlling the current is the series resistance and leakage reactance referred to the excited winding.

   Let V_s, I_s and P_s be the measured values of voltage, current and power.

   Then, $Z_s = |r_s + jx_s| = V_s/I_s$; $r_s = P_s/I_s^2$; $x_s = \sqrt{(Z_s^2 - V_s^2)}$.

We observe that both r_s and x_s are total resistance and leakage reactance in both primary and secondary referred to the winding to which the voltage source is connected in the short-circuit test. In all calculations involving transformers in a power system, there is no necessity for explicitly separating them into primary series impedance and secondary series impedance. The test yields the total and not the individual parts. If such a need arises, then the rule is that these two separate impedances are equal when referred to any one winding. By using the square of the turns ratio, it is easy to find their actual values. When using per-unit values, they are just made equal.

Example: A single-phase 50-Hz transformer forms one phase of a bank of 3 transformers with a total 3-phase rating of 30 MVA with voltages of 220 kV/13.2 kV, connected in *Y*/*Y* configuration of the 3 phases. Thus, each single-phase unit is rated 10 MVA, $(220/\sqrt{3}\ \text{kV})/(13.2/\sqrt{3}\ \text{kV})$. The turns ratio is $N_1/N_2 = 16.67$.

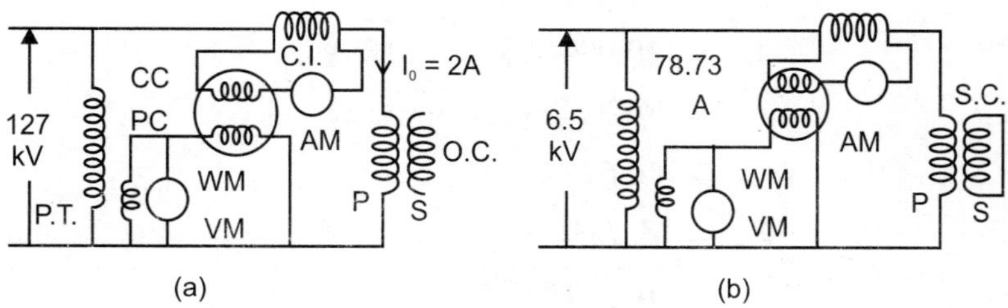

Fig. 12.5: O.C. and S.C. tests with instrument transformers.
(*a*) V_O = 127 kV, I_O = 2 A, P_O = 200 kW;
(*b*) V_S = 6.5 kV, I_S = 78.73 A, P_S = 102.27 kW.

The open-circuit and short-circuit test results are as follows, for each single-phase unit: See Figs. 12.5 (*a*) and (*b*):

(*i*) **O.C. Test:** L.V. winding open. H.V. winding excited.

$$V_o = 220/\sqrt{3}\ \text{kV},\ I_o = 2\ \text{A},\ P_o = 200\ \text{kW}$$

(*ii*) **S.C. Test:** L.V. winding short-circuited. Meters on h.v. winding.

$$V_s = 6.5\ \text{kV},\ I_s = 78.73\ \text{A},\ P_s = 102.27\ \text{kW}$$

Calculate the following from the equivalent circuit:

(*a*) The no-load-branch elements r_c and X_o;

(*b*) The series-impedance elements r_s and x_s.

(*c*) Taking base quantities as the rated voltage, current and/or power ($V_b = 220/\sqrt{3}$ kV, W_b = 10 MVA), draw the equivalent circuit in terms of per-unit values.

(*d*) At rated load of 10 MVA at 0.85 p.f. lag, calculate the efficiency of the transformer.

(*e*) At rated load of 10 MVA and at power factors of (*i*) Unity, (*ii*) 0.85 lag, and (*iii*) 0.85 lead, if rated voltage is maintained across the load of 1 p.u., calculate the regulation in p.u. and per cent.

Solution: The rated voltage, power, current and impedance of the two windings are first written down.

Primary:

$$V_{rp} = 220/\sqrt{3}\ \text{kV} = 127\ \text{kV}$$
$$W_{rp} = 10\ \text{MVA} = 10{,}000\ \text{kVA}$$

$$I_{rp} = 10{,}000/127 = 78.73 \text{ amp}$$

$$Z_{rp} = V_{rp}/I_{rp} = \text{kV}^2/\text{MVA}$$

$$= 127^2/10 = 1613\ \Omega$$

Also, $$Z_{rp} = V_{rp}/I_{rp} = 127/78.73 \text{ k}\Omega$$

$$= 1.613 \text{ k}\Omega = 1613\ \Omega$$

Secondary: $$V_{rs} = 13.2/\sqrt{3} \text{ kV} = 7.62 \text{ kV}$$

$$W_{rs} = 10{,}000 \text{ kVA}$$

$$I_{rs} = 10{,}000/7.62 = 1312 \text{ A}$$

$$Z_{rs} = 7620/1312 = 5.806\ \Omega$$

Also, $$a = 16.67$$

and $$Z_{rs} = Z_{rp}/a^2$$

$$= 1613/16.67^2$$

$$= 5.806\ \Omega$$

Also, $$Z_{rs} = 7.62^2/10 = 5.806\ \Omega$$

1. **O.C. Test:** $V_0 = 127$ kV, $I_0 = 2$ A, $P_0 = 200$ kW. At a current of 2 A in the primary, the copper loss in the winding is negligible. The wattmeter reading is entirely due to core losses at full rated voltage which has been used in this test.

$\therefore$ $$r_c = V^2/P_0 = (127 \times 10^3)^2/(200 \times 10^3)$$

$$= 80.65 \times 10^3 \text{ ohms}$$

$$= 80.65 \times 10^3/1613 \text{ p.u.} = 50 \text{ p.u.}$$

$$I_c = P_o/V = 200 \times 10^3/127 \times 10^3$$

$$= 1.575 \text{ amp}$$

$\therefore$ $$I_m = \sqrt{(I_0^2 - I_c^2)}$$

$$= \sqrt{(2^2 - 1.575^2)}$$

$$= 1.233 \text{ A}$$

$\Rightarrow$ $$X_0 = V_0/I_m = 103 \times 10^3\ \Omega$$

$$= 63.86 \text{ p.u.}$$

The resulting parallel equivalent circuit derived from the O.C. test results is shown in Fig. 12.6(a).

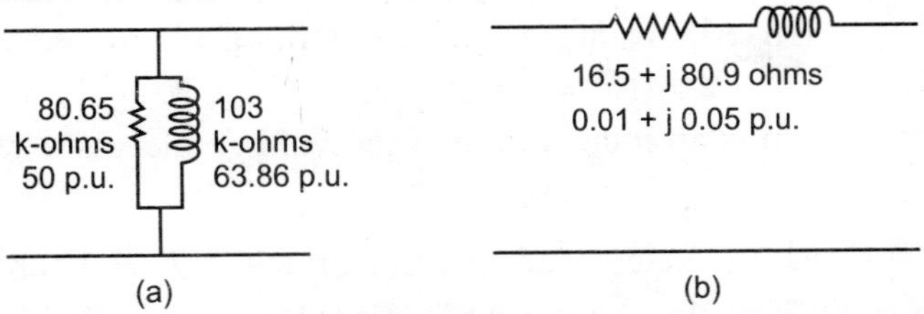

Fig. 12.6: Shunt and series impedance elements of transformer of example in Question 7. Actual ohmic values and per-unit values on bases of $V_b = 220/\sqrt{3}$ kV = 127 kV and W_b = 10 MVA.

2. **S.C. Test:**

$$V_s = 6.5 \text{ kV}$$
$$I_s = 78.73 \text{ A}$$
$$P_s = 102.27 \text{ kW}$$

Since the voltage is only about 5% of rated voltage of 127 kV, the core losses can be neglected. The wattmeter reading denotes the entire copper loss at the current used in the test, which happens to be the full rated current or 1 p.u.

$$\therefore \quad r_s = P_s/I_s^2 = (102.27 \times 10^3)/78.73^2$$
$$= 16.5 \text{ ohms} \approx 0.01 \text{ p.u.}$$
$$Z_s = V_s/I_s = 6500/78.73$$
$$= 82.56 \text{ ohms} = 0.051 \text{ p.u. (Base } \Omega = 1613)$$
$$\therefore \quad x_s = \sqrt{(Z_s^2 - r_s^2)} = 80.9 \text{ ohms} = 0.05 \text{ p.u.}$$

The resulting series impedance values are shown in Fig. 12.6(*b*).

(*c*) The equivalent circuit in per-unit values on transformer rated values as bases is also shown in Figs. 12.6(*a*) and (*b*).

(*d*) **Efficiency:** The power output is 10,000 × 0.85 = 8,500 kW. Core losses = open-circuit-test wattmeter reading = 200 kW. Copper losses in the two windings combined = S.C. test wattmeter reading = 102.27 kW.

∴ Total transformer losses for this loading = 302.27 kW

Efficiency, ε = output/input = output/(output + losses)
= 8,500/8,802.27
= 0.9657
= 96.57%

(*e*) (*i*) Current flowing in load, $I_L = 1.0(1 + j\,0)$ p.u. at unity p.f. voltage across load, $V_L = 1 + j\,0$ p.u. which is taken as reference phasor.

Series impedance, $z_s = 0.01 + j\,0.05$ p.u.

∴ Input voltage, $V_1 = 1 + j\,0 + (0.01 + j\,0.05)(1 + j\,0)$
$= 1.01 + j\,0.05 = 1.011\angle 2.83^\circ$ p.u.

Regulation = (Rise in voltage as load is thrown off)/Rated volts
= (1.011 – 1.0)/1.0 = 0.011 p.u. = + 1.1%

The positive sign is an indication that the voltage increases when the load is thrown off.

(*ii*) At 0.85 p.f. lag and rated load, current drawn is $I_L = 1(0.85 - j0.527)$.

Note: cos φ = 0.85, so that sin φ = 0.527. Since the current is lagging behind the voltage (the load being inductive), its j-part occurs with negative sign.

$\therefore$ Input voltage, $V_1 = 1 + j\,0 + (0.01 + j\,0.05)(0.85 - j\,0.527)$

$= 1 + 0.03485 + j\,0.03723 = 1.0355\angle 2.06°$

Regulation = (1.0355 – 1.0)/1.0 = + 0.0355 p.u. = + 3.55%

(*iii*) Current in load at 0.85 leading p.f., $I_L = 1(0.85 + j\,0.527)$

$\therefore \quad V_1 = 1 + j\,0 + 1(0.01 + j\,0.05)(0.85 + j\,0.527) = 0.983\angle 2.08°$

Regulation = (0.983 – 1.0)/1.0 = –0.017 p.u. = –1.7%

The input voltage is now lower than the rated voltage across the load. There will be a voltage drop of 1.7% when the load is thrown off across the secondary terminals. Note that the calculation has been performed by holding the secondary load voltage constant at 1.0 p.u. In normal practice, it is the primary input voltage that will be held constant irrespective of the secondary load. The reader can attempt a calculation under this assumption.

Q.8. What is the constructional detail and use of a 3-winding transformer? What is meant by the *Y*-equivalent circuit for one phase of a 3-phase 3-winding transformer?

Ans. Most large-rating h.v. transformers used at generating and receiving stations at the ends of a long transmission line have 3 windings or coils on the same magnetic core for each phase. They are called: (*a*) Primary; (*b*) Secondary; and (*c*) Tertiary windings. At a generating station, the usual generator output is 13.2 kV (line-to-line) in 50-Hz systems and 13.8 kV in 60-Hz systems. This voltage is stepped up in the secondary of the transformer to the required transmission voltage of the system, which could be 33 kV, 66 kV, 132 kV, 220 kV, 345 kV, 400 kV, 500 kV, 750 kV, 1000 kV, 1150 kV or even 1500 kV, which is the highest planned in the world. (There is a 1300-kV level in the U.K.). Usually, the primary and secondary windings are connected in Wye or Star. At the generating station, there will be a lot of auxiliary equipment such as motor-driven exciters for the generators, cooling fans, pumps, lighting and so on. To supply power to these loads a 3rd winding is necessary. This is the tertiary and its voltage may be 3.3 kV and it is usually connected in Delta. There are certain disadvantages in a *Y*/*Y*-connected transformer which are eliminated by having a Delta-connected tertiary on the same magnetic circuit. We will not discuss these problems, the most important of which is elimination of 3rd harmonic current in the lines (150 Hz or 180 Hz).

Similarly, at the receiving station, the primary in the 3-winding transformer is the highest-voltage winding. Loads or secondary transmission lines at lower voltages are fed from the secondary winding on the same core. At the receiving station, certain very special and important equipments are connected which are known variously as: (*a*) Synchronous Condensers; (*b*) Static VAR Systems or Thyristor-Switched Reactors; (*c*) Switched Capacitors. There are also station auxiliaries. These equipments are vital for the control of voltage at the receiving end of the transmission line. As has been shown in previous Questions, the load voltage varies with the power and its power factor so that voltage control is very essential by using suitable compensation equipment. All these are fed from the tertiary winding.

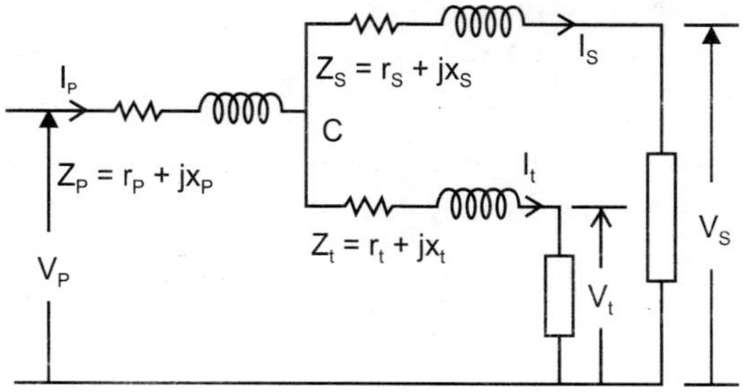

Fig. 12.7: Y-equivalent circuit for one phase in a 3-winding transformer.

The equivalent electrical circuit for one phase of a 3-winding transformer is represented in Fig. 12.7. There are 3 separate series impedances shown as Z_P, Z_S and Z_T connected in a Wye with the star-point shown as *C*. The input voltage V_P is fed to the primary. Loads can be connected to the secondary and the tertiary. If necessary, the no-load branches of r_c and X_0 can also be connected across the primary, but in normal network studies this can be ignored. All these three impedances are those referred to the same winding as was done in Question 7.

Q.9. What are the tests to be performed in order to determine Z_P, Z_S, Z_T and how are they calculated from the test results?

Ans. Three types of tests are necessary (short-circuit tests):

1. Keep the tertiary winding open. Short circuit the secondary. Apply a low voltage to the primary so that not more than rated current flows in the secondary. Note that the rating of the primary is equal to that of the secondary and tertiary combined so that when rated current flows only in the secondary, it will not exceed the primary rated current.

   Measure V_1, I_1 and P_1 on the primary side through proper PTs and CTs connected to the primary winding.

   Then,
$$Z_{PS} = Z_P + Z_S = V_1/I_1$$
$$r_{PS} = r_P + r_S = P_1/I_1^2$$
$$x_{PS} = x_P + x_S$$
$$= \sqrt{(Z^2{}_{PS} - r^2{}_{PS})}$$

2. Next, keep the secondary open. Short circuit the tertiary winding. Apply a low voltage to the primary so that not more than the rated tertiary current flows in that winding. Measure V_2, I_2 and P_2.

Then:
$$Z_{PT} = Z_P + Z_T = V_2/I_2$$
$$r_{PT} = r_P + r_T = P_2/I_2^2$$
$$x_{PT} = x_P + x_T = \sqrt{(Z^2_{PT} - r^2_{PT})}$$

3. Finally, keep the primary open. Short circuit the tertiary. Apply a low voltage to the secondary making certain that not more than the rated currents flow in either the secondary or the tertiary. Measure V_S , I_S and P_S on the secondary side. (Note that this test can also be performed with secondary short-circuited and meters used on tertiary side). Then convert these measured values of voltage, current and power to the primary side by the turns ratio and its inverse.

Let these be V_3, I_3 and $P_3 = P_S$
$$V_3 = (N_P/N_S)V_S$$
$$I_3 = (N_S/N_P)I_S$$
$$Z_{ST} = Z_S + Z_T = V_3/I_3$$
$$r_{ST} = r_S + r_T = P_3/I_3^2$$
$$x_{ST} = \sqrt{(Z^2_{ST} - r^2_{ST})}$$

We now have three equations for the impedances, all referred to the same winding, in this case to the primary:

1. $Z_{PS} = Z_P + Z_S$;
2. $Z_{PT} = Z_P + Z_T$;
3. $Z_{ST} = Z_S + Z_T$

We know the values of Z_{PS}, Z_{PT} and Z_{ST} obtained from the 3 short-circuit tests, and therefore we can solve for Z_P, Z_S and Z_T by calculation.

Now, $Z_{PS} + Z_{PT} + Z_{ST} = 2(Z_P + Z_S + Z_T)$

or, $Z_P + Z_S + Z_T = (Z_{PS} + Z_{PT} + Z_{ST})/2$

Consequently, $Z_P = (Z_{PS} + Z_{PT} + Z_{ST})/2 - Z_{ST}$

$Z_S = (Z_{PS} + Z_{PT} + Z_{ST})/2 - Z_{PT}$

and, $Z_T = (Z_{PS} + Z_{PT} + Z_{ST})/2 - Z_{PS}$

Under certain exceptional cases, an impedance represented in the Y-equivalent circuit for one phase has a chance of having a negative value. Physically, all impedances are positive quantities. As such they can be represented by physically realisable elements in a network study. In a Digital Computer calculation of the network using a 3-winding transformer, such difficulties do not pose problems.

Q.10. What is an auto-transformer? What are the relations between primary voltage, current, number of turns and secondary quantities as well as load quantities?

Ans. Refer to Fig. 12.8. An auto-transformer is called a "single winding" transformer. In practice, it does have a primary winding and a secondary winding on the same core but they are connected in series, while the load is connected to the secondary part of the transformer, drawing a current I_L. It is therefore, equivalent to one large winding tapped at some point to feed the load. The source is connected to the two windings in series and supplies the current shown as I_P. On account of the voltage applied, voltages are induced in both the primary and the secondary portions with turns N_P and N_S, respectively, which oppose the source voltage.

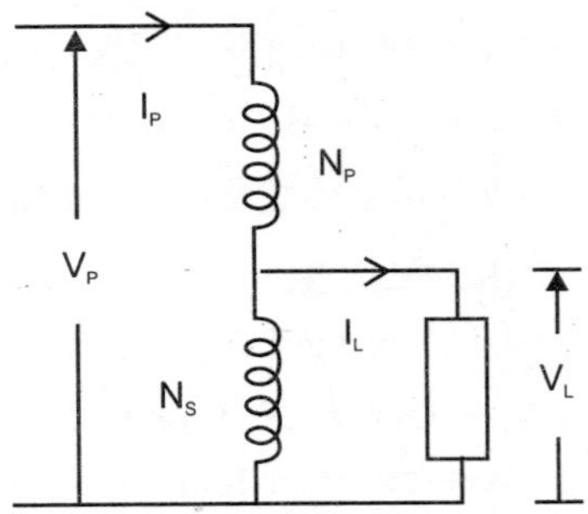

Fig. 12.8: Auto-transformer with load, source, common winding and electrical relations.

From the circuit diagram shown, the secondary is common to both the source and the load, and carries the current $(I_L - I_P)$. Thus, part of the current to the load is supplied directly by electrical conduction from the primary and the remaining part by induction from the secondary winding or the common winding. The two sides, namely the load side and the source side, are electrically connected so that insulation problems are very severe. An auto-transformer is used only when the ratio of source voltage to the load voltage does not exceed 3 for large powers and high voltages. It is, nonetheless, less expensive than using a 2-winding transformer where the primary and secondary are electrically isolated but linked by the magnetic circuit of the core. Therefore, they can be used for any ratio of voltages and insulation is available to separate the two windings adequately.

The electrical relations of the auto-transformer are as follows:

1. **Voltages:** $V_L/V_P = N_2/(N_1 + N_2)$

or, $V_L = V_P/(1 + N_1/N_2)$

2. **Currents:** Equality of power output and power input requires, that

$$V_P \cdot I_P = V_L \cdot I_L$$

$$\therefore \quad I_L = V_P \cdot I_P/V_L = (1 + N_1/N_2) \cdot I_P$$

The resulting current flowing in the common winding is:

$$I_C = I_L - I_P = I_P \cdot N_1/N_2$$

Also, ampere-turn balance requires:

$$N_2 \cdot I_C = N_1 \cdot I_P$$

so that, $$I_C = I_P \cdot (N_1/N_2)$$

Example: An auto-transformer has $N_1 = 2N_2$. The load is 10 MVA = 10,000 kVA at unity power factor and is supplied at a voltage of $132/\sqrt{3}$ = 76.2 kV. [The auto-transformer forms one phase in a 3-phase 132 kV system supplying a total load of 30 MVA]. Find the applied voltage V_P to the primary and the source current I_P.

Solution:

$$I_L = 10{,}000 \text{ kVA}/76.2 \text{ kV} = 131.23 \text{ A}$$

$$V_P = (N_1 + N_2)V_L = 3V_L$$

$$= 228.6 \text{ kV, line-to-neutral, (400 kV, 1–1)}$$

$$I_P = I_L \cdot N_2/(N_1 + N_2)$$

$$= I_L/3 = 43.74 \text{ A}$$

Common-winding current, $I_C = I_L - I_P = 87.5$ A

Rating of primary winding: Voltage of primary winding alone

$$= V_P \cdot 2/3 = 228.6 \times 2/3$$

$$= 152.4 \text{ kV}$$

∴ Rating $$= 152.4 \times 43.74 = 6{,}666 \text{ kVA}$$

Rating of common winding: $76.2 \times 87.5 = 6{,}666$ kVA

In a two-winding transformer, each winding would have been rated 10,000 kVA, which is equal to the load supplied.

Some of the largest transformers, up to 500 MVA 3-phase at voltages up to 1150 kV are auto-transformers, since they are intended to be connected to an existing 400-kV or 500-kV system. Also, transformers for very low voltages and powers called VARIACS or dimmer transformers are built as auto-transformers whose output can be varied from 0 to 100% voltage or more.

Q.11. What is a cascade-connected transformer and where is this used?

Ans. In order to test high voltage equipment at the very high voltages as specified by the Standards, normally exceeding 500 kV, two or more transformers are connected in series or "cascade". Also, when testing insulators under wet conditions and salt spray at high voltages, the current drawn is high. This will involve not only connecting testing transformers in series to obtain the very high voltages but also some units in parallel to supply the high current. Details of these testing transformers will be discussed in Chapter 16 under H.V. Laboratory Equipment and Layout.

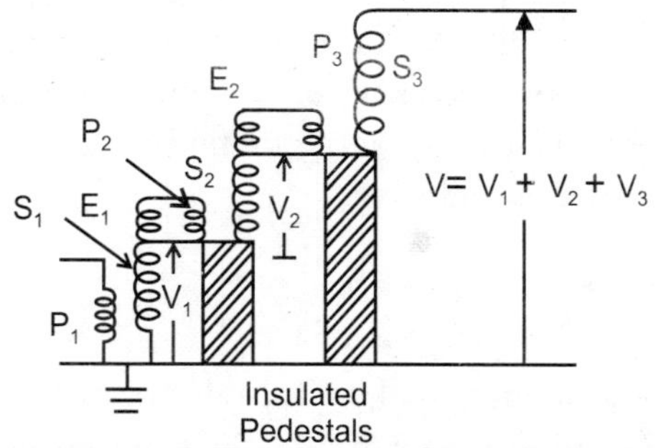

Fig. 12.9: 3-unit cascade-connected H.V. testing transformer for some million volts.

The basic circuit diagram of a 3-unit cascade-connected testing transformer unit is shown in Fig. 12.9. These testing transformers are very tall. The arrangement consists of a ground unit containing a low-voltage primary and high voltage secondary for up to perhaps 500 kV in the largest units or 600 kV, but not more. At the h.v. end of the secondary, a few turns are connected to the primary of the second unit. This third winding is called the "excitation winding", since it provides excitation to the primary of the next unit. In a similar manner, the second unit has its own secondary for very high voltage and an excitation winding to excite the primary winding of the next unit in the cascade; and so on. It is not common to find more than 3-units in cascade.

The secondary h.v. windings of all units are connected in series in order to yield a final voltage at the h.v. terminal of the cascade-connected units equal to the sum of the voltages of the individual units.

The most important characteristic to note is that since the bottom of the h.v. coil in unit 2 is at the high potential V_1, which is the high voltage generated by the ground unit, the entire second unit must be insulated from ground for the voltage V_1. It is usual to use high-strength porcelain insulators of the post type to support the heavy transformer with its windings, bushing and the tank filled with large quantity of oil or other liquid insulating and cooling medium.

In a similar manner, any 3rd unit must be insulated from ground for the voltage $(V_1 + V_2)$, which may be 1 million volts or more. This is the sum of the high voltage outputs of the two lower units. If the cascade-connected transformer unit is used indoors under the roof of a large hall of a laboratory, and they are protected from weather, less expensive fibre cylinders can support the transformers instead of porcelain insulators. But for outdoor units, porcelain must be used. Such high-strength fibre cylinders are as commonly found as porcelain supports. Another advantage of fibre cylinders is that they also serve as tanks in which the transformer coils and core can be immersed. But these are manufacturing details.

At the Indian Institute of Science H.V. Laboratory, 3-units of 350 kV each are connected in cascade to generate 1050 kV (1.05 million volts) at the h.v. terminal, (in 1950). Testing transformers for 750-kV equipment or higher voltages require voltages up to 6 MV, so that 1 million volts are not high in this game.

Another characteristic of such testing transformers is that since the equipment to be tested at high voltages is purely capacitive, the power factor is zero leading in contrast to power transformers which usually carry inductive loads and their power factors are lagging, usually of the order of 0.85 lag.

CIRCUIT BREAKERS

Q.12. What is the function of a circuit breaker in high voltage networks and other places?

Ans. A circuit breaker is a misnomer for what this piece of equipment does. It is a switch that can be used either to "make" or "break" a circuit, thereby either establishing the circuit current or breaking the circuit current. Since the duty imposed on this piece of equipment is most onerous or difficult while breaking or extinguishing the circuit current, it is generally known as the "circuit breaker" or CB.

When current in a circuit has to be interrupted, the breaker establishes a gap between two electrodes by having one or both electrodes move extremely rapidly. In most breakers, only one of the electrodes or contacts moves and is called the moving electrode while the other is known as a stationary electrode. But this need not be so in every case, as long as a gap is established in between the electrodes. In the process of parting of the contacts, a high-current arc is established which must be extinguished successfully. Circuit breakers differ in the mechanism by which the arc extinction occurs. Thus, the physics of arc phenomena, and its extinction process are basic sciences that must be carefully studied which are very involved as they depend on Plasma Physics and Aerodynamic properties of the medium used in the arcing chambers. These will not be discussed in this book.

Q.13. What are the different types of circuit breakers used in high voltage work?

Ans. There are as many types of circuit breakers as there are companies manufacturing them throughout the world. However, they can all be classified according to the insulating medium and the principle used for arc extinction. Broadly speaking, they can be classified as follows:

(*i*) Simple air-break disconnecting switches;

(*ii*) High-speed air-break switches for voltages up to 11 kV and not involving high inductance in the circuit;

(*iii*) Dead-tank circuit breakers using oil or compressed air;

(*iv*) Oil-minimum circuit breakers;

(*v*) Air-blast circuit breakers;

(*vi*) SF_6 circuit breakers with or without "puffer" type extinction; and

(*vii*) Vacuum circuit breakers.

Q.14. Describe the broad principles and distinctions between the above types of circuit breakers.

Ans. (*a*) **Compressed Air:** The arc space when subjected to a blast of compressed air allows the arc products to be blown out rapidly enough for arc extinction to occur. This is used where the voltage is not high but the current is high when arc splitter made of insulating barriers further help arc extinction. Such a breaker is known as a cross-blast compressed-air circuit breaker. For high voltages, the compressed air is blown through a nozzle. The air-pressure can go up to 50 or 80 atmospheres. Some schemes use a resistance in parallel with the arc during interruption which is placed in circuit by the parting of the contacts.

This type of breaker for very high voltages and currents is now being superseded by what is known as SF_6 "puffer breaker" which will be discussed later on.

(*b*) **Oil Circuit Breakers:** Dead tank or bulk Oil Circuit Breakers (OCB) use a large oil tank in which the arcing chambers are inserted. The arcing takes place inside a confined space called the arcing chamber which is filled with the mineral oil and is provided with passages for the gasified oil based on aerodynamic principles. When the oil becomes gaseous due to the heat generated by the arc, the gas develops sufficiently high pressure to blow the arc products out. The aerodynamic design provides effective passages for the gas products to cool the arc space. The metal tanks are at ground potential so that sufficient insulating clearance must be provided in the tank between the electrodes at high voltage and the grounded tank. Thus they have ample dimensions and look very large.

(*c*) **Oil-minimum Circuit Breaker:** In this type of breaker, the arcing chamber is surrounded by a closely-fitting oil-filled casing of insulating material. These are also known as "live-tank" circuit breakers and the arcing chamber is surrounded by a porcelain enclosure and the arrangement is electrically insulated for the high voltage from ground, usually by post-type insulators.

Oil circuit breakers are used for voltages up to 345 kV. Air-blast or compressed-air breakers are used for higher voltages. SF_6 can be used at all voltages.

(*d*) **SF_6 Gas-blast Circuit Breaker:** SF_6 is an electro-negative gas. See Chapter 3 for its properties in arc quenching. In early breakers, the SF_6 gas was used in place of air and used the compressed gas blast depending upon its electron-attachment and cooling properties for quenching the arc. Recently, the gas flow is piston-driven which forces the gas to blow the arc more forcibly. This "puffing action" is superior to the earlier breakers. Hence, the popular name "puffer breaker". This is an improvement over the earlier "oil pistons" used in oil circuit breakers.

(*e*) **Vacuum Circuit Breaker:** Vacuum environment has several advantages, namely, the absence of a material medium which breaks down dielectrically to create or sustain an arc. The use of vacuum was initially restricted to voltages up to 15 kV for distribution systems but has now been developed sufficiently well for transmission voltages of 132 kV and higher.

In a.c. circuit breakers, the arc is comparatively easier to quench (which is easier said than done!) than in d.c. systems. The reason is that the current passes through a natural zero every half-cycle and thus has no energy imparted to it at that instant. However, in d.c. work, the quenching is more difficult.

(*f*) **D.C. Circuit Interrupters:** For voltages of a few kV, air breakers provided with a coil in series with the arc, which generates a magnetic field and thus deflects the arc and lengthens it through insulated arc splitters or chutes is used. Thus, the air blast quenched the arc.

In modern high voltage d.c. transmission work, such a scheme is not practicable. The current through the arc must be brought to zero by auxiliary circuits. These take the form of: (*a*) Discharging a charged capacitor to force a current zero by having the capacitor discharge a current opposite in direction to the main circuit current; (*b*) To provide a capacitor in parallel with the two contacts so that when they part, the capacitor is placed in the local arc circuit. This creates an oscillatory circuit which behaves exactly as in an a.c. arc-quenching situation.

The former method (*a*) is preferred with oil-blast or vacuum circuit breakers. Method (*b*) is preferred for compressed-air blast, vacuum and SF_6 breakers since they depend upon the resistance characteristics of the arc. In certain arrangements, a non-linear lightning arrester type of material is also used to absorb the energy in the arc.

Q.15. What is the meaning of the following terms when referred to circuit breakers and circuit breaking?

(*a*) Arc voltage; (*b*) Restriking voltage; (*c*) Recovery voltage; (*d*) Natural frequency; (*e*) Rate of rise of restriking voltage, RRRV; and (*f*) Dielectric recovery strength.

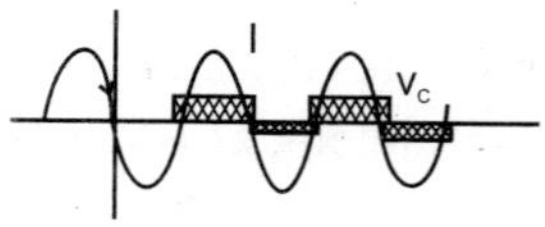

Fig. 12.10: Sinusoidal current waveform and arc voltage.

Ans. (*a*) **Arc Voltage:** An arc is the flow of highly ionised and electrically charged particles which are the products of ionisation of the arcing medium. Therefore, it has the properties of a pure resistance and the voltage across the contacts is in phase with the current. This is shown in Fig. 12.10, as well as the *I–V* characteristic. Because of increasing temperature and increase in ionisation level, the arc resistance reduces as the current increases. Arc temperatures can be as high as 20,000–25,000°C. Arc voltages are very low compared to the system voltage and is neglected in system studies but not in the study of the arc itself. It may be only 20–30 volts, most of the voltage drop occurring in thin layers in the anode and cathode electrode surfaces. Note that when the arc is extinguished the circuit finally opens and the voltage across the same contacts will reach the very high system voltage. Therefore, in the matter of a few milliseconds, the arc-space or space between the contacts rises from practically zero to hundreds of kilovolts. This is the entire problem in circuit breaking.

(*b*) **Restriking Voltage:** Refer to Fig. 12.11 where a series circuit consisting of *R–L* is connected to a source of direct voltage *E* through a breaker *B*. In the steady state when the switch is closed, the current flowing is $I_S = E/R$, since a pure inductance is a short-circuit for d.c. However, this current, nevertheless, builds up electro-magnetic energy in the magnetic field of the coil equal to $L \cdot I_S^2/2$ joules.

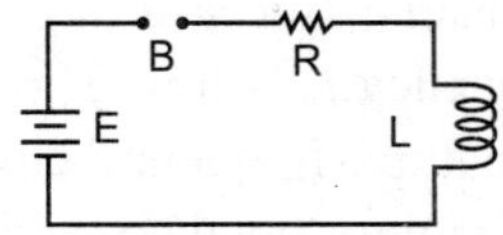

Fig. 12.11: Simple R–L series circuit across D.C. source with breaker.

When the breaker tries to interrupt the circuit, this energy must be dissipated in the same circuit consisting of resistance R and the arc resistance. The voltage generated by the collapse of the e.m. energy will be sufficient to sustain the arc until the quenching action has taken place and the current has reduced to zero. In case the arc medium has not recovered its dielectric strength sufficiently rapidly, this voltage can restrike the arc. This voltage is broadly called the restriking voltage.

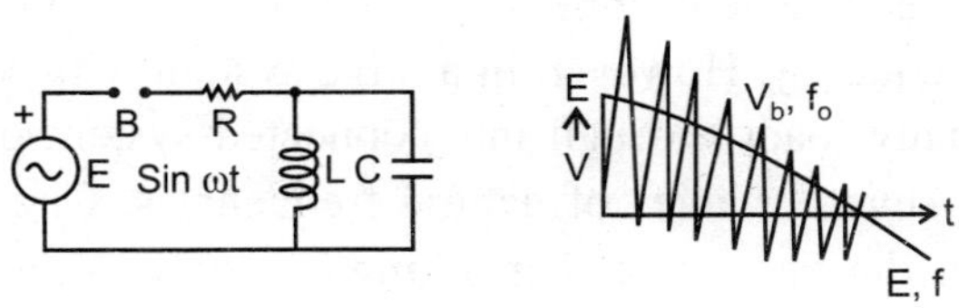

Fig. 12.12: Single-frequency network energised by a sinusoidal a.c. source; waveform of source voltage with high-frequency transient superimposed on it. Illustration of restriking voltage of circuit breaker.

However, in practice the restriking voltage has a deeper significance. All power systems contain capacitances in addition to inductances. These are also energised by sinusoidal voltages. This is shown in Fig. 12.12. Now, in addition to the power frequency f, the tank circuit (resonant circuit) consisting of parallel L–C elements has its own oscillatory frequency of $f_o = 1/2\pi\sqrt{(LC)}$, hertz. Without going through the differential equation for the circuit and its manner of solving it during the transient state, the voltage across the open contacts, one of which is connected to the source and the other to the tank, is given by $V_B = E\,(\cos 2\pi f t - \cos 2\pi f_o t)$, where E is the peak value of the system voltage. Usually, $f = 50$ or 60 Hz, and f_o can be a few kHz. Also, the series resistance helps damp the voltage. But, we observe that the voltage across the breaker contacts during the transient state can reach twice the system peak voltage. This can easily restrike the arc.

The resulting voltage is defined as the sum of the normal power-frequency voltage and the high-frequency voltage which is responsible for restriking the arc in case the quenching medium has not cooled the arc space sufficiently rapidly.

The restriking voltage is accompanied by its rate of rise. This is called the "Rate of Rise of Restriking Voltage", or RRRV. It is clear that the rate of rise of dielectric strength of the arc space must be higher than the RRRV for successful arc extinction to occur. In normal practice, small power systems have low value of inductance L and capacitance C. Thus, $f_o = 1/2\pi\sqrt{(LC)}$ is high. Oil circuit breakers can function very well as also other breakers. But in a large system, L and C are both high and the resonant or natural frequency f_o is low. Therefore, a voltage-zero does not occur as rapidly as for a higher natural-frequency circuit. This generally requires the use of air-blast, SF_6 and oil-minimum breakers which function better and more rapidly than normal air and oil circuit breakers.

(*c*) **Recovery Voltage:** When the contacts have finally parted and the arc has been quenched, the voltage appearing across the open gap in the circuit breaker is the full source voltage of the system. The circuit has now fully recovered from the transient shock. This voltage is called the Recovery Voltage.

(*d*) **Natural Frequency:** The frequency $f_0 = 1/2\pi\sqrt{(LC)}$ is known as the Natural Frequency. It refers to the connected system parameters L and C. A small power system usually feeds a load through one transmission line. There is, thus, only a single natural frequency. However, in a large system with several generating stations connected to many loads through interconnected system of h.v. transmission lines, there will exist several values of natural frequencies. Circuit breakers are designed and selected on the basis of either a single natural frequency or multiple natural frequencies.

Of even more serious concern is when a short transmission line is interposed between the source and the fault location. This is known as a "short-line fault" or SLF. The travelling wave created on this stub line will affect the restriking voltage and behave as a second natural frequency. This type of fault has been considered to be very serious from the point of view of operation of a circuit breaker, since the RRRV is extremely high.

(*e*) **Dielectric Recovery Strength:** When the arc is in the process of being quenched, the gap between the contacts acquires a steadily increasing dielectric strength (volt/m) which has to sustain the restriking and recovery voltages during and after full opening of the contacts. If the quenching medium does not cool the arc products or blow them away sufficiently rapidly, the dielectric strength is not achieved rapidly enough. Successful arc interruption has been described as the race between rate of rise of dielectric strength and rate of rise of restriking voltage, the former being higher than the latter. Restriking establishes the circuit continuity once again and superimposes the new transient over the previously-existing one and can damage the circuit breaker. This is the most usual cause for explosion of circuit breakers.

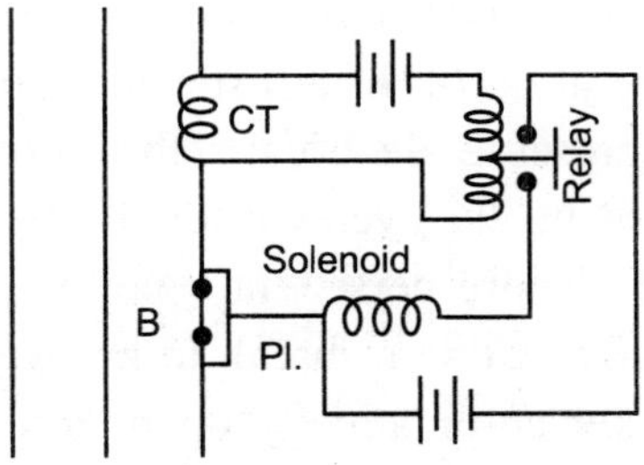

Fig. 12.13: Sequence of operation of a circuit breaker. Battery in relay circuit is optional for providing d.c. bias.

Q.16. What is the sequence of operation of a circuit breaker protecting a power system from the time of initiation of a fault and the opening of the contacts? Draw a single-line diagram.

Ans. See Fig. 12.13. The following sequence of operations is representative of the protection scheme. Details vary with the systems.

(*i*) A fault (short circuit of various types) generates a current value which exceeds normal rated current carried by the system. The fault current is thus an over current which is harmful.

(*ii*) The overcurrent is sensed by a CT and is applied to the operating coil of a relay.

(*iii*) The relay can be of the electro-mechanical type, or, in modern systems of the solid-state type.

(*iv*) The aim of the relay is to close an electrical contact which completes an electric circuit, which drives a larger current into a solenoid magnet.

(*v*) The solenoid, when energised, moves a heavy slug or iron piece inside the coil which is called an "armature". This armature in turn hits the operating link or mechanism and "trips" the breaker mechanical operating system. This could be in the form of a simple rod or the release of a compressed spring from its latch, or open a valve to release compressed air or gas.

(*vi*) The tripping mechanism then moves the mechanical linkage to operate the breaker contacts to open them. Normally, one of them is stationary while the other moves rapidly; but in other cases, both contacts could be made to move apart.

(*vii*) To close the contacts from an open position, the reverse generally takes place although now there is no need for the relay to operate. The closing operation merely energises the operating solenoid which then acts on the mechanical linkage to close the contacts. In the case of normal operation of a open-close circuit breakers, the closing operation is carried out by a manual switch which is closed by the human operator from the control switchboard. Modern circuit breakers also incorporate an open-close-open sequence. The breaker opens on a fault through the over current relay and automatically recloses after a delay of a few cycles.

If the fault has been very short lived such as an insulator flashover on a tower due to lightning or operation of an arrester at a station, then there is necessity to wait long. In order to preserve electrical stability of the system, the circuit breaker can be made to reclose in 2½ cycles after the first operation of opening on the fault. If the fault is of a more persistent type in nature such as arcing at a tower or a permanent short-circuit due to break in an overhead conductor and falling on a tower, then the breaker automatically opens after the reclosing operation. Such faults are known as "bolted faults". The time elapsed between the reclosing and second opening or the delay time is called "dead time" and can be arranged to be up to 10–15 seconds and not a mere few cycles. All this depends on proper system study.

Figure 12.13 is a skeleton of the sequence mentioned before. The operating rod is shown as being the "armature" inside the solenoid. But in general there are other mechanical parts which require much larger and speedier mechanical operation which the solenoid only initiates.

Q.17. Draw the schematic diagram of the following types of circuit breakers showing essential parts: (*a*) Dead-tank circuit breaker; (*b*) Live-tank circuit breaker; (*c*) Air-blast circuit breaker; (*d*) SF_6 breaker; and (*e*) Vacuum circuit breaker or interrupter.

Ans. (*a*) **Dead-tank Circuit Breaker:** This is so named because the tank containing the oil or fluid or pressure air is located on ground and therefore is electrically "dead". It is also known as bulk-oil circuit breaker when oil is the quenching medium. Figure 12.14 shows the overall view of an Oil Circuit Breaker (OCB) and the arc inside the arcing chamber inside the oil. It is usual to have two breaks in series in two separate arcing chambers, and for very high voltages up to four breaks in series.

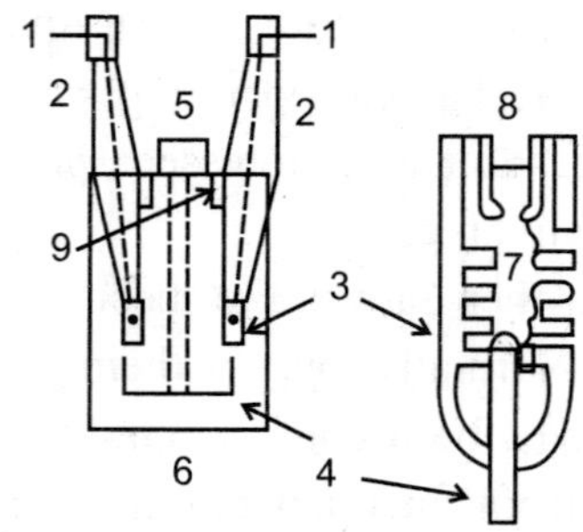

Fig. 12.14: Dead-tank OCB with details of arcing chamber.

1. H.V. leads; 2. H.V. bushing; 3. Arcing chamber; 4. Movable contact; 5. Box containing operating mechanical parts and insulated operating rod; 6. Grounded oil tank; 7. Arc and arc passages in chutes; 8. Fixed contact; 9. Bushing CTs on "Line side" and "Bus side".

The current transformer for sensing the over current can be located inside the h.v. bushing at nearly ground potential. The secondary coil surrounds the h.v. lead which is threaded inside the coil. The dead-tank breaker permits easy mounting of the CT so that the lead-in to the operating relay at the switchboard is easy. In other schemes, there could be a separate CT for doing the same job.

(*b*) **Live-tank Circuit Breaker:** In this type of breaker, the interrupting chambers are located at high voltage. The operating mechanism or mechanical link or compressed-air pipe must be insulated from ground through porcelain insulators. The entire breaker assembly resembles a network of porcelain insulators as shown in Fig. 12.15.

Since there is no bushing leading the h.v. conductor to nearly ground level, the CT must be located at the h.v. terminal. The lead-in problem is severe for the switchboard and relay. However, this is circumvented by using a separate CT unlike the bushing CT for the dead-tank arrangement. The h.v. conductor is led into the CT through a bushing on the CT tank and the secondary is near ground potential.

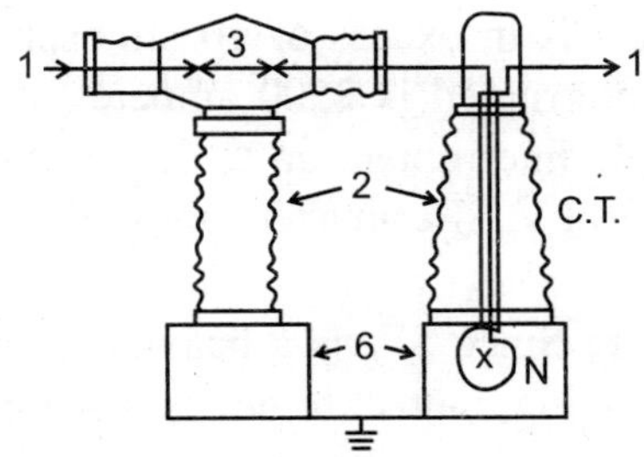

Fig. 12.15: Live-tank OCB with separate CT. 1. H.V. leads; 2. H.V. pedestal bushings; 3. Arc interrupters; 4. Grounded oil tank for C.B. and CT.

(*c*) **Air-blast Circuit Breaker:** Figure 12.16 shows schematics of the overall arrangement of the interrupters at high voltage, the compressed-air tank at ground level, as well as the air flow inside the interrupter in a simplified form. Details of course vary from manufacturer to manufacturer. When the compressed-air tank is at ground level, it also acts as a support for the rest of the breaker consisting of porcelain insulators. However, this arrangement requires a longer pipe for guiding the compressed air to the interrupter at some height above the tank and thus entails delay in operating the breaker contacts in the arcing chamber. In other arrangements, the compressed-air tank is also kept at high voltage or else the arc interruption can occur inside that tank itself, which entails no delay in guiding the air into the interrupter. In very cold climates (where winter minimum temperature can reach below –50°C), moisture condensation will block operating valves. Therefore, the compressed-air tanks are buried below ground and they may also be provided with heaters to avoid condensation.

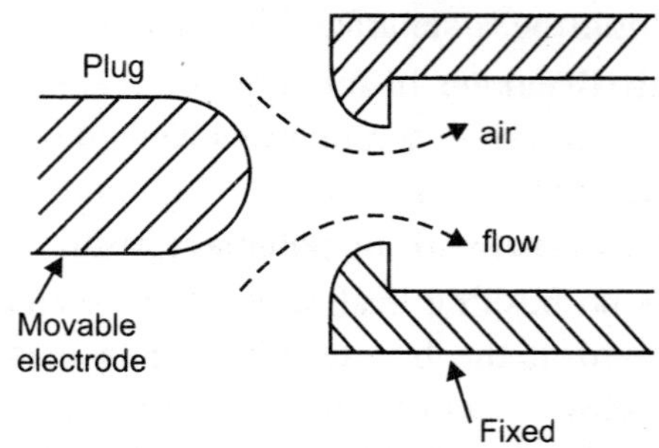

Fig. 12.16: Air-blast CB. See also Fig. 12.15 for arrangement of Compressed-air tank at ground level and separate CT.

The air-pressure used is in excess of 10 atmospheres and may reach 50 – 80 atm. The usual current interrupted is 2500 amperes at all voltage levels from 11 kV to 1150 kV, using single interrupters up to 33 kV and invariably 2 or 4 or even more for voltages up to 330 kV. SF_6 breakers are replacing air-blast breakers for very high voltages.

(*d*) **SF_6 Circuit Breaker:** Since SF_6 is a heavier gas than air, a mechanical piston can be coupled to the moving contact to act as a pump or "puffer" to force the gas into the arc space. This requires a little extra power to operate the piston of the puffer cylinder than merely opening a nozzle valve. But it has such superior arc-quenching properties that nearly all SF_6 breakers manufactured in the world are puffer-type breakers.

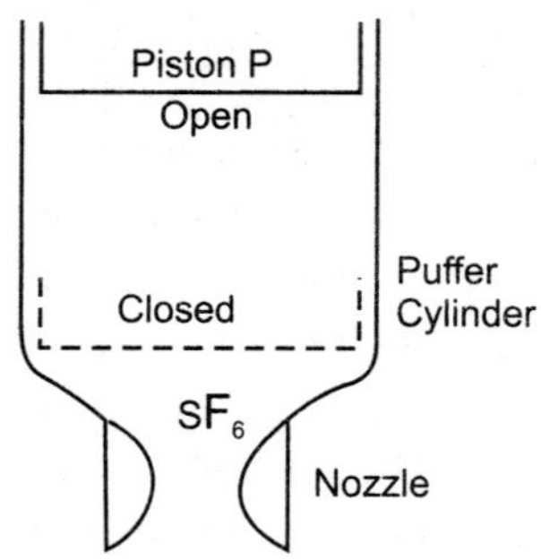

Fig. 12.17: Details of SF_6 puffer cylinder, piston and nozzle. Overall circuit breaker and CT schematic is similar to Fig. 12.15.

Figure 12.17 shows the schematic of the puffer cylinder, piston and nozzle for one type of breaker. In this figure, P = puffer piston shown in full open position (and shown dotted in full closed position), N = interrupting nozzle and stationary contact, C = puffer cylinder. The SF_6 gas is stored in a tank and let into the interrupter chamber through suitable valves by the operation of the solenoid mechanism which in turn is energised by the over current relay and battery power supply. At the same instant, the puffer piston is also actuated and blows the high-current arc out. (A similar oil-piston was also developed for OCBs to pump cool oil into the arcing chamber in order to lower the temperature of the arc).

The pressure used is in the range of 5–20 atmospheres, the most often preferred pressure being 15–17 atm.

(*e*) **Vacuum Circuit Breaker or Interrupter:** Vacuum interrupters have been developed for voltages up to 132 kV at the present time and for currents up to 2000 amperes. The contacts in this type of breaker have many shapes and sizes. The simplest is a plain butt contact of 5 cm diameter. But for larger currents, they take the shape of segments or petals as in a flower which are also known as spirals. These are shown in Fig. 12.18. For high voltages, there are many breaks in series, a typical number being 14 interrupting gaps per phase for 750 kV. Thus, single vacuum breakers operate at lower voltages than other types of circuit breakers.

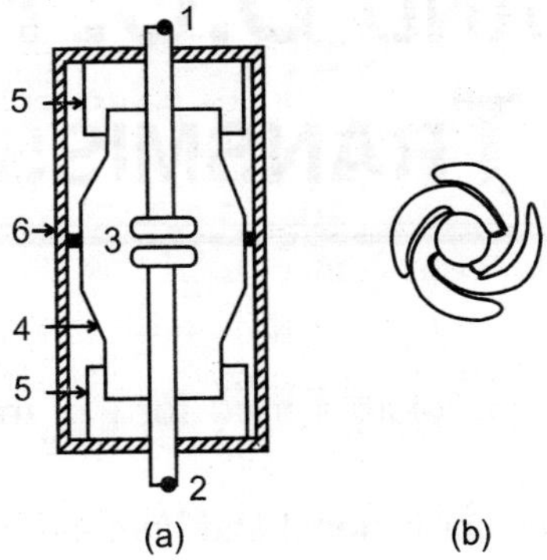

Fig. 12.18: Vacuum interrupter. (*a*) Butt electrode; (*b*) Spiral electrode. 1. Fixed contact; 2. Movable contact; 3. Butt electrodes; 4. Vapour condensation shield; 5. End shields; 6. Porcelain insulating envelope. Diameter for 15 kV, 20 kA interrupter is only 10 cm.

❑❑

CHAPTER 13

VOLTAGE CONTROL OF A.C. AND D.C. HIGH VOLTAGE TRANSMISSION SYSTEMS

Q.1. What are the standard high voltages used for a.c. transmission lines and high voltage equipment?

Ans. These are specified by relevant national and international standards. The following values are accepted as standards:

Nominal kV	11	33	66	132	220	345	400	500	750	1000	1150
Maximum kV	12	36	72.6	145	245	362	420	525	800	1050	1200

These are line-to-line r.m.s. voltages in 3-phase configuration.

Q.2. Why is there a maximum kV for each level of nominal kV?

Ans. Referring to Chapter 7, Q.9, we observed that because of Ferranti Effect, the charging current of a long transmission line, when flowing through the series inductance of the line, can make the voltage at the receiving end higher than the voltage at the source end. Therefore, equipment must be able to withstand any rise in voltage along the line. This rise must also be limited to the maximum allowable voltage permitted by the standards. The insulation of all equipment must be designed based on maximum power-frequency voltage of the system and not on the nominal system voltage.

Q.3. Since the Ferranti Effect depends upon the load condition also, how is the voltage controlled or made to stay at specified constant values at the sending end and receiving end of a long transmission line as the load varies from no-load condition to full-load condition?

Ans. Let E_S = r.m.s. value of line-to-line voltage at the sending end and E_r = voltage at the receiving end. During the entire range of operating condition from no-load to full-load, the magnitudes of E_S and E_r will be held at the specified constant values by voltage regulators. At the generator terminals, this is achieved by control of d.c. excitation current to its field winding on the rotor. But at the receiving end, certain types of compensating equipment must be connected so as to vary the reactive component of the current flowing in the line, since the resistive component is fixed by the load requirement. By varying the VARS consumed at the load or receiving end of the line, the voltage E_r is held steady. This VAR control is achieved by a voltage regulator which operates on the compensating equipment.

Q.4. What are the different types of compensating equipments used in practice at the receiving end?

Ans. These supply or consume pure reactive power and operate at zero power factor lead or lag. Leading VARS are supplied by:

(*a*) Synchronous Condensers, and (*b*) Switched Capacitors.

Lagging VARS are supplied by: (*c*) Synchronous condensers in under-excited state of operation of the d.c. field winding, or, (*d*) Thyristor-Controlled (Switched) Static Reactors called Static VAR Systems (SVS).

At low loads, where the charging current of the transmission line tends to yield high voltage at the receiving end, zero-power factor inductive VARS are supplied to compensate for the capacitive nature of current. At high loads, where the current drawn by the load is normally highly inductive, the compensating equipment provides zero-power factor leading VARS.

Examples are provided in order to explain these points through theoretical and numerical procedures.

Q.5. What are the equivalent representations of a transmission line having conductor resistance, series inductance and shunt capacitance to ground?

Ans. We have discussed the methods of calculating the inductance and capacitance of overhead transmission lines in Chapter 8. For calculating the voltage conditions and power flow, we consider here that the line parameters are lumped. Every line has distributed parameters and unless the problem warrants it, the distributed nature will be ignored for now. As such, line representation is known as short-or medium-length line. The equivalent circuits utilised in practice are shown in Fig. 13.1. These are known as (*a*) the π-representation and (*b*) *T*-representation. A line with total length *X* between the sending-end and receiving-end has distributed resistance of *r* ohm/km, inductance of l_u henry/km, and shunt capacitance to ground of c_u farad/km. These are positive-sequence values since in the steady state, the generator at the sending end generates only positive-sequence voltages, namely, $E_a = E\angle 0°$, $E_b = E\angle -120°$ and $E_c = E\angle 120°$. We will assume that *E* is the r.m.s. value of line-to-neutral voltage in all calculations to follow.

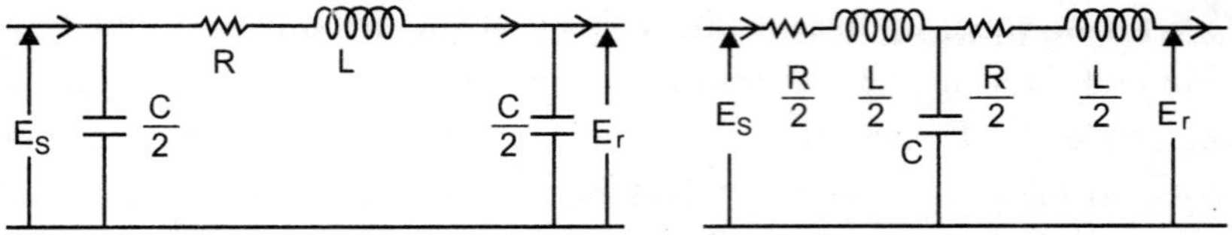

Fig. 13.1: π and T-equivalent circuits of short-and medium length a.c. transmission lines.

Then, the total series resistance, inductance and shunt capacitance of the line will be: $R = r \cdot X$, $L = l_u \cdot X$, and $C = c_u \cdot X$. As shown in Fig. 13.1(*a*), the shunt capacitance is split into two equal halves $C/2$ each and connected at the entrance and exit of the line. The series branch between E_S and E_r consists of the entire R and L. Such a representation is called the π-representation of a lumped-parameter line.

Figure 13.1(*b*) shows an alternate representation called the *T*-representation or circuit. Here, the series R and L are split into two halves while the total capacitance C is retained as such and connected in the middle of the line.

Example: A transmission line has a conductor resistance of $r = 0.08$ ohm/km, inductance $l_u = 1$ mH/km, and shunt capacitance of $c_u = 10$ nF/km. The line is 200 km long. Find the parameters of the line in (*a*) π-representation, and (*b*) *T*-representation. Give also the series reactance and shunt susceptance of inductance and capacitance at 50 Hz.

Solution:

$$R = 0.08 \times 200 = 16 \text{ ohms}$$
$$L = 10^{-3} \times 200 = 0.2 \text{ henry}$$
$$X_L = 2\pi \times 50 \times 0.2 = 62.8 \text{ ohms}$$
$$C = 10 \times 10^{-9} \times 200 = 2\ \mu\text{F}$$
$$S_c = 2\pi \times 50 \times 2 \times 10^{-6} = 628 \times 10^{-6} \text{ mho or siemen.}$$

The resulting equivalent circuits are shown in Figs. 13(*a*) and (*b*).

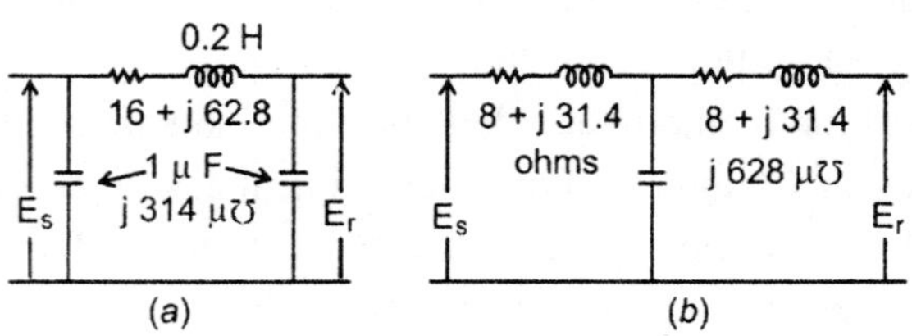

Fig. 13.2: Equivalent circuits for Q. 5.

Q.6. With no load connected at the receiving end, the voltage there is 127 kV, line-to-neutral ($220/\sqrt{3}$ kV). Calculate the magnitude and phase angle δ_0 of the voltage at the sending end, using the numerical values of line constants of Fig. 13.2.

Ans. We take E_r as reference and its phasor value $E_r = E_r \angle 0°$. The corresponding sending-end voltage is represented as $E_S = E_S \angle \delta_0$. Use numerical values of line parameters of previous question.

(*a*) **π-representation:** Current in $C/2$ at receiving end is:

$$I_r = j2\pi f\,(C/2)\,E_r,$$

which leads E_r by 90°.

∴ Voltage drop in series *R-L* will be:

$$(R + j2\pi fL)I_r = (R + jX_L)\cdot j(S_C/2)E_r$$
$$= [(-S_C\cdot X_L/2) + j(S_C\cdot R/2)]\cdot E_r$$

$$\therefore \quad E_S\underline{|\delta_0} = E_r\,[(1 - S_C\cdot X_L/2) + jS_C\cdot R/2]$$
$$= E_r[(1 - S_C\cdot X_L/2)^2 + (S_C\cdot R/2)^2]^{1/2}\ \underline{|\tan^{-1}\{S_C\cdot R/(2 - S_C\cdot X_L)\}}$$

Using $R = 16$, $X_L = 62.8$, $S_C = 628\times10^{-6}$,

we obtain:

$$E_S = 127[(1 - 628\times10^{-6}\times62.8/2)^2 + (628\times10^{-6}\times16/2)^2]^{1/2}$$
$$= 127\,(0.961 + 25\times10^{-6})^{1/2}$$
$$= 127\times0.98$$
$$= 124.5 \text{ kV}$$

Note that the sending-end voltage is lower than the receiving-end voltage by 2%. This is the Ferranti Effect for this line at no load.

$$\delta_0 = \tan^{-1}[628\times10^{-6}\times16/(2 - 628\times10^{-6}\times62.8)]$$
$$= \tan^{-1} 0.005$$
$$= 0.294°$$

The sending-end voltage E_S is leading E_r.

Note that the effect of *R* is negligible. With $R\approx 0$, $\delta_0\approx 0$

(*b*) **T-representation:** Since there is no current flowing in the series branch $(R/2 + jX_L/2)$ on the right at no load, $E_C = E_r$ at the middle of the line.

$$\therefore \quad I_C = jS_C\cdot E_r$$

$$\text{and} \quad E_S\underline{|\delta_0} = E_r + (R/2 + jX_L/2)\cdot jS_C\cdot E_r$$
$$= E_r\,[\,(1 - S_C\cdot X_L/2) + jS_C\cdot R/2]$$

This is the same as for the π-representation.

Q.7. With no load at the receiving end, it is required to raise the voltage E_S at the sending end to 110% of E_r in magnitude. Using the π-representation, calculate the nature and magnitude of reactive VARS to be provided at the receiving end.

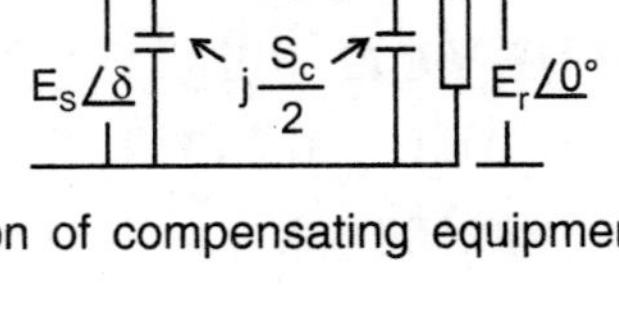

Fig. 13.3: Connection of compensating equipment. No-load condition.

Ans. Figure 13.3 shows the condition of the network with a compensating equipment connected across E_r. Let the current drawn by this equipment be $j \cdot I_C$. If the value of I_C turns out to be positive, then the compensating equipment is supplying VARS, that is, it is purely capacitive in nature since the current I_C leads the voltage $E_r \underline{|0°}$ by 90°. If it is negative, then the compensating equipment will be purely inductive. The final calculation will show this nature. Since the half of the line-capacitance at the sending end does not affect the conditions in the line for calculating E_S, we ignore it.

Total current drawn at the receiving end:

$$I_r = j(S_C/2)E_r + jI_C$$

Voltage drop in series branch

$$= (R + jX_L)\cdot j(I_C + S_C \cdot E_r/2)$$
$$= -X_L(I_C + S_C \cdot E_r/2) + jS_C \cdot R\cdot E_r/2$$

The problem requires that $|E_s| = 1.1\ |E_r|$

Using
$$E_r = 127 \text{ kV}$$
$$R = 16 \text{ ohms}$$
$$X_L = 62.8 \text{ ohms}$$
$$S_C = 628 \times 10^{-6} \text{ mhos,}$$

and keeping I_C in kilo-amperes, we have all voltages in kV.

Thus,
$$|1.1 \times 127| = |127 - 62.8(I_C + 628 \times 10^{-6} \times 127/2)| + j314 \times 10^{-6} \times 127|$$
$$= |(124.5 - 62.8\ I_C) + j\ 0.638|$$
$$\therefore (124.5 - 62.8\ I_C)^2 + 0.638^2 = (1.1 \times 127)^2$$
$$= 139.7^2$$

This gives:
$$124.5 - 62.8\ I_C = [139.7^2 - 0.638^2]^{1/2} \approx 139.7$$

or,
$$I_C = (124.5 - 139.7)/62.8$$
$$= -0.242 \text{ kA}$$

$\therefore$ I_C is inductive.

Therefore, the compensation equipment draws a current of 0.242 kA at 127 kV. The corresponding rating is 127 kV × 0.242 kA = 30.74 MVAR. Such an equipment would take the form of an underexcited Synchronous Condenser or Static Reactors. They would normally be connected to the low-voltage winding of the receiving-end transformer at a voltage of 3.3 kV to 13.2 kV.

With no compensation equipment, we observed that the sending-end voltage had to be 124.5 kV. With the compensating equipment connected, the current flowing at the receiving end of the line is

$$I_r = jS_C \cdot E_r/2 + jI_C$$
$$= j\ 0.04 - j0.242$$
$$= -j0.202 \text{ kA}$$

Example: If in the above numerical example, the sending-end voltage also is to be kept at 127 kV, find the nature and magnitude of compensating VARS to be connected at the receiving end, and the total current in line after compensation.

Solution: The equation for I_C will now be:

$$(124.5 - 62.8I_C)^2 + 0.638^2 = 127^2$$

We note that the second term on the right, namely 0.638^2, is caused by line resistance R and is negligible. Therefore, for voltage-control calculations, line resistance can be neglected. But as will be seen next, it must be retained for calculating line losses and efficiency of transmission.

$$\therefore \qquad I_C = (124.5 - 127)/62.8 = -0.04 \text{ kA}$$

This gives the VARS in the compensating equipment as still inductive and of magnitude (127 kV × 0.04 kA) = 5.08 MVAR. The total current in the line is now zero, since $(j\, S_C\, E_r/2)$ is also j 0.04 kA.

The reader should be able to work out these two examples with the T-representation of the line of Figs. 13.1(b) and 13.2(b).

Q.8. The above line is now carrying full load of 100 MVA at 0.9 lag. Determine the nature and amount of compensating equipment to be connected at the receiving end across $127\underline{|0°}$ kV, if the sending-end voltage is to be held constant at:

(a) $|E_s| = 1.1 \times 127$ kV, and (b) $|E_S| = 127$ kV.

Calculate also the power-angle δ, the angle between the phasors E_S and E_r. Use the π-representation again.

Ans. Let $W_L = P_L + jQ_L$, the total MVA with P_L = real power and Q_L = reactive power of load. For inductive loads, convention requires that Q_L be positive. However, the current flowing in an inductive load must lag behind the voltage. Denoting the complex conjugate of a complex quantity with (*), the load current will then be:

$$I_L = (W_L/E_r)^* = (P_L - jQ_L)/E_r = I_R - jI_X$$

With W in MVA and E_r in kV, the current I is in kA. Then, the receiving-end current with the compensating equipment will be, as shown in Fig. 13.4:

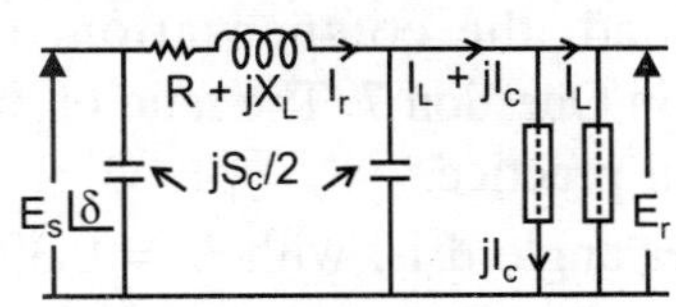

Fig. 13.4: Connection of compensating equipment. On-load condition.

$$\begin{aligned} I_r &= j\, S_C \cdot E_r/2 + I_R - jI_X + jI_C \\ &= I_R + j(S_C \cdot E_r/2 + I_C - I_X) \end{aligned}$$

$\therefore$ The sending-end voltage will be:

$$E_S\underline{|\delta} = E_r\underline{|0^\circ} + (R + jX_L)[I_R + j(S_C \cdot E_r/2 + I_C - I_X)]$$
$$= [E_r + R \cdot I_R - X_L(S_C \cdot E_r/2 + I_C - I_X)] + j\,[X_L \cdot I_R + R(S_C \cdot E_r/2 + I_C - I_X)]$$

We can calculate the magnitude of the complex quantity on the right-hand side and equate this to the given $|E_s|$. Since all quantities except I_C are known, the resulting equation can be solved to yield I_C, the current flowing in the compensating equipment.

Numerical Example:

(*a*)

$$|E_s| = 1.1 \times 127 = 139.7$$
$$E_r = 127$$
$$R = 16$$
$$X_L = 62.8$$
$$S_C = 628 \times 10^{-6}$$
$$I_R - jI_X = (0.9 - j0.436) \times 100/127$$
$$= 0.709 - j0.343 \text{ kA}$$

$$\therefore \quad E_S\underline{|\delta} = 139.7\underline{|\delta}$$
$$= (157.37 - 62.8I_C) + j\,(36.68 + 16I_C)$$

The resulting quadratic equation for I_C is obtained by equating the squares of the magnitudes on both sides, thus:

$$139.7^2 = (157.37 - 62.8I_C)^2 + (36.68 + 16I_C)^2,$$

which yields

$$I_C^2 - 4.427I_C + 1.575 = 0.$$
$$\therefore \quad I_C = [4.427 \pm \sqrt{(4.427^2 - 4 \times 1.575)}]/2$$
$$= 4.037 \text{ or } 0.39 \text{ kA}$$

The two values for compensating MVAR are: 4.037 × 127 = 512.7 MVAR, or, 0.39 × 127 = 49.53 MVAR, both being purely capacitive. Since the lower value would be the more reasonable of the two, we must select I_C = 0.39 kA giving 49.53 ≈ 50 MVAR for the compensating reactive power, capacitive.

Note that at no load, the compensation to be provided came out to be 30.7 MVAR, inductive, in Question 7. The ratio of the two is 30.7/49.53 = 0.62, which is a reasonable value in practice.

The resulting power angle d is, with I_C = 0.39 kA,

$$\tan\delta = (36.68I_C + 16I_C)/(157.37 - 62.8I_C)$$
$$= 42.92/132.88$$
$$= 0.323$$
$$\therefore \quad \delta = 17.9^\circ$$

The sending-end voltage E_S leads the receiving-end voltage E_r by 17.9°.

$\therefore\ E_S = 139.7\underline{|17.9°}$ when $E_r = 127\underline{|0°}$.

(*b*) If $|E_s| = |E_r| = 127$ kV, the equation for I_C now becomes:

$$127^2 = (157.37 - 62.8I_C)^2 + (36.68 + 16I_C)^2,$$

or, $I_C^2 - 4.427I_C + 2.3815 = 0,$

$\Rightarrow I_C = 1.253$ kA as the more reasonable value.

$\therefore$ Compensating MVAR = 1.253 × 127 = 159 MVAR capacitive.

$$\tan\delta = (36.68 + 16 \times 1.253)/(157.37 - 62.8 \times 1.253)$$
$$= 0.721$$

$\therefore$ $\delta = 35.8°$

giving $E_S = 127\underline{|35.8°}$

for $E_r = 127\underline{|0°}$

[As a note, we might point out that the limit for δ, the power angle or phase difference between the sending- and receiving-end voltages is normally kept at 30° from consideration of system stability. Keeping this in mind, the engineer must select the proper value of sending-end voltage E_S. But this is beyond the scope of this treatise in High Voltage].

Q.9. In Question 8, calculate the line ohmic losses at full load for both $|E_s| = 1.1 \times 127$ kV and 127 kV. Then, calculate the efficiency of transmission based on real powers sent and received.

Ans. (*a*) $E_S = 139.7$ kV,

$I_L = 0.709 - j\,0.343$ kA in the load

Current in line capacitance at receiving end

$$= S_C \cdot E_r/2$$
$$= j\,0.04 \text{ kA}$$

and current in compensating equipment is $I_C = j\,0.39$ kA

$\therefore$ Current flowing in the line π-representation of the line will be:

$$I_r = 0.709 - j\,0.343 + j\,0.04 + j\,0.39$$
$$= 0.709 + j\,0.087 \text{ kA}$$
$$= 0.7143\underline{|7°}$$

$\therefore$ $I_r^2 \cdot R = 0.7143^2 \times 16$

$$= 8.16 \text{ MW}$$

Load power = 100 × 0.9 = 90 MW

$\therefore$ Efficiency of transmission, ε = Load power/source power

$$= 90/(90 + 8.16) = 0.917$$
$$= 91.7\%$$

Line losses = 8.3%

(*b*)

$$E_S = 127 \text{ kV}$$

$$I_C = j\,1.253 \text{ kA}$$

$$I_r = 0.709 - j\,0.343 + j\,0.04 + j\,1.253$$

$$= 0.709 + j\,0.95$$

$$= 1.1854\underline{|53.26°} \text{ kA}$$

$$\therefore \quad I_r^2 \cdot R = 1.1854^2 \times 16 = 22.48 \text{ MW}$$

and $$\varepsilon = 90/(90 + 22.48) = 0.8 = 80\%$$

Line losses now amount to 20% of source power. Thus, by selecting different values for the sending-end voltage, the engineer can control line losses and select the optimum value.

Q.10. Summarise the results of the previous calculations and give your comments. Which voltage value for E_S is the better of the two from system performance point of view?

Ans.

$E_S =$		$\frac{\lvert E_s\rvert=1.1\lvert E_r\rvert}{139.7\,\text{kV}}$	$\frac{\lvert E_s\rvert=1.0\lvert E_r\rvert}{127\,\text{kV}}$
(a) No-load condition:			
Compensating current,	I_C :	0.241 kA, inductive	0.04 kA, ind.
Compensating MVAR	Q_C :	30.6 MVAR, ind.	5.08 MVAR, ind.
Power angle,	δ :	≈ 0	≈ 0
(b) Load = 100 MVA at 0.9 p.f. lag = 90 MW + j 43.4 MVAR			
Compensating current,	I_C :	0.39 kA, capacitive	1.253 kA, cap.
Compensating MVAR,	Q_C :	50 MVAR, cap.	159 MVAR, cap.
Power angle,	δ :	17.9° leading E_r	35.8° leading E_r
Line losses		8.2 MW	22.5 MW
Efficiency,	ε :	91.7%	80%

Comments or Conclusions

1. By changing the amount and nature of compensation at the receiving (or load) end, the sending-end voltage and its phase angle can be selected suitably for a given receiving-end voltage. When both are selected for the best performance of the system, they are held constant by voltage regulators. The system voltages cannot change at either end from no-load to full rated load conditions.
2. If the voltage magnitudes at both ends are equal, the compensation MVAR at no-load is only 5 MVAR which is negligible. However, at rated voltage and load, it requires 159 MVAR capacitive which is much higher than even the load MVAR of 100 MVAR. The phase-angle difference δ between E_S and E_r, i.e., the power angle, is 35.8° with E_S leading E_r. The line losses amount to 22.5 MW and the efficiency of transmission is only 80%.

The result is a very poor performance on account of the following reasons:

(*a*) Cost of compensating equipment will be very high, in fact unreasonably high. Ratio of inductive compensation required at no-load to the capacitive compensation required at full load = 5/159 ≈ 1/32. This ratio should be nearly 0.7 as will be explained under Synchronous Condenser Compensation.

(*b*) The line current is also high at 1.1854 kA with full-load current being 100 MVA/127 kV ≈ 0.8 kA.

(*c*) The power angle δ is nearly 36°. For reasons of system stability, it is usual practice to limit this angle to 30°, since otherwise the system will perform poorly during faults caused by short-circuits. This aspect of system performance, called System Stability, cannot be discussed here.

Raising the sending-end voltage E_S to be 10% higher than E_r gives the following advantages and improved performance:

(*d*) Compensation required at no-load = 30.6 MVAR inductive, and at full load it is 50 MVAR, capacitive. The ratio of the two is about 0.6. This is close to 0.7 and therefore the design is nearly optimum.

(*e*) Line losses are reduced to 8.2 MW from 22.5 MW as in the previous case when $E_S = E_r = 127$ kV. Then efficiency of transmission is now 91.7% instead of 80%.

(*f*) The power angle δ is 18° which is very desirable from system stability point of view. The only other check is whether for this voltage level, namely 220-kV line-to-line, 10% excess voltage is permissible. The maximum allowable voltage for nominal voltage of 220 kV is 245 kV and so 220 × 1.1 = 242 kV is within the limit.

By selecting various magnitudes for E_S, the sending-end voltage, an engineer can calculate: (*a*) Compensation required at no-load; (*b*) Compensation required at full rated load; (*c*) Resulting line losses and efficiency ε of transmission; (*d*) The power angle δ.

After such calculations are carried out, which are very easy on a Digital Computer if one is available, the best decision for the choice of E_S can be made. There are also certain graphical procedures available. They are not discussed here. [It is of course expected that the reader will verify all calculations given here].

Q.11. What is meant by "Series-Capacitor Compensation"? What are its advantages and disadvantages in a h.v. system?

Ans. When a transmission line has voltages $E_S\underline{|\delta}$ and $E_r\underline{|0°}$ at its two ends, we have seen in Chapter 7, Question 10, that the real power P that can be transmitted or handled can be written as the tie-line equation $P = E_S \cdot E_r \cdot \sin\delta / Z_S$. In general, $Z_S \approx X_S$, since resistance R of the line is negligible. Therefore, $P = E_S \cdot E_r \cdot \sin\delta / X_S$. The value of the power angle δ is limited to 30° for reasons of stability of the system during faults. Thus, $P \approx E_S \cdot E_r/2X_S$. It is evident that with given magnitudes of E_S and E_r and the power angle δ, the power that can be transmitted or handled by a line increases if X_S is lowered.

Since the series reactance X_S of the line is inductive, the total or effective reactance can be lowered by connecting a high voltage capacitor in series with the inductive reactance. This capacitor is usually lumped whereas the inductive reactance is distributed throughout the length of the line. However, from performance point of view, in the π- or *T*-representation, the series capacitor can be seen to reduce the lumped representation of the line inductance. For long lines and high voltages, one capacitor at the middle of the line is connected in series. This is true for voltages from 400 kV to 750 kV. For higher voltages of 1000 kV, 1150 kV or 1500 kV, and very long lines, two lumped capacitors can be connected at 1/3 points. With the series capacitor in, the total series reactance of the line then becomes $X_S = X_L - X_C$ with one capacitor in the middle. This is shown in Fig. 13.5.

The ratio of X_C/X_L is known as the "degree of compensation". This is usually 40–50%, and is then known as 40% series compensation or 50% series compensation, etc.

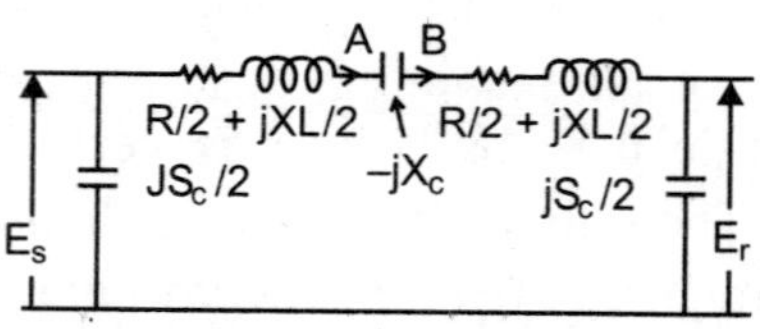

Fig. 13.5: Series-capacitor compensation for a long line for Question 10.

One of the serious disadvantages accruing out of this series compensation is evident by a look at Fig. 13.5. Let there be 50% series-capacitor compensation. Then, if a short-circuit occurs at the receiving end of the line, the total reactance limiting the fault current from the source is only $0.5X_L$ and thus is twice the current without the series capacitor. It is still more dangerous if the fault were to occur on the load-side terminal of the series capacitor shown as point *B*. The total reactance into the fault is now zero! The only element restricting the fault current is the small line resistance *R*/2 up to the fault. This condition is the most serious to reckon with in a 50% series-capacitor compensated line.

Thus, schemes must be adopted so as to remove the capacitor out of the circuit instantaneously when such faults occur. This will increase the reactance up to the fault, if it were to occur beyond the capacitor on the load-side of the transmission line. This is known as "capacitor flashover." A small sphere gap is provided across the capacitor terminals which is allowed to spark over the instant a short-circuit occurs and the voltage across the series-capacitor terminals rises above normal.

A second scheme utilises SiC or ZnO arrester across the capacitor to absorb the surge or overvoltage caused across the capacitor due to the high fault current, since the voltage across it is $V_C = X_C \cdot I_f$, where X_C = reactance of the capacitor at power frequency f, and I_f = current flowing through it during the fault.

One other serious phenomenon occurring with series-capacitor compensated line is known as "Sub-Synchronous Resonance." Because an inductance L of the line and the series capacitor C are in series, they form a resonant circuit. The resonance frequency cccurs,

when $2\pi f_r \cdot L = 1/2\pi f_r \cdot C,$

giving $f_r = 1/2\pi\sqrt{LC}$

If $X_C = 0.5X_L$ (50% compensation) at power frequency f,

then, $1/2\pi fC = 2\pi fL/2$

$\therefore$ $\{(1/2\pi\sqrt{(LC)}\}^2 = 0.5f^2,$

giving $f_r = \sqrt{0.5} \cdot f$

$= 0.707f$

which is less than f, or sub-synchronous. Thus, the sub-synchronous resonance frequency is f_r = 35.5 Hz for 50-Hz systems and 42.42 Hz for 60-Hz systems. Any disturbance in the line giving rise to currents at these frequencies will kick the voltage across the capacitor to practically infinite value, but is limited by the resistance R of the line. Remedies against the Sub-Synchronous Resonance (SSR) condition are very important. We will not discuss them here.

Q.12. How is a Synchronous Condenser used for controlling the voltage at the receiving end of a transmission line from no-load to full-load? What is meant by the V-curve of a synchronous condenser or phase modifier?

Ans. We have observed in Question 9 that the operating condition of a transmission line can reach optimum values (reasonable amount of compensating VARS, minimum line losses, maximum efficiency, reasonable value of power angle δ), if the magnitude of the sending-end voltage E_S is selected suitably for a given receiving-end voltage E_r. This has necessitated that at no-load, inductive VARS must be connected at the receiving-end load, and at full-load condition with usually lagging power factor, capacitive VARS must be supplied by the compensating equipment. For the numerical example considered there, the ratio of the two values of VARS (inductive/capacitive) was about 0.6.

A synchronous condenser can provide both types of zero-power-factor reactive powers, see Chapter 7, Question 16. When the d.c. field excitation is varied, taking normal excitation to be that value at which the a.c. armature current is minimum, when the field is under-excited the armature draws current from the line at zero power factor lagging. This means that it is drawing inductive VARS which are necessary at no-load and low-load conditions. If on the other hand, the d.c. field current is increased beyond the normal excitation value, that is, the machine is operating in the over-excited condition, it supplies capacitive VARS to the system. The voltage regulator at the receiving end acts to change the field current and hence the nature and amount of VARS drawn or supplied to the system vary easily and smoothly.

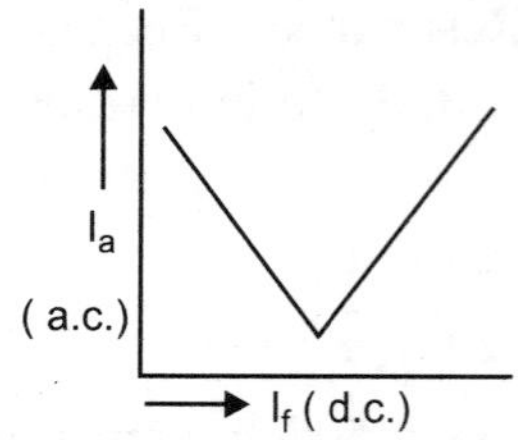

Fig. 13.6: V-Curve of an unloaded synchronous motor. (Synchronous Condenser or Phase Modifier).

Figure 13.6 shows the graph of variation of armature a.c. current drawn when the d.c. field current is varied. This has the form of a *V* and is known as the famous *V*-curve of an unloaded synchronous motor or the synchronous condenser or synchronous phase modifier.

One serious restriction about using a synchronous condenser to supply both types of VARS is that if the field is under-excited to a very low value so as to draw a large inductive current, the magnetic field becomes very weak and any disturbance on the machine cannot generate restoring e.m. forces to make the machine keep running at synchronous speed. The machine "hunts" or oscillates very badly and may loose synchronism, thereby creating more disturbances to the system. It is for this reason that the inductive VARS have a certain maximum value. From practical experience the ratio (inductive VARS/capacitive VARS) supplied by the synchronous condenser is limited to 0.7. Therefore, the magnitude of voltage E_S at the sending-end must be so selected that this ratio is obtained by the method of calculation described before. This is a major design problem for a system engineer.

Q.13. What are Static VAR Systems? Give examples of Thyristor-Switched Reactors and Capacitors.

Ans. A synchronous condenser is a rotating VAR system. Its maximum capacity is limited and cannot be used for very long lines. These are now being replaced by non-rotating or static equipment which will smoothly-vary VARS. For supplying inductive VARS, reactors are used while for supplying capacitive VARS, capacitor banks are used. However, capacitors cannot be varied smoothly in value and both types are necessary to adjust the VAR requirements. The availability of high voltage high-current high-speed or fast-switching Thyristors or Silicon-Controlled Rectifiers (SCR) has made this possible. The voltage regulator generates suitable pulses to the third electrode or GATE of the SCR which then controls the firing angle and voltage output of the suitable switching circuits and the reactance of the reactor. Thus, overall improvement is obtained in power-transfer capability and so increased economy.

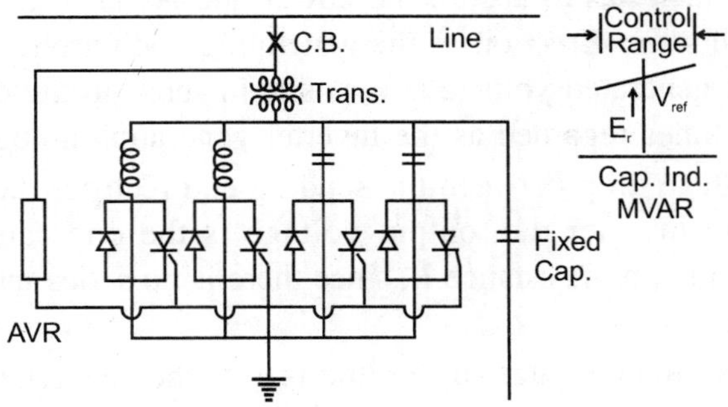

Fig. 13.7: General circuit of Static VAR Systems with controlled-rectifiers and reactors and capacitors. Also shown is control-range exercised by switched reactors and capacitors.

Examples of Static VAR compensating systems called SVC schemes are broadly the following, and shown in Fig. 13.7 for one type which is basic. Also shown is the range of control obtainable with these schemes.

The current flows in the form of pulses, or portions of the sine wave, and generates undesirable harmonics which will be injected into the line. These harmonics interfere with signalling and communication circuits and must be eliminated by using properly-tuned filters. These are not discussed here.

It can be shown that the reactive power generated at any firing angle α is:

$$Q = 1.35[\pi - \alpha - 0.5 \sin 2(\pi - \alpha)] \cdot I \cdot V_{ph}, \text{ VAR 3-phase,}$$

where I = peak value of fundamental component of current,

when, $\alpha = 90°$

and, V_{ph} = r.m.s. value of phase voltage

for, $\alpha = \pi/2$

$$Q = 2.12\, I\, V_{ph} = 3 I_{ph} V_{ph}$$

where I_{ph} is now the r.m.s. value of current:

$$I_{ph} = I/\sqrt{2}$$

Q.14. What is the voltage pattern of a d.c. transmission link between the source and load through a long h.v. line?

Ans. 1. d.c. transmission line is only a link between two a.c. systems.

2. At both ends, there are a.c. sources generating and using sinusoidal voltages. (**Note:** Inverters at load end "generate" sine waves of voltage);
3. Power can be arranged to flow from any end to the other end. Thus, the a.c. system at one end can be made to function as a source of power and the other end can be made to function as a source of absorbing the power, or in other words, as a load;
4. At the end which sends the power or current into the transmission line, a rectifier station is provided with a series smoothing reactor to absorb ripples in the rectified a.c. voltage;

5. At the end which has to absorb the power and act as a load, while maintaining the current in the same direction as the generating end current, the converter acts as an inverter. The generated voltage is opposite in sense to the direction of current flow. This is sometimes regarded as the inverter generating a negative d.c. voltage value;
6. The only voltage drop between the sending-end rectifier d.c. output voltage and the receiving-end inverter d.c. output voltage is the one caused by the line current I_d flowing in the line-resistance R, since there is no series inductance involved in d.c. transmission;
7. The voltage at the entrance to the line is V_{dS}, the subscript d standing for d.c. and s for sending end or source. The voltage at the receiving-end of the line is V_{dr}, the subscript r standing for the receiving end. Thus, $V_{dS} = V_{dr} + I_d \cdot R$;
8. The d.c. voltage output of the rectifier bridge at the source end is higher than V_{dS} by the voltage drop in the series reactor.

   The rectifier output voltage will be shown to be

$$V_{rS} = (3\sqrt{3}/\pi)\ V_m \cos\alpha$$
$$= 1.654\ V_m \cdot \cos\alpha,$$

   where V_{rS} = d.c. output of rectifier bridge,

   V_m = peak value of a.c. phase voltage,

   and α = firing angle or the angle on the a.c. waveform at which a rectifier (SCR) connected to any phase commences to conduct; α is counted from a voltage zero on the sine wave of phase voltage while passing from negative to positive value.

   The voltage drop in the series reactor is proportional to the current I_d. This drop is $3X_S \cdot I_d/\pi$, where X_S = reactance of smoothing or series reactor at power frequency, 50 or 60 Hz;
9. Thus, the voltage at the entrance to the line is:

$$V_{dS} = (3\sqrt{3}/\pi) \cdot V_m \cdot \cos\alpha - (3/\pi) \cdot X_S \cdot I_d$$

   We note that if V_{LL} = r.m.s. value of line-to-line voltage on the a.c. side in a 3-phase Y-connected secondary of the transformer feeding the rectifier bridge,

   then, V_m = peak value of phase voltage = $(\sqrt{2}/\sqrt{3})\ V_{LL}$

   ∴ The d.c. output voltage of the rectifier at the source end can be written as

$$V_{rS} = (3\sqrt{3}/\pi) \times (\sqrt{2}/\sqrt{3})\ V_{LL} \cdot \cos\alpha$$
$$= (3\sqrt{2}/\pi) \cdot V_{LL} \cdot \cos\alpha$$
$$= 1.35\ V_{LL} \cdot \cos\alpha$$

   Some authors prefer to use this expression.

   When $\alpha = 0$, which is called the no-load value of the firing angle, the d.c. output voltage is:

$$V_{rO} = (3\sqrt{3}/\pi)\ V_m$$
$$= 1.654\ V_m$$
$$= (3\sqrt{2}/\pi)\ V_{LL}$$
$$= 1.35\ V_{LL}$$

10. These three voltages, namely, V_{rO}, V_{rS} and V_{dS} are shown pictorially in Fig. 13.8 at a given d.c. line current I_d and SCR firing angle α.

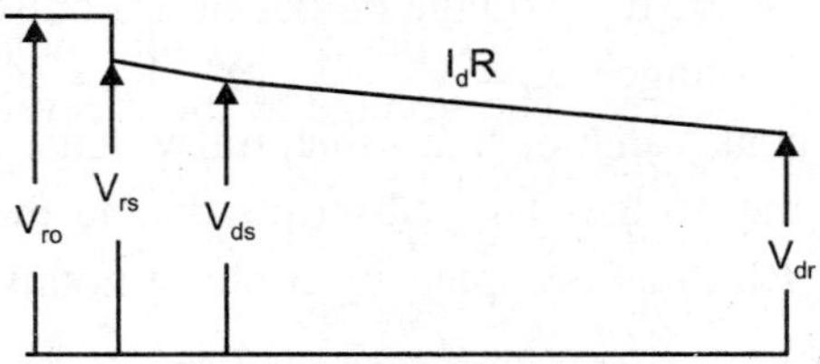

Fig. 13.8: Sending-end voltage conditions of a d.c. link and voltage drop along the line.

11. The voltage drop along the line is $R \cdot I_d$, and therefore at the entrance to the inverter at the receiving end of the line, the d.c. voltage is $V_{dr} = V_{ds} - R \cdot I_d$. This is also shown in Fig. 13.8.
12. At the receiving end, similar voltage conditions exist. The series reactor is connected between the line and the inverter bridge. Here, the d.c. input is converted to a.c. so that the reactor stores energy from the incoming current I_d and supplies it to the load as a.c. On account of this, there is a voltage rise from the line entrance to the inverter input of the value $(3/\pi)X_S \cdot I_d$. Normally, the reactor values are equal at both ends. Let the voltage input from the line to the inverter be V_{iS}.

$$\therefore \qquad V_{iS} = V_{dr} + (3/\pi)X_S \cdot I_d$$

13. This voltage is equal to the output voltage of the inverter. Here, the angle that determines the voltage output of the inverter is denoted as the "extinction angle" and not firing angle. This angle is $\beta = \pi - \alpha$, where α = firing angle counted from voltage zero. We will show that for $\alpha > 90°$ and $< 180°$, the d.c. voltage output of a controlled rectifier opposes the current flow.

$$\therefore \qquad V_{iS} = (3\sqrt{3}/\pi)\ V_m \cdot \cos\beta$$
$$= (3\sqrt{2}/\pi)\ V_{LL} \cdot \cos\beta$$

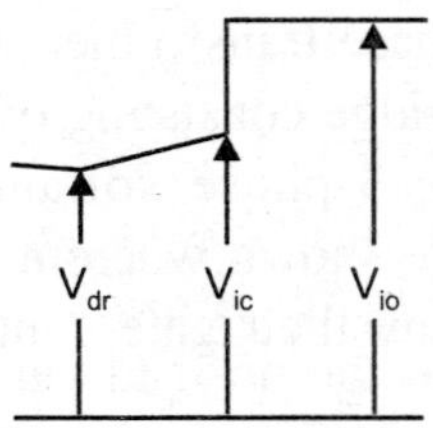

Fig. 13.9: Receiving-end voltage conditions at inverter station.

14. At no-load, $\beta = 0$, and this voltage is:

$$V_{iO} = (3\sqrt{3}/\pi)V_m = (3\sqrt{2}/\pi)V_{LL}$$

We note that V_{iO} = a d.c. voltage,

V_m = peak value of phase voltage,

and V_{LL} = r.m.s. value of line-to-line voltage of the 3-phase a.c. system.

These voltages are shown in Fig. 13.9.

Q.15. Derive the expression for the d.c. voltage output of a 6-pulse Graetz Bridge when excited by a 3-phase source of voltages $V_a = V_R = V_m\underline{|0^\circ}$, $V_b = V_Y = V_m\underline{|-120^\circ}$ and $V_c = V_B = V_m\underline{|120^\circ}$, where V_m = peak value of line-to-neutral voltage output of the h.v. transformer winding connected to the bridge. The subscripts denote the phase-markings (a, b, c) or colours (Red, Yellow, Blue) as accepted by 3-phase notation.

Ans. It will be shown that the average or d.c. output voltage is $V_{dc} = (3\sqrt{3}/\pi)V_m \cdot \cos\alpha$, where α is the firing angle or delayed angle of commutation or transfer of conduction from one thyristor to the next in the sequence of firing of gates arranged in the sequence.

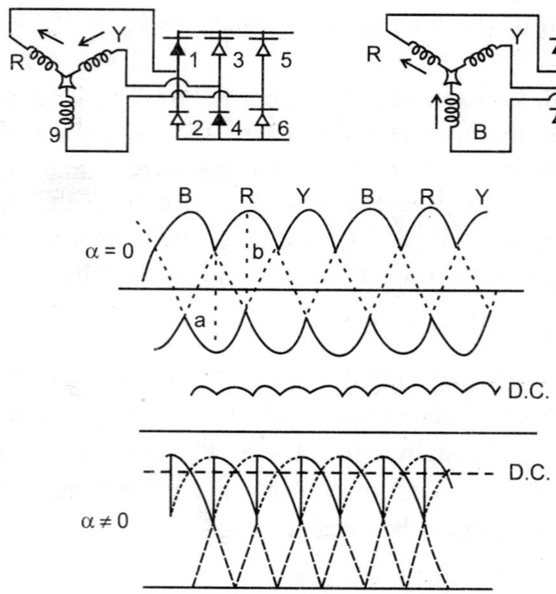

Fig. 13.10: 6-Pulse Graetz Bridge converter configuration and a.c. voltages.

Figure 13.10 (top) shows a 3-phase transformer winding connected in star or Y feeding into a 6-pulse Graetz rectifier-bridge consisting of 6 thyristors or controlled rectifiers. In the lower parts are shown the 3-phase voltages drawn as sine waves with phase sequence R, Y, B. Conditions are shown wherein phases R and Y are conducting due to the rectifiers 1 and 4 are fired by their gate controls.

Let conduction start in the normal manner without delay ($\alpha = 0$). From the lower figure we observe that at instant *a*, the anode of rectifier 1 connected to phase *R* has the highest positive voltage. This happens when the voltage of phase *R* is passing at 30° from its voltage zero where phase *B* voltage is crossing phase *R* voltage but is falling. At this instant, phase *Y* is most negative of all 3 phase voltages and rectifier 4 is in the conduction state since its cathode is most negative of all three rectifiers connected to the negative pole of the d.c. output. Conduction of 4 will stop when phase *B* voltage intersects phase *Y* voltage and becomes more negative than phase *Y* voltage. This rectifier 4 will therefore stop conducting 60° later at which instant phase *R* voltage is passing through its positive peak value and phases *Y* and *B* are passing through their negative half-peak values.

With Natural Commutation: The average of the output voltage between + and – poles of the bridge will be:

$$
\begin{aligned}
V_{RY} = V_{SO} &= (3/\pi)V_m \Big|_{(\pi/2)}^{(\pi/6)} [\sin\theta - \sin(\theta - 120°)] \cdot d\theta \\
&= (3/\pi)V_m [\cos\theta - \cos(\theta - 120°)] \Big|_{(\pi/2)}^{(\pi/6)} \\
&= (3/\pi)V_m [(\cos \pi/6 - \cos \pi/2) - (\cos \pi/2 - \cos \pi/6)] \\
&= (3/\pi)V_m \cdot 2 \cdot \cos \pi/6 \\
&= (3\sqrt{3}/\pi)V_m \\
&= 1.654 V_m
\end{aligned}
$$

With Delayed Commutation: By feeding properly-timed gate-firing pulses to the thyristors, the conduction of the diodes can be delayed by any desired angle α from the instants of natural commutation. This is done in order to vary or control the output d.c. voltage. Thus, with delay angle α of conduction or firing, we obtain:

$$
\begin{aligned}
V_{dc} = V_{SC} &= (3/\pi)\, V_m \Big|_{(\pi/2)+\alpha}^{(\pi/6)+\alpha} [\sin\theta - \sin(\theta - 120°)] \cdot d\theta \\
&= (3/\pi)\, V_m [\cos\theta - \cos(\theta - 120°)] \Big|_{(\pi/2)+\alpha}^{(\pi/6)+\alpha} \\
&= (3\sqrt{3}/\pi)\, V_m \cdot \cos\alpha
\end{aligned}
$$

For $0 < \alpha < \pi/2$, V_{dc} is positive, since $\cos \alpha > 0$. But if the angle of firing is delayed beyond 90° or $\pi/2$, i.e., $90° < \alpha < 180°$, then $\cos \alpha < 0$ and the polarity of output voltage is reversed. The bridge acts as an inverter now. It becomes a voltage sink when connected in the circuit. It can draw a current through its negative pole and pump it out of the positive pole. If, therefore, a rectifier ($0° < \alpha < 90°$) is connected in series with an inverter ($90° < \alpha < 180°$), the current I_d flows in the same direction.

Since the inverter firing angle is between 90° and 180°, we define the angle $\beta = 180° - \alpha$ as the "extinction angle". Thus, β is an acute angle lying between 0° and 90°.

The next step is to know how the rectifier and inverter voltages should be controlled in response to variations in system current I_d flowing in the line in response to variations in load.

Q.16. How is the control of voltage and current for d.c. transmission line arranged as load variations occur at the receiving end?

Ans. This is a very complicated and specialised problem in electronic circuitry, but the principles in a simplified form of the overall picture can be provided here. Like all advances in control circuitry, it involves most up-to-date advances in d.c. operational amplifiers, optical-fibre transmission of signals, etc.

There are two possible ways of effecting control of power flow:

1. By varying the voltage at both ends, or keeping the voltage constant at one end and varying it at the other end; or,
2. Keep the current I_d constant and vary the voltages suitably. Both types are used in practice. As can be realised, if the current is held constant and voltages are adjusted to suit the load, then line losses will be constant. This will lower the efficiency at low loads. Therefore, the choice of type of control will vary from system to system. It is usual to keep the voltage at the receiving end or the inverter output constant by keeping the extinction angle constant. This is known as "Constant Extinction Angle" control or CEA-control. We consider variation in I_d for illustrative purposes.

We observe that as current I_d in the line varies in response to load changes, the only voltage drop is, see Fig. 13.8, equal to that in the line resistance, thus

$$V_{dS} = V_{dr} + R \cdot I_d,$$

or,

$$I_d = (V_{dS} - V_{dr})/R$$

Since the line resistance R is very low, very little change in V_{ds}, the output d.c. voltage of source, is necessary to effect a large change in current I_d. Thus, by having a control system operating on the rectifier bridge, the output voltage V_{ds} can be varied. This is the main principle of voltage-control of the line. We also note that

$$V_{dS} = V_{rS} - (3/\pi)\, X_S \cdot I_d,$$

where

$$V_{rS} = (3\sqrt{3}/\pi)\, V_m \cos \alpha$$

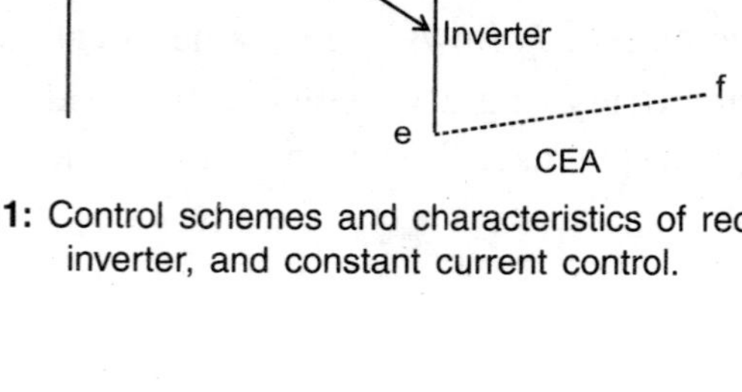

Fig. 13.11: Control schemes and characteristics of rectifier and inverter, and constant current control.

Figure 13.11 shows on a *V-I* diagram the type of variation of voltages and control characteristics that can be achieved. The voltages V_{rO} and V_{iO} are open-circuit voltages ($\alpha = 0$ and $\beta = 0$) of the rectifier and inverter. V_{rC} and V_{iC} are the voltage outputs at rectifier and inverter terminals under conduction control, i.e. $\alpha \neq 0$ and $\beta \neq 0$. The voltage drops $(3/\pi) \cdot X_S \cdot I_d$ show the voltage drops in the smoothing reactors due to current I_d. There is a region between which the current in the line can be held constant by sensing the d.c. current and feeding a small voltage to an operational amplifier which exercises control on the firing and extinction angles of the bridges. Thus, several types of control are possible both in theory and practice.

We note that when load requirements change, we can either hold the current constant or the voltages constant. Keeping voltages constant utilises the insulation levels better, since it is designed for maximum operating voltage levels. In such a case, the control resembles very nearly the control of voltages on an a.c. transmission line.

When both α and β are to be changed in response to load changes, a tele-communication channel should be available. All these several types of control have been adopted in practice, in addition to tap-changing of transformers at both ends to change the input voltages to the rectifier and inverter suitably. However, this is a slow process compared to other modes of control.

Q.17. What are the approximate power levels which one circuit of an a.c. transmission line and a d.c. transmission line can handle at various voltage levels?

Ans. (*a*) **A.C. Transmission Line:** We observed in Question 11 that the power in MW that an a.c. line can handle is given by the tie-line equation $P = E_S \cdot E_r \cdot \sin \delta / X_S$

where, E_S = magnitude of sending-end voltage in kV,

E_r = magnitude of receiving-end voltage in kV,

X_S = series inductive reactance of line in ohm,

and, δ = power angle

= phase difference between E_S and E_r.

From system-stability point of view, the angle δ can be pushed to 30° as limit. If E_S and E_r are in kV, line-to-neutral, and X_S is in ohms per phase, then the power P in MW per phase. But if E_S and E_r are in kV line-to-line, then P = total 3-phase power in MW. We will use this notation and keep E_S and E_r in kV line-to-line.

We have already seen in Chapter 8 how to calculate the line reactance X_S. For preliminary estimates for the power-handling capacity of a single 3-phase circuit, we will assume that $E_S = E_r$ i.e., equal sending-end and receiving-end voltages. The engineer can change these values of P, if necessary when actual values of E_S, E_r and δ are decided from voltage-control point of view using compensating reactors and capacitors.

Then, the following Table shows approximate values of X_S for given voltage levels of transmission and power P in MW/circuit when $\delta = 30°$. Series-capacitor compensation is not considered, so that X_S is the actual value of line inductive reactance in ohms.

The average dimensions used for these calculations are given in the following Table immediately below. The reader should verify all these calculations. Only S/C horizontal configuration for the lines has been used.

Voltage, kV	**132**	**220**	**345**	**400**	**500**	**750**	**1000/1200**
Line height, m	10	11	12	15	15	18	21
Phase spacing, m	3.5	5	7.3	12	9	15	18/21
Conductor No.	1	1	2	2	2	4	6/8
Diameter, m	0.0254	0.028	0.043	0.032	0.045	0.03	0.046
Bundle spacing, m			0.4572				
Bundle diameter, m				1.2/1.2			

These lines follow practices in different countries so that the reader might not be able to find any regularity or uniformity in dimensions all over the world.

Voltage, kV 132	220	345	400	500	750	1000	1200	
X_S (50 Hz) (Ω/ph/km)	0.36	0.38	0.29	0.33	0.29	0.27	0.23	0.23
P (for $\delta = 30°$) 3-phase MW: $P = E_S \cdot E_r/2X_S$.								
Line Length, km								
150	160	430	(3-phase MW)					
200	120	320	1020					
250	100	250	820					
300		210	680	900	1400			
400			510	670	1080	2860	6000	8625
500			410	540	860	2300	4800	6900
600				450	720	1900	4000	5750
800				335	540	1425	3000	4310
1000				270	430	1140	2400	3450
1200				225	360	950	2000	2875

(*b*) **D.C. Transmission Line:** The power-handling capacity of a bipolar circuit is based here on a uniform current-carrying capacity of 2000 amps per pole at all voltage levels. There exist no standard voltages presently worldwide for d.c. transmission voltages. Therefore, a set of possible or probable voltages is selected and the power per bipole is calculated and shown in the following Table.

kV	±100	±150	±200	±250	±300	±400	±500	±600
I_d, amp.	←	- - - - - -	- - - -	2000	- - - -	- - - -	- - - -	→
P, MW	400	600	800	1000	1200	1600	2000	2400

A table of major d.c. h.v. cable transmission systems has been given in Chapter 6, Question 9. Some representative overhead bipolar lines are given below here. (See book *H.V. Power Transmission Systems* by K.R. Padiyar, New Age International (P) Ltd., New Delhi, 1990, for an extensive table of h.v.d.c. power systems all over the world).

No.	System	Voltage	Length	Power, (MW)	Curr. (kA)
1.	Skagerrak (Denmark–Norway):	±250 kV	113 km	500	I = 1 kA
2.	Cabora Bassa–Apollo (Africa)	±533 kV	1414 km	1920	1.8
3.	Square Butte (USA)	±250 kV	750 km	500	1
4.	Nelson River (Canada)	±250 kV	930 km	1000	2
		±500 kV	930 km	2000	2
5.	Itaipu (S. America):	±300 kV	800 km	1575	2.6
		±600 kV	800 km	3150	2.6
6.	Gezhouba–Nan Qiao: (China)	±500 kV	1080 km	1200	1
7.	Rihand–Delhi (India)	±500 kV	1000 km	1000	1

❑❑

CHAPTER 14

WAVE PHENOMENA IN HIGH VOLTAGE SYSTEMS

Q.1. How do wave phenomena occur in high voltage systems? What is the basic mechanism of generation and propagation of an electro-magnetic (e.m.) wave?

Ans. Wave phenomena take place in h.v. systems whenever an electric network with distributed inductance and capacitance is present. These situations occur in: (*a*) Overhead transmission lines; (*b*) Cables; (*c*) Transformer windings; and (*d*) Generator and motor windings.

Ever since James Clerk Maxwell postulated in the 19th century that light is an e.m. wave and Heirich Hertz demonstrated experimentally its existence, we know that all e.m. signals must propagate as waves with a definite velocity. In all cases, three quantities are present in a wave: (1) An electric field, E; (2) A magnetic field, H; and (3) Energy or power P. In an e.m. wave in free space (called an unguided wave), these three form orthogonal vectors in 3 mutually-perpendicular directions in space. But in high voltage systems, the waves are guided by the distributed inductance and capacitance network.

In h.v. systems, waves are initiated when a severe disturbance to the system occurs suddenly. For example, when lightning strikes a line, the voltage and current in the stroke can travel as waves and reach another point on the line. Similarly, when lightning hits a transformer terminal or a generator terminal, the energy travels on the line or generator conductors. These guide the wave at characteristic velocities, and the ratio of voltage to current on the line or windings is known as the characteristic or surge impedance. Similarly, for cables.

The basic mechanism of propagation of a guided wave is explained as follows: At the point struck, by say lightning on a line, the voltage in the stroke charges the capacitance network, namely the shunt capacitance to ground and the mutual capacitance between the conductors. Thus, Electrostatic Energy is first built up. This e.s. energy is then transferred to neighbouring sections of capacitance and in doing so the charging current flows through the series inductance network and e.m. energy is built up in the inductance network. There takes place a continuous transfer of e.s. energy and e.m. energy so that the voltage and current propagate as waves along the capacitance—inductance network.

Consider a single conductor above ground to begin with.

Let, c_u = shunt capacitance of conductor to ground per metre

$= 2\pi\varepsilon_0/\ln(2H/R_{eq})$ farad/m

and l_u = series inductance per metre

$= \mu_0 \cdot \ln(2H/R_{eq})/2\pi$ henry/ m

We will show later on that the following relations hold:

Velocity of wave propagation, $v = 1/\sqrt{(c_u \cdot l_u)}$, metres/second;

and characteristic impedance of wave, $z = \sqrt{(l_u/c_u)}$ ohm.

In all cases, the ground forms the return conductor for the energy in the wave so that we should consider that the wave is "guided" between the conductor and ground. If we neglect the inductance of ground,

Then, $v = 3 \times 10^8$ m/s

= velocity of e.m. wave propagation in free space,

and,
$$z = \sqrt{(l_u/c_u)} = \sqrt{(\mu_0/\varepsilon_0)} = 377 \text{ ohm}$$

Referring to Question 3 in Chapter 8, we observed the same relations. We also note the following important relations among l_u, c_u and v and z:

$$z = v \cdot l_u = 1/(v \cdot c_u)$$

since,
$$v \cdot l_u = l_u/\sqrt{(l_u \cdot c_u)} = \sqrt{(l_u/c_u)} = z$$

and,
$$v \cdot c_u = c_u/\sqrt{(l_u \cdot c_u)} = \sqrt{(c_u/l_u)} = 1/z$$

or,
$$z = 1/(vc_u)$$

Note that the characteristic impedance z and velocity of propagation v are independent of line height and conductor dimensions for an overhead line in air, even though the inductance and capacitance per unit are functions of line dimensions. z and v depend solely on the properties of the medium, namely, μ_0 and ε_0. The above relations between z, v, l_u and c_u of an overhead transmission line in air are known as the "light-velocity theory" relations. There is no explicit need to calculate both l_u and c_u, since if we calculate one quantity the other can be calculated by using light velocity. This is extremely convenient when we have to calculate the line constants for a multi-conductor line.

Q.2. What are the equivalent circuits for wave propagation of: (*a*) Overhead transmission lines; (*b*) Cables; (*c*) Transformer windings; and (*d*) Generator windings. Neglect resistances in all cases.

Ans. Figure 14.1(*a*) shows the equivalent distributed-constant network of an overhead line and a cable. They are described by distributed series inductance l_u and shunt capacitance c_u (or c_g) per unit length. At one end, a surge such as lightning or due to switching originates and propagates to the farther end where the line or cable may be terminated by some equipment, chiefly a transformer an lightning arrester or an entire receiving station. In the case of cable, because of losses in the dielectric, there would normally be a resistance or shunt conductance in parallel with c_g.

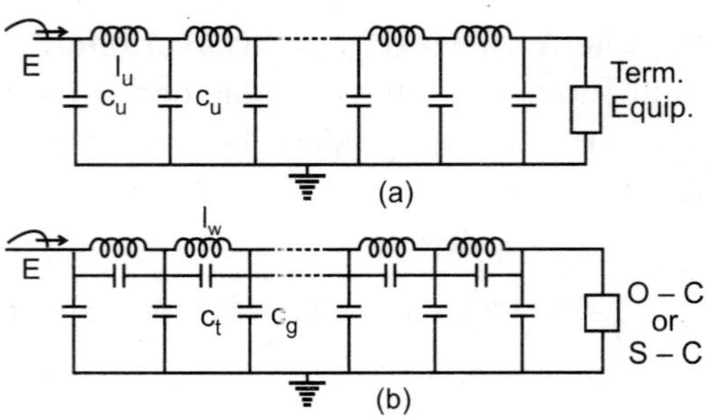

Fig. 14.1: (*a*) Equivalent distributed-parameter network of transmission line and cable. (*b*) Equivalent distributed-parameter network of transformer winding and generator of winding resistances neglected in both cases.

Figure 14.1(*b*) is the normally-accepted distributed-parameter equivalent circuit of a transformer winding or generator winding. The winding has shunt capacitance c_g to ground, a series inductance l_w and in addition a turn-to-turn capacitance c_t. All these are distributed and described as being per unit length of the winding. At one end, called the "entrance", a surge impinges on the winding, usually limited in magnitude by a lightning arrester. At the far end, which is the neutral point of the 3-phase *Y*-connected windings, the neutral may be solidly grounded or may be left open in order to connect protective equipment such as current transformers, neutral reactors, vacuum breakers, etc. This holds for both transformers and generators.

Q.3. A transmission line or cable with distributed series inductance and shunt capacitance to ground is energised suddenly by a constant source of voltage *E* at the source-point *S*. Neglect all resistances. Describe the events taking place when the voltage *E* and accompanying current $I = E/z$ travel on the line or cable. The line length is *L* and the velocity of travel of the voltage and the current surges is $v = 1/\sqrt{(l_u \cdot c_u)}$. Take the travel time for one line length to be $\tau = L/v$. (Note: In a cable because of material medium other than air, velocity v is not generally equal to light velocity $c \cdot v = c/\sqrt{\varepsilon_r}$. For example, for plastic insulation, $\varepsilon_r = 2.25$ so that $v = c/\sqrt{2.25} = c/1.5 = (2/3)c$.

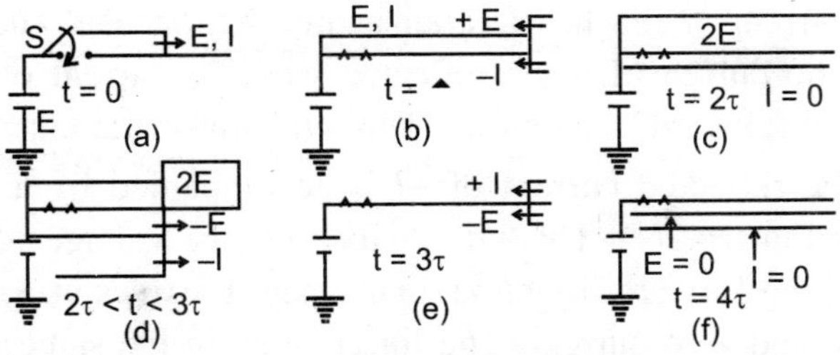

Fig. 14.2: Open-ended transmission line energised suddenly by constant voltage source E (shown as a battery). Sequence of travelling waves of voltage and current. Periodic time = 4 L/v.

Ans. Refer to Figs. 14.2 (*a*) to (*f*). The sequences are as follows for the propagation of the waves:

1. When switch S is closed on an unenergised line of length L, the voltage E and current $I = E/z$ which accompanies it travel on the line at the velocity v. At a time $\tau = L/v$, they reach the far end O. The voltage and current are assumed to be travelling in the "forward" direction from source to the far end O which is open for now, Fig. 14.2(*a*).
2. At the open end, there can be no current. Therefore, whatever current reaches O, reflects back to the source as $-I$. This is known as a "backward" travelling wave of current, see Fig. 14.2(*b*).
3. We will show from theoretical considerations later on that a backward travelling wave of current is accompanied by a backward travelling wave of voltage, but their polarities are opposite, see Fig. 14.2(*b*).
4. Therefore, the negatively-reflected current is accompanied by a positive-polarity voltage $+E = -I(-z)$. In this sense, one may imagine that the ratio of voltage to current for the backward travelling voltage and current is negative, or in other words, the surge impedance or characteristic impedance is $(-z)$.
5. When the wave $-I$ reaches the source after a time interval of $2\tau = 2L/v$ from the instant of initiation of the source voltage onto the line, the current in the line everywhere is zero. But the voltage is $+2E$ everywhere, see Fig. 14.2(*c*).
6. We have neglected attenuation of voltage and current magnitudes caused by the presence of ohmic losses, such as conductor resistance, ground-return resistance and shunt conductance such as the presence of corona losses or insulation losses in cable.
7. When the voltage $2E$ and current $-I$ reach the source end S, the source keeps the voltage at that end at E. Therefore, a negative wave of voltage $-E$ with accompanying current $-I$ will originate at S and travel down the line towards the open end in the "forward" direction. The voltage of $+2E$ combined with $-E$ will

make the voltage of the line equal to $+E$. But the current flowing in the line is $-I$, see Fig. 14.2(d).

8. When the current $-I$ reaches O after a time 3τ, because the end is open and there can remain no current there, a reflected wave of current of $+I$ flows back from O to S in the "backward" direction. This will make the current zero on the line.
9. However, the reflected current of $+I$ is accompanied by a voltage $-E$ of opposite polarity, see Fig. 14.2(e). This will "wipe out" the voltage $+E$ on the line which was travelling in the forward direction, and when it arrives at S, the entire line now has zero voltage and zero current. The line is now in the same initial unenergised state as it was at time $t = 0$, see Fig. 14.2(f).
10. Once again the source energises the "dead" line and the sequence repeats itself all over again. The phenomenon of voltage and current distribution due to travelling waves of voltage and current repeats with a periodic time of $4\tau = 4L/v$.

Thus, when all resistances are neglected, the voltage of the line never stabilises itself. The waves keep going forth and back indefinitely, theoretically for ever or for infinite time. In practice, however, the travelling waves decay due to dissipation of energy and the line acquires a steady-state voltage E, but with no current flow in an open-ended line.

Q.4. A transmission line of length L has distributed series inductance l_u henry/metre and shunt capacitance to ground of c_u farad/m. It is energised suddenly at one end by a constant voltage source E while the other end is terminated in a general impedance Z. ($Z = \infty$ for an open-ended line and $Z = 0$ for a short-circuited line to ground).

Derive the travelling-wave equations for the line for voltage and current. Solve the differential equations on the line. Show that the phenomenon consists of a "forward" set of voltage and current and a "backward" set of voltage and current at any point x on the line and time t. Also show that the ratio of voltage/current for the forward travelling waves is $+z = \sqrt{(l_u / c_u)}$ and for the reverse or backward travelling waves is $-z$.

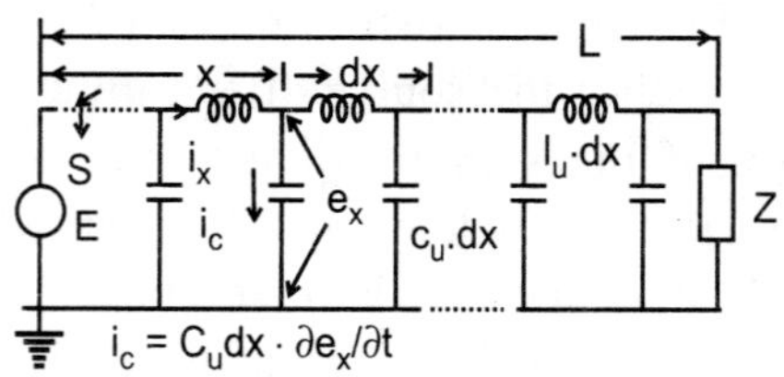

Fig. 14.3: Voltage and current relations over a differential length dx of a distributed-parameter transmission line or cable. Derivation of "transmission-line" equations. Resistance (losses) omitted.

Ans. This is a lengthy derivation but straightforward and the reader must be patient. In Fig. 14.3 is shown the source of voltage E, a switch S, terminating impedance Z for a line of length L and distributed constants l_u and c_u. We have $v = 1/\sqrt{(l_u / c_u)}$ and $z = \sqrt{(l_u / c_u)}$ · The travel time for one length is $\tau = L/v$.

Consider a differential length dx of the line at a distance x from the source end. The series inductance of this section is $l_u \cdot dx$, while the shunt capacitance is $c_u \cdot dx$. The current at x is I_x while the voltage is e_x to ground. The changes in voltage and current over the small length dx will be $(\delta e/\delta x) \cdot dx$ and $(\delta i/\delta x) \cdot dx$, respectively. We apply Kirchhoff's laws at x for the voltage and current as follows:

(*a*) The change in voltage is caused by the voltage drop in the series inductance.

Thus: $$[e_x + (\delta e/\delta x) \cdot dx] - e_x = -(l_u \cdot dx) \cdot \delta i/\delta t,$$

or, $$\delta e/\delta x = -l_u \cdot \delta i/\delta t \qquad ...(1)$$

[The negative sign indicates that as distance x is increased from the source end, the voltage e decreases].

(*b*) Similarly, the change in current is caused by the current drawn by the shunt capacitance.

∴ $$[i_x + (\delta i/\delta x) \cdot dx] - i_x = -(c_u \cdot dx) \cdot \delta e/\delta t$$

or, $$\delta i/\delta x = -c_u \cdot \delta e/\delta t \qquad ...(2)$$

Here, $\delta/\delta x$ and $\delta/\delta t$ are partial derivatives with respect to x and t, respectively.

(*c*) Differentiate (1) with respect to x, and (2) with respect to t:

$$\delta^2 e/\delta x^2 = -l_u \cdot \delta^2 i/\delta x \cdot \delta t \qquad ...(3)$$

and, $$\delta^2 i/\delta x \cdot \delta t = -c_u \cdot \delta^2 e/\delta t^2 \qquad ...(4)$$

Combining (3) and (4), we obtain:

$$\delta^2 e/\delta x^2 = l_u\, c_u \cdot \delta^2 e/\delta t^2$$
$$= (1/v^2) \cdot \delta^2 e/\delta t^2 \qquad ...(5)$$

Similarly, $$\delta^2 i/\delta x^2 = l_u\, c_u \cdot \delta^2 i/\delta t^2$$
$$= (1/v^2) \cdot \delta^2 i/\delta t^2 \qquad ...(6)$$

In order to solve a partial differential equation of this type, we assume a solution for e to consist of the product of two functions: one of them is a function of x only and the other a function of t only. Assume that $e(x, t) = X(x) \cdot T(t)$, where $X(x)$ is a function only of distance x, and $T(t)$ is a function of time t only. Substitute this in equation (5). Then,

$$\delta^2 e/\delta x^2 = T(t) \cdot \delta^2 X/\delta x^2$$

and $$\delta^2 e/\delta t^2 = X(x) \cdot \delta^2 T/\delta t^2$$

∴ $$T(t) \cdot \delta^2 X/\delta x^2 = (1/v^2) \cdot X(x) \cdot \delta^2 T/\delta t^2$$

or, separating the function of x only and t only, we have:

$$(1/X) \cdot \delta^2 X/\delta x^2 = (1/v^2) \cdot (1/T) \cdot \delta^2 T/\delta t^2$$

Since the left side is a function of x only and the right side is a function of t only, they must then be each equal to a constant K^2, say. Thus, the partial derivatives can be converted to total derivatives. We then obtain:

$$(1/X) \cdot d^2X/dx^2 = K^2$$

and $(1/v^2)\cdot(1/T) \cdot d^2T/dt^2 = K^2$

We can solve these two equations independently of each other:

$$d^2X/dx^2 = K^2 \cdot X$$

$\therefore$ $X = C_1 \exp(Kx) + C_2 \exp(-Kx);$

Also, $d^2T/dt^2 = K^2 \cdot v^2 \cdot T$

$\therefore$ $T = C_3 \exp(Kvt) + C_4 \exp(-Kvt)$

C_1, C_2, C_3 and C_4 are constants which will be determined by boundary conditions in x and initial conditions in t.

$\therefore$
$$\begin{aligned} e(x, t) &= X(x) \cdot T(t) \\ &= C_1C_3 \cdot \exp\{K(x + vt)\} + C_2C_4 \cdot \exp\{-K(x + vt)\} \\ &\quad + C_1C_4 \cdot \exp\{K(x - vt)\} + C_2C_3 \cdot \exp\{-K(x - vt)\} \end{aligned}$$

We now observe that the voltage e at any distance x and at any time t from the initiation of the voltage on the line consists of two types of terms:

1. A function of $(x + vt)$ and (2) A function of $(x - vt)$. We can combine the terms and write the complete solution as: $e(x, t) = f_1(x - vt) + f_2(x + vt)$.

   We now provide a physical interpretation of what these terms mean. They form travelling waves in the "forward" direction and "backward" direction, as seen from the explanation given below:

   Consider the term $f_1(x - vt)$. When time t is increased i.e., as time progresses, the function remains constant in magnitude only when $x - vt$ = constant. This means that as time t is increased, x must also be increased. The velocity is $dx/dt = + v$. This function is therefore travelling in the "forward" direction.

   Next consider the term $f_2(x + vt)$. When time t is increased, the constant value is found at a negative value of x in order to maintain $x + vt$ = constant,

   or,

   $$dx/dt + v = 0$$

   $\therefore$ $dx/dt = -v$

   The velocity of propagation is negative, or, the wave is travelling backwards.

   Therefore, we have arrived at the property that the behaviour of voltage at any point on the line and at any time t is a combination of a forward-travelling component $f_1(x - vt)$ and a backward-travelling component $f_2(x + vt)$ with velocities $+v$ and $-v$, respectively.

   In a similar manner, the current can also be written as:

   $$i(x, t) = g_1(x - vt) + g_2(x + vt)$$

   where g_1 and g_2 are the forward-and backward-travelling components of current. The functions g_1 and g_2 are related to f_1 and f_2 of the voltage and are not independent, as shown below:

Consider equation (1): $\delta e/\delta t = -l_u \cdot \delta i/\delta t$

$\therefore \quad f_1(x - vt) + f_2(x + vt) = -l_u \{-v \cdot g_1(x - vt) + v \cdot g_2(x + vt)\}$

Equating corresponding components on the two sides, we have:

$$f_1(x - vt) = l_u \cdot v \cdot g_1(x - vt),$$

or, $$f_1/g_1 = l_u \cdot v = \sqrt{(l_u / c_u)} = +z$$

$$f_2(x + vt) = -l_u \cdot v \cdot g_2(x + vt),$$

or, $$f_2/g_2 = -\sqrt{(l_u / c_u)} = -z,$$

upon using the relation $v = 1/\sqrt{(l_u / c_u)}$. Thus, we have shown that:

(*a*) The forward-travelling component of voltage is accompanied by a forward-travelling component of current. Their ratio is $+z$.

(*b*) The backward-travelling component of voltage is also accompanied by a backward-travelling component of current. But their ratio is $-z$.

The voltage and current can now be written as:

$$e(x, t) = f_1(x - vt) + f_2(x + vt),$$

and $$i(x, t) = (1/z) \cdot f_1(x - vt) - (1/z) \cdot f_2(x + vt)$$

Q.5. On a real transmission line or cable, the conductor has a series resistance of r ohm/metre. Show that the voltage and current waves, in travelling a distance x or a time $t = x/v$, attenuate by a factor $\exp\{-(r/2l_u)t\} = \exp\{-(r/2z)x\}$.

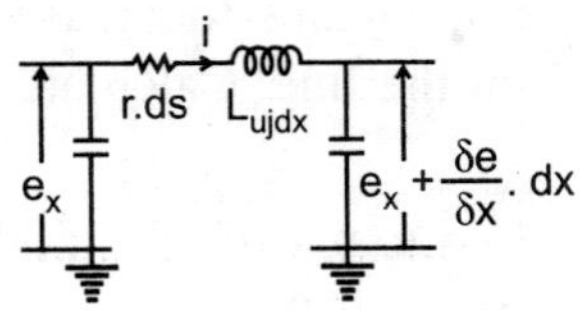

Fig. 14.4: Line section of length dx. Voltage and current relations. Resistance of line considered.

Ans. With resistance included, the equivalent circuit of the line over a differential length dx is shown in Fig. 14.4. The differential equations for voltage and current will now become:

$$\delta e/\delta x = -r \cdot i - l_u \cdot \delta i/\delta t$$

and $$\delta i/\delta x = -c_u \cdot \delta e/\delta t$$

$\therefore$ $$\delta^2 e/\delta x^2 = -r \cdot \delta i/\delta x - l_u \cdot \delta^2 i/\delta x \cdot \delta t$$

$$= rc_u \cdot \delta e/\delta t + l_u c_u \cdot \delta^2 e/\delta t^2$$

Once again, let $e(x, t) = X(x) \cdot T(t)$.

Then we obtain two separate equations for $X(x)$ and $T(t)$ to be:

$$(1/X) \cdot d^2X/dx^2 = K^2 = r \cdot c_u \cdot (1/T) \cdot dT/dt + l_u \cdot c_u \cdot (1/T) \cdot d^2T/dt^2$$

$\therefore$ $$X(x) = C_1 \cdot \exp(Kx) + C_2 \cdot \exp(-Kx), \text{ as before.}$$

Then, $d^2T/dt^2 + (r/l_u) \cdot dT/dt = (K^2/l_u \cdot c_u) \cdot T = K^2 v^2 T$

Denoting $d/dt = D$, we obtain the auxiliary equation:

$$D^2 + (r/l_u)\cdot D - K^2v^2 = 0.$$

Its two roots are:

$$D_1, D_2 = -(r/2l_u) \pm [(r/2l_u)^2 + K^2v^2]^{1/2}$$

When $(r/2l_u) << K^2v^2$,

we can write $D = -(r/2l_u) \pm Kv$

Then, $T(t) = \exp\{-(r/2l_u)\, t\} \cdot [C_3 \cdot \exp(Kvt) + C_4 \cdot \exp(-Kvt)]$

This is similar to the solution obtained in Question 4 earlier when r was not taken into consideration. Therefore, we can write:

$$e(x, t) = \{\exp(-r/2l_u)t\}/[f_1(x - vt) + f_2(x - vt)]$$

These are still travelling waves but attenuated by the factor

$$\exp(-r/2l_u)t$$

Now, $t = x/v$

$\therefore$ $(r/2l_u)t = (r/2l_u) \cdot x/v$

$= (r/2z)x$

since $z = l_u \cdot v$

$\therefore$ The attenuation factor $\alpha = \exp(-r/2l_u)t = \exp(-r/2z)x$

Example 1: A travelling-wave attenuates (decreases in amplitude) to 80% of its initial value after a travel of 200 km on the line. Calculate the attenuation factor per km length of line. What are its units?

Solution: $E_{200}/E_0 = 0.8 = \exp\{(-r/2z) \cdot 200\}$

$\therefore$ $\exp\{(r/2z) \cdot 200\} = 1.25 = e^{0.223}$

This gives $200 \cdot r/2z = 0.223$

and $r/2z = 1.1157 \times 10^{-3}$ Nepers/km

If the surge impedance of the line is $z = 400$ ohms, calculate the line resistance per km.

$$r = 2 \times 400 \times 1.1157 \times 10^{-3}$$

$$= 0.8926\ \Omega/\text{km}$$

Example 2: A 400-kV transmission line has a resistance of $r = 0.034$ ohm/km and inductance of $l_u = 1$ mH/km. Calculate the attenuation of a surge on the line after travelling 600 km. Assume light velocity. What is the surge or characteristic impedance z of the line?

Solution: $\alpha = \exp\{(-r/2l_u)t$

and $t = x/v = 600/(3 \times 10^5) = 2$ ms

$\therefore$ $\alpha = \exp\{-(0.034/0.002)\times 2 \times 10^{-3}\}$

$= e^{-0.034}$

$= 0.9666$

The voltage surge attenuates to 0.9666 or 96.66% of its initial value after a travel of 600 km.

$$z = v \cdot l_u = 3 \times 10^5 \times 0.001 = 300 \text{ ohms}$$

The resistance given above is a typical value at 50 Hz and the attenuation of the wave is very low. The surge attenuates by less than 4% over a travel of 600 km. This is typical of a switching surge or a travelling wave initiated by a fault on the line since their frequencies are low. However, lightning surges rise to crest value in about 2 μs, which can be considered as one-quarter cycle with periodic time of 8 μs and equivalent frequency of ($1/8 \times 10^{-6}$) = 125 kHz. The skin-effect resistance of the conductor will be very high compared to the d.c. or power-frequency value of resistance so that the lightning surge will attenuate to low values. However, a more important contributing factor is the corona loss since the lightning voltage will generally be much higher than the corona-inception voltage for the conductors. These problems are not discussed here.

Q.6: A wave of voltage is travelling on a line with surge impedance Z_1 when it encounters another line with surge impedance Z_2. $Z_1 \neq Z_2$. Explain what happens to the voltage at the junction of the two lines. What are: (*a*) Reflection Coefficient, and (*b*) Refraction or Transmission Coefficients in terms of Z_1 and Z_2 for the voltage and current?

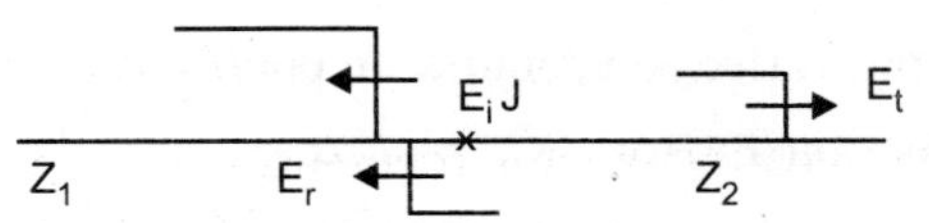

Fig. 14.5: Transmission line with surge impedance Z_1 joined to line with surge impedance Z_2. Incident, reflected and transmitted voltages at junction J.

Ans. Refer to Fig. 14.5. The incident voltage E_I meets the junction J, and since $Z_1 \neq Z_2$, a part of the voltage E_r is reflected back on the incoming line of surge impedance Z_1 and the remaining part E_t travels forward on the outgoing line of surge impedance of Z_2. The voltage at the junction is E_t, which is equal to the transmitted part.

We have observed that the voltage at any point of a line is the sum of a forward-travelling wave and a backward-travelling wave. Also, the ratio of voltage/current for the forward-travelling wave is $+z$, whereas for the backward-travelling waves it is $-z$. Using these properties, we can set up the following equations:

1. $E_i/I_I = Z_1,\ E_r/I_r = -Z_1,$

   and $E_t/I_t = +Z_2,$

   where the subscripts I, r, and t stand for incident, reflected and refracted or transmitted quantities.

2. $E_I + E_r = E_t;$

   $I_I + I_r = I_t;$

   or, $E_i/Z_1 - E_r/Z_1 = E_t/Z_2$

3. ∴ $E_I - E_r = E_t \cdot (Z_1/Z_2)$

This gives:

$$2E_I = (1 + Z_1/Z_2) \cdot E_t$$
$$= \{(Z_1 + Z_2)/Z_2\} \cdot E_t$$

Consequently, the junction voltage as well as the voltage which travels onwards on the line with surge impedance Z_2 is:

$$E_t = E_i \cdot 2Z_2/(Z_1 + Z_2), \text{ the transmitted voltage.}$$

Substituting this value in (2), the reflected voltage is

$$E_r = E_t - E_I$$
$$= E_i \cdot (Z_2 - Z_1)/(Z_2 + Z_1)$$

Similarly, $I_r = -E_r/Z_1$

$$= -(1/Z_1) \cdot (Z_2 - Z_1) \cdot Z_1 \cdot I_i/(Z_2 + Z_1)$$
$$= I_i \cdot (Z_1 - Z_2)/(Z_1 + Z_2)$$

Also, $I_t = E_t/Z_2$

$$= (1/Z_2) \cdot 2Z_2 \cdot Z_1 \cdot I_i/(Z_1 + Z_2)$$
$$= I_i \cdot 2Z_1/(Z_1 + Z_2)$$

We now obtain the reflection factors or coefficients as follows:

(*a*) Voltage Reflection Factor: $K_{re} = E_r/E_I$

$$= (Z_2 - Z_1)/(Z_2 + Z_1);$$

Current Reflection Factor: $K_{ri} = I_r/I_I$

$$= (Z_1 - Z_2)/(Z_1 + Z_2) = -K_{re};$$

(*b*) Voltage Transmission Factor: $K_{te} = E_t/E_I$

$$= 2Z_2/(Z_2 + Z_1)$$

Current Transmission Factor: $K_{ti} = I_t/I_I$

$$= 2Z_1/(Z_2 + Z_1)$$

These coefficients or factors have very great importance and utility in high voltage work. We will now consider their applications through several important practical examples. Note that these same factors apply if the line with surge impedance Z_2 is replaced by a lumped resistance, $R_2 = Z_2$.

Q.7. A transmission line is open at the far end and a travelling wave impinges on the line at the other end. What are the reflection and refraction coefficients or factors for voltage and current at the open end? What are the junction voltage and current?

Ans. Refer to Fig. 14.6(*a*). For the open end, $Z_2 = \infty$.

∴

$$K_{re} = \{(Z_2 - Z_1)/(Z_2 + Z_1)\}\Big|_{Z_2 = \infty} = +1;$$

$$K_{te} = \{2Z_2/(Z_2 + Z_1)\}\Big|_{Z_2 = \infty} = +2$$

$$K_{ri} = -K_{re} = -1$$
$$K_{ti} = \{2Z_1/(Z_1 + Z_2)\}\Big|_{Z_2 = \infty} = 0$$

Conclusions: At an open circuit

1. The junction voltage, (at the open circuit), is twice the incident voltage;
2. The reflected voltage is equal to the incident voltage and of the same polarity. $E_t = E_I + E_r = 2E_I$.
3. The current at the open circuit = 0;
4. The reflected current at the open circuit is the negative of the incident current: $I_r = -I_I$. $\therefore$ $I_t = I_I + I_r = 0$.

This happens when there is a broken conductor or the line is terminated in the high voltage winding of a transformer whose high value of inductance presents an open-circuit at the instant of incidence of a lightning or other type of surge.

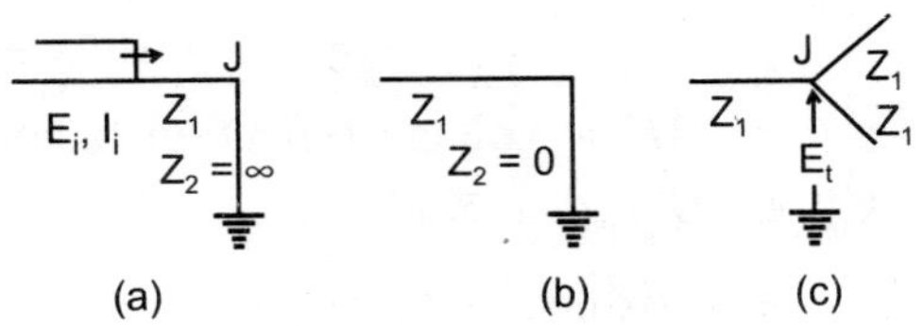

Fig. 14.6: (*a*) Open-circuited line; (*b*) Short-circuited line; (*c*) Transmission line connected to two similar lines at junction J. To obtain relations between E_I, E_r, E_t and I_I, I_r, I_t.

Q.8. What are the conditions of voltage and current if there is a short-circuit at the junction *J*?

Ans. Refer to Fig. 14.6(*b*). For a short circuit, $Z_2 = 0$.

$$\therefore \quad K_{re} = \{(Z_2 - Z_1)/(Z_2 + Z_1)\}\Big|_{Z_2 = 0} = -1;$$
$$K_{te} = \{2Z_2/(Z_2 + Z_1)\}\Big|_{Z_2 = 0} = 0$$
$$K_{ri} = -K_{re} = +1;$$
$$K_{ti} = \{2Z_1/(Z_1 + Z_2)\}\Big|_{Z_2 = 0} = +2.$$

Conclusions: At a short circuit

1. The voltage is zero;
2. Current in short circuit = 2 × current in the incident wave;
3. The reflected current is equal to the incident current and of the same polarity.

$$I_r = I_I$$
$$\therefore \quad I_t = I_r + I_I = 2I_I$$

4. The reflected voltage at a short circuit is equal to the incident voltage but of opposite polarity.

$$E_r = -E_I$$

$$\therefore \quad E_t = E_r + E_I = 0$$

This situation occurs when a conductor breaks and touches a tower and falls on ground. Travelling waves are initiated at the grounded fault which travel back to the source where there are protective relays. This is the modern method of protection which operates circuit breakers to open the line. This is known as "Travelling-Wave Protection" scheme.

Q.9. A transmission line of surge impedance Z_1 has a junction from which two similar lines branch out. What are the reflected and transmitted voltages and currents in the three lines?

Ans. Refer to Fig. 14.6(*c*). The incident voltage E_I and current I_I are travelling towards the junction J on a line with surge impedance Z_1. At the junction, two similar lines each with surge impedance Z_1 are connected. Thus, $Z_2 = Z_1/2$, since they are in parallel.

$$\therefore \quad K_{re} = \{(Z_2 - Z_1)/(Z_2 + Z_1)\} \quad \Big| \quad = (0.5 - 1)/(0.5 + 1)$$

$$= Z_2 = Z_1/2$$

$$= -1/3 = \text{voltage reflection factor}$$

$$K_{te} = 2Z_2/(Z_1 + Z_2)$$

$$= Z_1/(Z_1 + 0.5Z_1) = 2/3$$

Also,

$$K_{te} = 1 + K_{re} = 2/3$$

$$K_{ri} = (Z_1 - Z_2)/(Z_1 + Z_2)$$

$$= 1/(2 + 1) = 1/3$$

$$= -K_{re}$$

$$K_{ti} = 2Z_1/(Z_1 + Z_2)$$

$$= 2/(1 + 0.5)$$

$$= 4/3$$

$$= 1 + K_{ri}$$

The junction voltage has fallen to 2/3 the incident voltage while the reflected voltage is (–1/3) times the incident voltage.

$$E_t = E_I + E_r = (1 - 1/3)E_I = (2/3)E_I$$

The current at the junction is 4/3 times the incident current while the reflected current is 1/3 of the incident current and of the same polarity. One half of the transmitted current flows in each of the connected lines beyond the junction J.

Such instances occur at a receiving station where there may be only one incoming line but several outgoing lines of the same geometry. This brings the station bus-bar voltage to a value lower than the incident voltage. It offers a certain amount of built-in protection against high lightning voltages. However, suitable lightning arresters will always be present at receiving stations and at generating stations.

Q.10. A surge of 1000 kV peak value is travelling on an overhead transmission line with $z = 300$ ohms into a receiving station where it is connected to a high voltage cable having a surge impedance of 30 ohms. The other end of the cable is connected to the high voltage terminal of a transformer winding which can be considered as an open circuit ($Z_T = \infty$). What are the voltages at the entrance junction of the cable and at the junction of the cable and transformer?

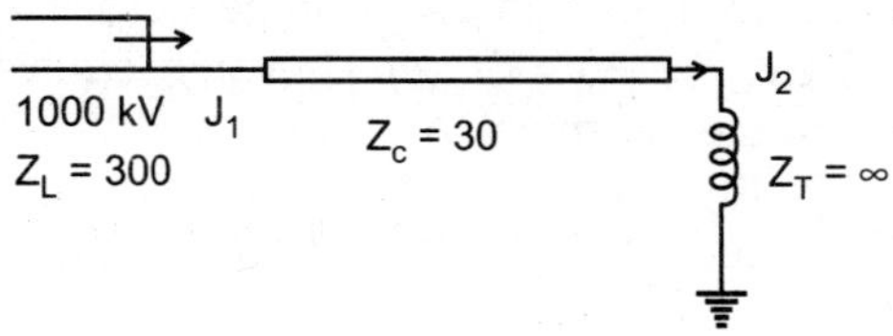

Fig. 14.7: Transmission line of 300 ohms surge impedance connected to high voltage terminal of transformer winding ($Z_T = \infty$), through a length of cable with $Z_C = 30$ ohms.

Ans. Refer to Fig. 14.7.

1. At junction $J_1 : E_t = E_i \cdot 2Z_C/(z + Z_C)$

$$= 1000 \times 2 \times 30/(300 + 30)$$

$$= 182 \text{ kV}$$

2. This transmitted voltage into the cable travels to junction J_2 where it meets $Z_T = \infty$, an open circuit. Therefore, the incoming voltage of 182 kV doubles at the open circuit. Therefore, voltage at transformer terminal = 364 kV.

Q.11. In the previous problem, suppose there is no cable interposed between the line and the transformer. What then would be the voltage at the high voltage terminal of the transformer winding?

Ans. Since $Z_T = \infty$, the incoming voltage of 1000 kV on the line will become double at the transformer terminal.

$\therefore$ voltage = 2000 kV

Note that with the 30-ohm cable in, the voltage at the transformer terminal was only 364 kV. Such protection schemes are used in many situations to reduce transformer insulation levels.

Q.12. A surge of 900 kV is travelling on a line with $z = 300$ ohms. It encounters a lightning arrester at a station which holds the voltage across itself at a constant value of 600 kV. What is the current flowing through the arrester?

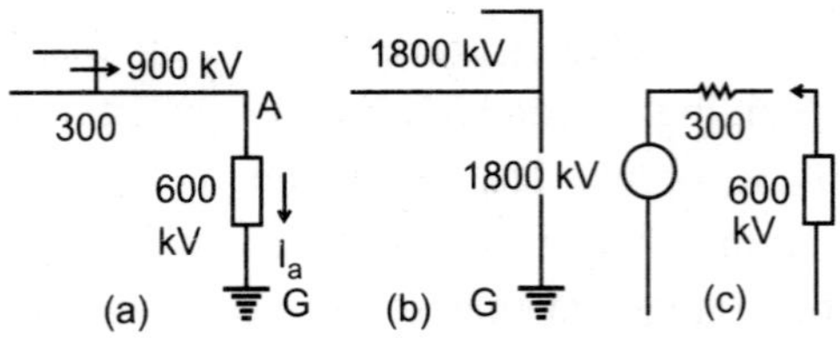

Fig. 14.8: (*a*) Arrester with constant voltage of 600 kV connected to a transmission line with z = 300 Ω. (*b*) Thevenin o.c. voltage across arrester terminals A – G · $V_{th} = 2V_L$ = 1800 kV (*c*) Thevenin equivalent circuit across arrester terminal A – G · $I_A = (2V_L - V_A)/z$.

Ans. We use Thevenin's Method for solving this problem. Refer to Fig. 14.8. The following steps have to be used:

Step 1: Open the arrester terminals *A–G*, in which we wish to calculate the current flow. Fig. 14.8(*b*).

Step 2: Determine the voltage across the opened terminals.

This is 2 × 900 kV = 1800 kV.

Step 3: Looking through the open terminals *A–G*, the impedance of the line = 300 ohms.

Step 4: The resulting Thevenin equivalent circuit is shown in Fig. 14.8 (*c*). It consists of a voltage source of 1800 kV connected to an impedance of 300 ohms in series up to *A–G*.

Step 5: Now connect the arrester across *A–G*. The current flowing through the arrester is then:

$$I_A = (1800 - 600)/300 \text{ kA} = 4 \text{ kA}$$

[**Note:** At this instant, the arrester material has a resistance of 600 kV/4 kA = 150 ohms. We can then see that the current coming on the line through 300 ohms is I_L = 900/300 = 3 kA. The current-transmission factor into the arrester is:

$$\begin{aligned} K_{ti} &= 2Z_L/(Z_L + Z_A) \\ &= 2 \times 300/(300 + 150) \\ &= 4/3 \end{aligned}$$

∴ Current in arrester, $I_A = K_{ti} \cdot I_I$ = (4/3)× 3 kA = 4 kA, as before.]

In general, let V_L = voltage travelling on line,

V_A = constant value of voltage maintained by arrester, and

z = surge impedance of line.

Then, with arrester removed, open-circuit voltrage = $2V_L$.

Thevenin equivalent impedance looking through the arrester terminals = z

∴ Current flowing in arrester $I_A = (2V_L - V_A)/z$

❑❑

CHAPTER 15

PROTECTION OF HIGH VOLTAGE SYSTEMS

Q.1. What are the types of protection required in high voltage systems?

Ans. These can be categorised as follows:

(*a*) Overcurrent protection; (*b*) Overvoltage protection; (*c*) Short-circuit protection; (*d*) Ground-fault protection; (*e*) Surge protection; (*f*) Cathodic protection; and (*g*) Overload or over-heating protection. Some of these may overlap.

Q.2. What are the types of equipment and lines for which protection has to be offered?

Ans. (*a*) Generators; (*b*) Cables; (*c*) Transformers; (*d*) Communication systems; (*e*) Transmission lines.

Q.3. What are the different components forming protective schemes?

Ans. They are very numerous but can be categorised as follows:

(*a*) Current transformers; (*b*) Potential transformers; (*c*) Relays; (*d*) Circuit-breakers; (*e*) Arresters or Surge Absorbers; (*f*) Thermocouples for temperature detection and hot spots; (*g*) Carrier Protection Equipment; (*h*) Travelling-wave Detectors; (*i*) Fibre optical cables.

Q.4. A generator (or a generating station with several generators operating in parallel) is connected to a bus to which is connected a transformer (or several transformers). The h.v. winding(s) is (are) connected to the h.v. bus from which a transmission line (or lines) is (are) connected. What are the types of protective schemes adopted? Show them on a single-line diagram.

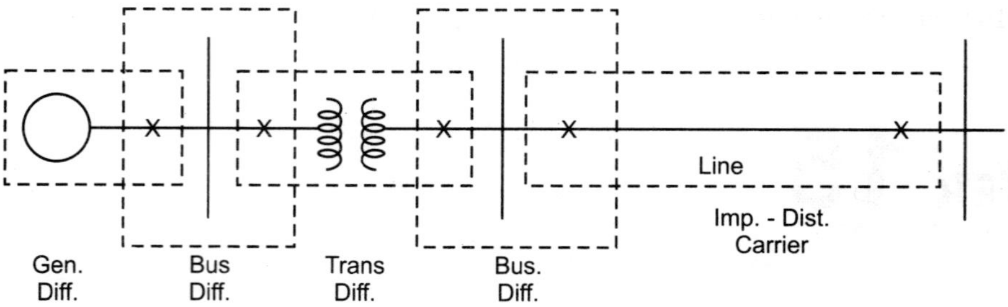

Fig. 15.1: Schematic diagram of basic types of protection used in a simple system. (1) Generator Differential Protection; (2) L.V. Bus Differential Protection; (3) Transformer Differential Protection; (4) H.V. Bus Differential Protection at Generating Station; (5) H.T. Line: Impedance-DistanceProtection. x-Circuit Breakers.

Ans. Refer to Fig. 15.1. Except for the long high voltage transmission line, the generator low-voltage bus, transformer, h.v. bus are protected by differential schemes. Here, essentially, CTs are connected on both sides of the winding or bus. Any internal fault is indicated by difference in current flow on the two sides. This is sensed by a relay which operates to close electrical contacts and operate the proper circuit breaker.

For a fault on the long line, the change of impedance caused to normal impedance is sensed and relay operation occurs. Additionally, signals can also be sent to carrier equipment on the high voltage line-conductors themselves. The relays operate by a stoppage of the signals caused by the fault. The most modern method is to use the travelling waves initiated by the fault and detecting them suitably at the two ends of the line.

Q.5. What are the types of protection used for generators?

Ans. These are:

1. Differential-current protection;
2. Over-voltage protection;
3. Under-voltage protection;
4. Under-or over-frequency protection;
5. Over-speed protection;
6. Loss of field protection;
7. Current-reversal protection;
8. Over-load or temperature-rise protection;
9. Surge protection. All of these may not be adopted at any one generator but must be provided for to some degree.

Q.6. How are the above protective schemes carried out?

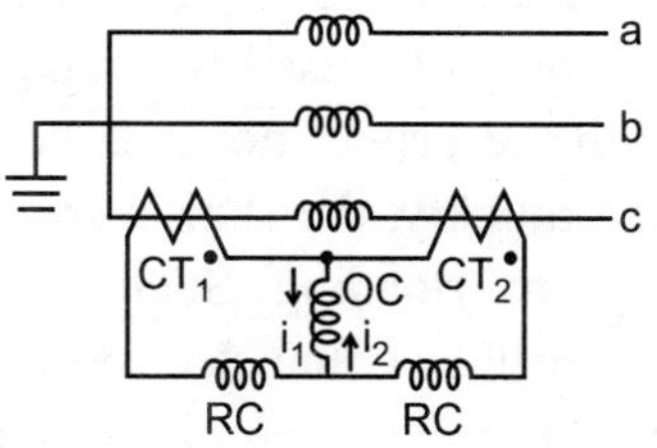

Fig. 15.2: Differential Protection of Generator. OC–Operating coil; RC–Restraining coil for percentage differential protection.

Ans. 1. **Differential-current Protection:** Refer to Fig. 15.2. Two CTs are connected on the two sides of a phase winding. Only one phase is shown in the Figure. Normally, if the currents on the two sides of the phase-winding are equal, the operating coil has no current flowing through it, if it is connected to the midpoint of the coil shown as "restraining coil", RC. If there is a breakdown of insulation inside the winding to ground, the currents in the two CTs become unequal so that a current flows through the operating coil. The voltage across it is used to operate a relay which opens or "trips" a generator circuit breaker taking the machine out of the system.

Some schemes use a "percentage differential protective scheme". In this scheme, the operating coil is connected to the restraining coil with unequal resistance values for the two sides. In such a scheme, a prescribed percentage difference in current on the two sides will cause sufficient current flow through the operating coil to trip the breaker. When no percentage differential is used it is known as a "straight-through" differential protection.

In the case where the neutral of the generator is not grounded, a PT is inserted between the neutral point and ground. Under balanced and healthy state of operation of the generator, there exists no voltage between neutral and ground. However, under an internal fault in one winding, the neutral voltage rises above ground potential which is sensed by the PT and this is used for operating a relay to trip the circuit breaker. Most large generators are solidly grounded for reasons of stability and surge protection so that they use differential CT protection only.

2. **Overvoltage and Undervoltage:** This type of operation is harmful to the system. The PTs on the generator terminal used for voltage measurement are arranged to feed proper relays to open the circuit breaker.

3. **Frequency Protection:** Over-frequency protection is provided along with over-speed protection described below. However, frequency-measuring meters can be arranged to provide operating signals to circuit breakers. Under-frequency operation is very dangerous to a machine or a system since this is caused by severe overloading condition. A generator or a generating station connected to a large system is controlled for its load condition from a central Load-Dispatching Station or Centre

for the system as a whole. In case of normal operation of the system, load requirements change and a generating station is required to take its share of the load, then its frequency must be raised by changing the speed-droop characteristics of the generator (through the prime mover governor). Its frequency must be raised. If no control is provided quickly, the station-frequency is lowered when load comes on. This usually leads to overloading. Such control methods are known as "Load-Frequency Control" of an inter-connected system or a "grid".

4. **Overspeed Protection:** Modern electronic frequency counters can be used to open circuit breakers under overspeed condition which raises the frequency of the generator. This may occur when the mechanical governor of the turbine or prime mover fails. But in most machines a centrifugal switch is provided inside a hollow shaft. A small weight is attached to the shaft in a hollow which will fly outward from its normal position under overspeed. This can be made to close an electrical circuit to remove the generator from the system.
5. **Loss-of-Field Protection:** In a single generator connected to a bus, any loss of d.c. field excitation due to inadvertent operation of the field switch usually causes loss of voltage. An under-voltage relay can therefore be used to cut the machine out of the bus bar. However, where there are several machines operating in parallel, the loss of d.c. field of one machine makes that machine operate as an induction motor causing reversal of current in the generator stator winding. A directional relay can therefore detect this and protect the machine by removing the machine from the bus.
6. **Current Reversal Protection:** The output of the CT on the line is connected through a rectifier and to a relay. Current reversal can be detected easily and protection offered.
7. **Overload or Temperature-rise Protection:** The main CTs can be used to operate overcurrent relays if the generator current exceeds a certain prescribed value of overload. Normally, the synchronous reactance of a generator is nearly 1 per unit so that on a 3-phase fault, the current is still not in excess of the rated current of the machine in the steady state. Thus, this type of fault will not affect the overload relay and protection scheme. A 3-phase symmetrical fault can be protected against only by under-voltage relays since the output voltage of the generator is zero at its terminals.

   Overload on a machine for some length of time results in excessive rise in temperature of the windings. Thermocouples are embedded in some of the coils inside the generator which measure the temperature under all conditions of operation. These are used for providing overload protection also.
8. **Surge Protection:** Lightning arresters or surge absorbers are used for surge protection under lightning in the same manner as is provided for a transformer or

transmission line. These arresters for generator protection are designed for low voltages for the voltage levels prescribed under Standard Specifications.

Q.7. What are the types of protection offered to transformers? Describe their operation.

Ans. These are: (1) Overload protection; (2) Fault detection; and (3) Surge protection.

1. **Overload Protection:** A thermal relay immersed in transformer oil is used. This relay is energised by the output of a CT which detects the overcurrent and operates a circuit breaker or else sound an alarm, since in most cases the time-constant for heating of the transformer is quite long. Transformers use various types of cooling methods. These are:
   (*a*) Natural cooling for small transformers where sufficient amount of oil or insulating fluid is used and the tank has sufficient area to cool the insulant.
   (*b*) Forced cooling where radial fins are attached to the tank in the form of pipes. These provide a natural circulation for the oil and offer sufficiently large area for cooling. The oil can also be led into a heat-exchanger where a water jacket surrounds the pipe. Water pumps must be used and any fault in the circulation pumps must be made to sound an alarm.
   (*c*) For larger transformers very large fins or radiators are attached in which the oil circulates and cooling fans are attached which constantly blow air on the radiators to cool the oil. Any fault in the cooling system must be made to sound an alarm.
2. **Internal Fault Detection:** These are of both electrical and mechanical types.
   (*a*) **Differential Protection:** This is similar in principle to generator differential protection, but will vary according to transformer-winding connections. They depend upon whether they are connected in *Y*-*Y*, *Y*-Delta, or Delta-Delta. High voltage high-power transformer are mostly connected in *Y*-Delta for neutral stabilisation by eliminating 3rd harmonic currents. The Delta-connected winding provides a short-circuit for the 3rd harmonic current and eliminates them in the lines as well as on the *Y*-side.

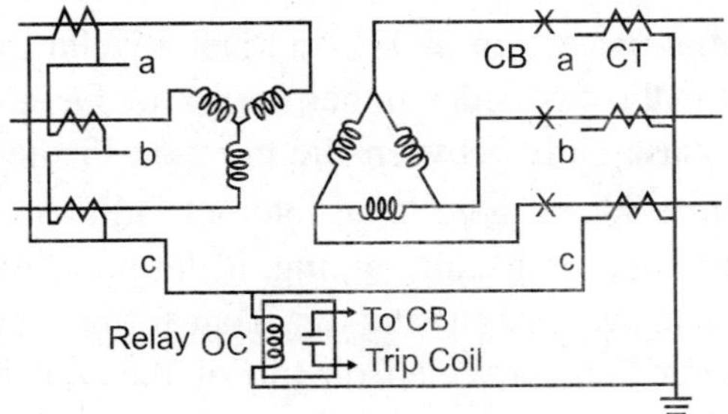

Fig. 15.3: Differential protection for Y-Delta transformer. OC-Operating Coil.

Figure 15.3 shows complete circuit diagram of the differential protection scheme of a *Y*-Delta transformer. Because there is a phase difference of 30° existing between line currents on the two sides, they must be brought back in phase in the secondaries of the CTs connected to the operating coil of the relay. Only one phase is shown for the operating coil connection of the secondaries of the CTs in the Figure. Note that on the Delta-side, the three secondaries of the CTs are connected in *Y*. Also, the secondaries of the CTs on the *Y*-connected winding of the main transformer are connected in Delta.

A percentage differential protection can be used, just as for the generator differential protection scheme, by connecting the operating coil of the relay in each phase to a restraining coil RC, which is not shown in Fig. 15.3. Only straight-through differential arrangement is shown.

In India, very serious faults occur to the insulation of transformers quite frequently in both urban distribution and long-distance transmission as well as to lightning arresters. We cannot discuss the reasons for all these failures here. In urban distribution up to even 66 kV, when power is led from receiving stations on the outskirts of a major city to distribution substations inside the metropolis, because of the need to eliminate overhead lines strung on poles, the entire network has gone "underground" involving very long lengths of high voltage and medium-voltage cables. These have much higher capacitance than overhead lines. In case the circuit-breaker poles, when switching these cables (or what is popularly known as "load shedding") do not close simultaneously, the cable capacitance forms a series circuit with the transformer inductance causing series resonance giving rise to dangerously high voltages across the insulation of both the transformer inductance and the cable capacitance. These situations are very common in everyday experiences in large cities like Mumbai, Delhi, Kolkata and in very large steel plants using hundreds of kilometres of cable. On long transmission lines, switching of synchronous condensers at receiving stations has also caused severe damages to transformer insulation to whose winding (usually the low-voltage tertiary of 3-winding transformer) the large synchronous motor is connected. Thus, transformer insulation failures are caused by improper operation of circuit breakers. So also the cable insulation. Thus, circuit breakers must be maintained in first-class condition.

Differential protection of *Y-Y* and Delta-Delta transformers have the secondaries of the CTs for the two sides either in *Y-Y* or Delta-Delta and not in *Y*-Delta, since there is no phase shift between the currents in the lines on the two sides.

(*b*) **Gas Detection:** When there is an internal fault to the insulation of the windings of a transformer, there is usually arcing at the point of failure which gives rise to gas from the oil. This gas rises to the top. Transformers not provided with external coolers have a breather tank provided on top of the transformer tank where the oil level changes due to temperature changes. The gas bubbles collect in this breather tank since they rise above the oil. A float valve using mercury contacts is tilted from its normal position to sound an alarm or operate a relay to open a circuit breaker. This is called the Bucholtz Relay or system of protection.

3. **Surge Protection:** Lightning arresters are used for protecting transformer windings, as has been described in detail in Chapter 9 to which the reader is referred. These are not repeated here.

Q.8. What are the main characteristics of an induction-type overcurrent relay?

Ans. This type of electro-mechanical relay is slower-acting than a solid-state relay using transistors. The relay has an aluminium disc capable of rotation on a vertical shaft or spindle with two e.m. coils one on top of the other with a gap separating the two. The disc rotates in this gap between the poles of two coils. A permanent magnet is also provided in whose gap the disc rotates whose function is to adjust the operating speed of the disc and hence the operating time of the relay. The disc is normally prevented from rotating by the force exerted by a restraining spring. As soon as the current in the coils exceeds a desired value, as set by the tension on the spring and the position of the permanent magnet, the disc develops sufficient torque to rotate against the restraining force. An armature attached to the shaft can be made to close a normally-open pair of electrical contacts or open a normally-closed pair. These contacts then control the current flow into the operating solenoid or other operating mechanism to trip the circuit breaker. Once the relay has operated, it has to be manually reset by bringing the disc back to its original position for the next sequence of operation.

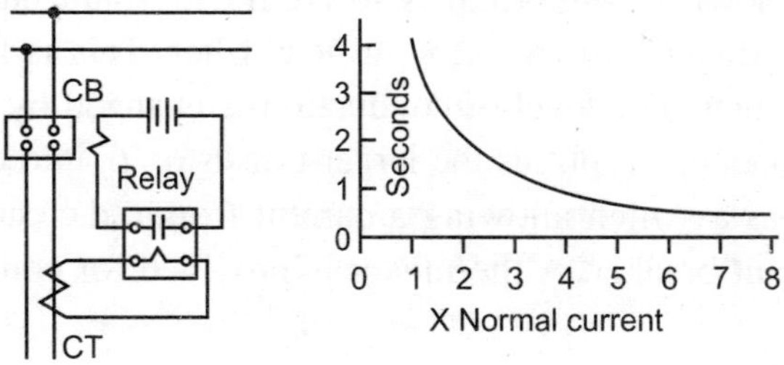

Fig. 15.4: Inverse-time characteristic of a typical induction-type electro-magnetic overcurrent relay.

The most important electrical characteristic of this type of relay is known as the "inverse-time" characteristic. This is shown in Fig. 15.4 from which one can observe that as the current in the operating coil (which is proportional to the fault current in the system as sensed by CTs either in the circuit-breaker bushings or separately located) becomes higher, the operating time becomes shorter. Thus, more severe the fault, quicker will the relay clear the fault. However, a minimum time is required for all such electro-mechanical relays to operate even if the current is increased to very high values. The inverse-time characteristic can be varied by changing the mechanical position of the braking magnet, i.e., the permanent magnet. Thus, similar relays can be used at different locations for coordinating operating times of circuit breakers.

Q.9. How does a solid-state overcurrent relay operate?

Ans. The e.m. induction-type relay takes several cycles to operate. Fast relays can reach 0.5 second (25 cycles in 50-Hz systems) to operate while circuit breakers can operate in 10 cycles normally. But now 2.5-cycle breakers have come into use, that is, the contacts are opened within 2.5 cycles from the instant of initiation of the fault. This requires relays which can operate in 1 cycle leaving 1.5 cycles for breaker operation. Such fast-acting timings are possible only by using solid-state relays using very fast-switching transistors. It takes but a millisecond (1/20-th of a cycle in 50-Hz systems) for the relay to operate, since the instant of sensing the overcurrent to the instant of closing or opening contacts.

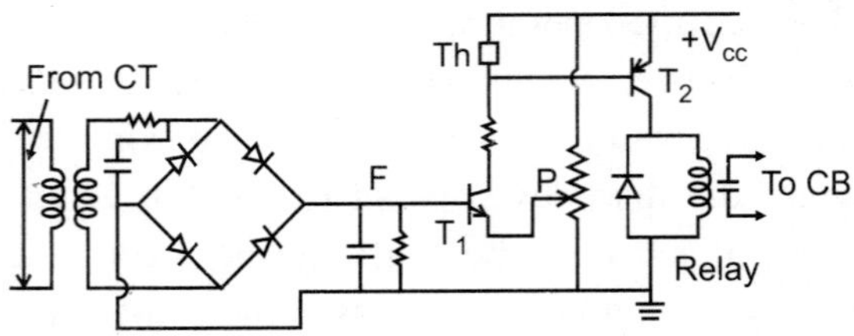

Fig. 15.5: Schematic of a solid-state overcurrent relay. Line current is rectified and fed to Base of transistor T_1.

The scheme, shown in Fig. 15.5, is to rectify the fault current sensed by the CT by a full-wave rectifier and smooth it by an R-C filter. This is fed to the base of a transistor to get it to conduct. The level of conduction is changed by a potentiometer *P*. Once this transistor T_1 conducts, it places the larger transistor T_2 into conduction and this switching transistor acts as a switch allowing a current from a d.c. supply to flow through the trip coil of the circuit breaker. A thermistor is provided for control of run-away condition of T_2.

Q.10. What are the different types of protective schemes adopted for transmission-line protection?

Ans. Transmission lines form an inter-connected system with several lines connected to a single bus, with some of them bringing power towards the bus and others taking power away from the bus. Therefore, current directions are different so that directional discriminating relays are always necessary. However, we cannot discuss such vast problems. Here, we will only discuss methods of protection applied to a single line and a single line-section with circuit breakers at the two ends. Depending upon the importance and extent of inter-connection, the method adopted will vary. Transmission-line protective schemes can be classified broadly as follows:

1. Impedance-distance relaying which is known as either ohm-distance or mho-distance relaying;
2. Carrier protection schemes;
3. Pilot-wire relaying in which an existing telephone line can carry the operating signals from one end to the other;

4. Travelling-wave protection schemes.

   Recently, optical fibres have come into use which are replacing auxiliary channels such as telephone circuits.

Q.11. What is an Impedance-distance relay and how does the protective scheme function with this relay in the system?

Ans. Figures 15.4 and 15.5 illustrated simple overcurrent relays whose operating coils were fed by a CT. It depended entirely on the level of overcurrent. It is clear that closer the fault is to the breaker location, larger will be the current and lower will be the voltage of the bus at the circuit breaker. If the fault occurs at a larger distance from the bus sending the power, the fault current will be low and the voltage dip will be less, that is, the voltage at the bus will be higher. In other words, the fault current and the voltage at the bus depend upon the impedance of line up to the fault. The ratio of voltage to current is the impedance in ohms while the inverse or current/voltage is in mhos or siemens. The impedance-distance relay uses this property in the following manner: Two electromagnets are used in the relay, one of them excited by the bus voltage through a PT and the other by the line current through a CT. A magnet beam is supported over the pole pieces in such a way that the forces due to the two coils oppose each other. The voltage-excited coil acts as a restraining force acting against the force of attraction due to the current-excited coil. In the normal position, the beam is balanced and so the relay is also known as a "Balanced-beam relay".

As soon as the current exceeds the value as set by the protective scheme, (say 125% overload), the current-excited coil exerts sufficient force to tip the beam thereby operating electrical contacts to operate the circuit-breaker trip mechanism. The larger the current, the lower also will be the voltage so that the operating time is reduced since the operating force is large and restraining force becomes low. Thus, a kind of inverse-time characteristic is built in. The operating time is therefore directly proportional to the impedance in ohms or inversely proportional to the ratio of current/voltage or mhos.

Figure 15.6 shows schematically the balanced-beam relay and the time-to-operate versus the distance to fault or impedance up to the fault. Note that when a long line is sectionalised, and a fault occurs to the right of the circuit breaker 2, the relay at breaker 2 has shorter operating time than the relay shown next to breaker 1 in the previous line section. Therefore, only the breaker connected to the faulted section will operate to trip the line.

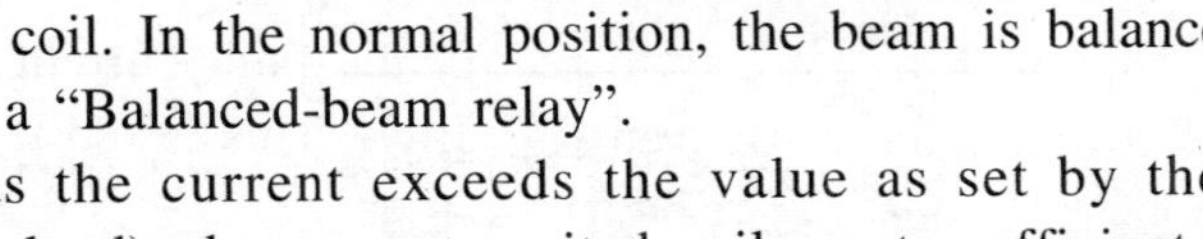

Fig. 15.6: Balanced-beam impedance-distance (or mho-distance) relay. Time-distance (or impedance) setting of relay.

In order to change the operating characteristic i.e., to change the voltage/current ratio for operating the relay, it is usual to change the current flowing in the voltage coil by connecting a variable resistance in series with it. Modern solid-state relays use fast-response current comparators to provide operating signals to the transistors. But the principle is the same as the e.m. impedance-distance relay.

Q.12. What are the principles of Power-line Carrier-current relaying and protection?

Ans. In an overcurrent relay and impedance relay, the overload current was used for operating the relay coils. In the case of impedance-distance relay scheme, several line sections could be selected by selective tripping of only the faulted section of line by setting the impedance-distance-time characteristic properly as shown in Fig. 15.6.

As systems have grown in size and with extensive interconnections, this type of arrangement becomes even more sophisticated so that simple high-speed impedance-distance relay application becomes too complicated or even impossible if improved systems are not available. Fortunately, the power-line carrier protection scheme has improved protective nature of schemes and thereby stability limits also to keep pace with power-flow requirements.

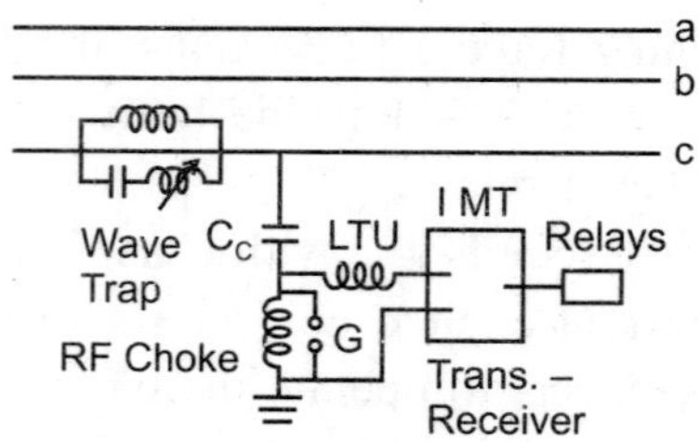

Fig. 15.7: Power-line Carrier protection scheme for line sections. WT: Wave Trap; C_C: H.V. Coupling Capacitor; L: R.F. Choke; L_t: Tuning Choke; IMT: Impedance-Matching Transformer; RR: Receiver Relay; X-R: Transmitter-Receiver; IDR: Impedance-Distance Relay; G: Protective Spark Gap.

The basic principles and components of carrier protection are as follows. Refer to Fig. 15.7.

(*a*) Each line section between major buses at the ends of lines is provided with a high-frequency transmitter and receiver operating between 50 and 200 kHz. The signal from the transmitter is coupled to the overhead-line conductor at high voltage by the coupling capacitor C_C so that the signal propagates on the h.t. line itself and acts as the channel for the carrier signal. It is received by the receiver at the other end of the line section.

(*b*) Wave traps (consisting of parallel *L–C* tank-circuit) are connected at each end of the line section. They are located at the h.v. end of the line and are designed to carry the full power-frequency current as well as the short-time heavy short-circuit current in any fault (for about 2 seconds or until the fault is cleared). An idea of these currents is up to 4000 amps steady-state and 60,000 amps short-circuit for 2 seconds.

(*c*) At each end is located a relay RR called the Receiver Relay. The contacts of this relay are normally open and operate the impedance-distance relay of the high-speed type of normal design described earlier.

(*d*) The transmitter sends a signal continuously while the receiver receives it if there is no fault on the line and the line is healthy. The relay RR keeps the contacts which supply power to the breaker trip-coil open. When there is no fault in the line section, power is flowing into the line section at one end and flows out at the other end. This also happens when there is a fault in the next line section and the fault current has a continuous passage. The carrier transmitter is provided with a directional sensor such that when power is flowing out of this line section, it is transmitting its signal to the receiver at the other end over the power-line conductors.

(*e*) However, when a fault develops within the line section between the two transmitters and receivers, which is called an "Internal Fault", power flows into the fault from both sides. The directional relay senses this and stops the signal being transmitted. When this happens, the receiver relays at both ends operate to close their contacts and place the operating contacts of the impedance-distance relay into active position. A CT and a PT energise the two coils of the relay continuously as described before. Thus, the function of the carrier signal is only to keep the relay RR open when the line section is healthy and close it when a fault develops by stoppage of the transmitted h.f signal.

(*f*) The circuit breakers at both ends of the line section are tripped or opened simultaneously. Thus, the carrier protection scheme is selective as to line sections protected.

Some of the advantages of the carrier-protection scheme over simple impedance-distance relaying can be summarised as follows:

1. Each line section protected is independent of other line sections. This improves system stability since unnecessary tripping is avoided.
2. Total time from the occurrence of fault to energising the trip coil of the circuit breaker is usually less than 3 cycles. In order to obtain even higher speeds, such as less than 1 cycle, static electronic relays have to be used instead of electromagnetic type of relays.
3. Opening of both ends of a faulted line section is necessary also from the point of view of quick re-closing of the breakers. Single-pole reclosing maintains stability. Fast clearing also lessens danger from shock to the system such as voltage dips and removal of loads which cause power surges.

Referring to Fig. 15.7, we observe that the carrier signal transmitter-receivers are coupled to the overhead power line through coupling capacitors C_C. These are outdoor porcelain-housed capacitors of about 1 nF (10^{-9} F) capacitance and drop the entire high voltage across themselves. (Reactance at 50 Hz is $X_C = 1/2\pi fC = 10^9/314 = 3.2 \times 10^6$ ohm. At 100 kHz, the carrier signal frequency, $X_C = 10^9/2\pi \times 10^5 = 1600$ ohms). At the ground end, two chokes are provided. The one shown as R.F. Choke drops the high tension to a low value so that the transmitter-receivers do not experience a voltage of say no more than 20 volts at 50 Hz. For example, for a 500-kV line (line-to-ground voltage = $500\sqrt{3}$ kV), its reactance should be nearly $20\sqrt{3} \times 3.2 \times 10^6/500 \times 10^3$ = 1150 Ω. Thus, its inductance is 1150/314 = 3.65 henry at 50 Hz. The second inductor shown as LTU is called the Line-Tuning Unit. It is in series with the coupling capacitor in the transmitter-receiver ($X - R$) circuit and offers a series-resonant circuit with C_C at carrier frequency. For 100 kHz, and $C_C = 10^{-9}$ F, we must then have $10^5 = 1/2\pi\sqrt{(LC)}$ giving L = 2.5 mH. This is nearly a short circuit at 50 Hz. The gap G protects the entire l.v. circuit in the event of the R.F. choke developing an open circuit, placing high voltage across the transmitter-receiver. An impedance-matching transformer is necessary to match the impedance of the X–R to the surge impedance of the line (about 300–400 ohms).

Telephone circuits for carrier voice communication and measurement of system conditions, called Telecommunication and Telemetering can also be coupled to the h.v. line. This is not discussed here since even under fault conditions on the h.v. line they must be functioning to be useful.

We also note that the cost of carrier relay system does not depend on the length of line protected. In practice, this system is used for long line sections.

Q.13. What is meant by Pilot Wire Relaying?

Ans. In the carrier-relaying scheme described before, the high-frequency signal from the transmitter to the receiver at the other end under healthy line condition was transmitted over the h.t. line conductor. In the event of an internal fault in that line section, the signal is blocked and the receiver relay RR acts to energize the breaker trip-coil-circuit. For a long line, the cost of the terminal equipments at the two ends is justified for the high speed of operation and maintaining stability.

However, for lines of less than 20 km, the expense for carrier equipment proves prohibitive. Simple over-current protection will suffice. In addition, if a telephone line is already available nearby or some such communication channel, the fault condition can be suitably transmitted to the other end and will function exactly the same way as the carrier protection scheme. However, since the fault occurs at high voltage, suitable isolating transformers must be utilised at both ends. Also the telephone or pilot wire must be removed at a sufficiently large distance so that electro-statically-induced voltages from the h.v. line do not disturb the main fault signals.

Q.14. How are optical fibres utilised in protection schemes?

Ans. When pilot-wire relaying is used over telephone channels the signals are affected by noise, i.e., undesired signals, because the telephone wires are exposed and pick up extraneous disturbances. However, modern low-attenuation optical fibres are replacing the communication channels for the protection signals. They also have a metallic outer shield which can be grounded, so that no interfering noise or voltage from the h.t. line can be induced in the cores, even if the cable is threaded over the h.t. towers and stay closer to the h.t. conductors than is permissible for open telephone lines. Furthermore, many channels of communication can be used simultaneously in both directions by suitable frequency multiplexing. Amplifiers can also be used for further enhancing efficiency of signal communication. This requires d.c. power supplies. It is expected that unattended solar-cell power packs will be used for this purpose.

Q.15. What are the principles of Travelling-wave protection of long transmission lines?

Ans. Much progress has been achieved in protection schemes for fast clearing of line faults by using Digital Processors, Microprocessors, fibre-optical links and high-speed communication, as well as using static relays and fast-clearing circuit breakers which can operate in 1 cycle. One scheme that has been developed since 1978 is based on sensing the properties of travelling waves generated by the fault itself which travel on the line conductors at high voltage. These travelling waves are generated at the location of the fault and travel to both sides as forward-travelling and backward-travelling voltage and current waves as was discussed in detail in Chapter 14, Question 4.

We have derived that the voltage and current at any point x and time t on a line can be written as:

$$e(x, t) = f_1(x - vt) + f_2(x + vt)$$

and,

$$i(x, t) = (1/z) \cdot f_1(x - vt) - (1/z) \cdot f_2(x + vt)$$

The second equation can also be written as:

$$z \cdot i(x, t) = f_1(x - vt) - f_2(x + vt)$$

This allows us to write the functions $f_1(x - vt)$ and $f_2(x + vt)$ as

$$f_1(x - vt) = (1/2)\ [e(x, t) + z \cdot i(x, t)]$$

the forward-travelling wave,

and

$$f_2(x + vt) = (1/2)\ [e(x, t) - z \cdot i(x, t)]$$

the backward-travelling wave.

These are known as D'Alembert solutions and can be used for discriminating or separating the forward-travelling component from the backward-travelling component in the voltage.

In the protective scheme used, discriminators are provided which function as follows and which are shown in Fig. 15.8.

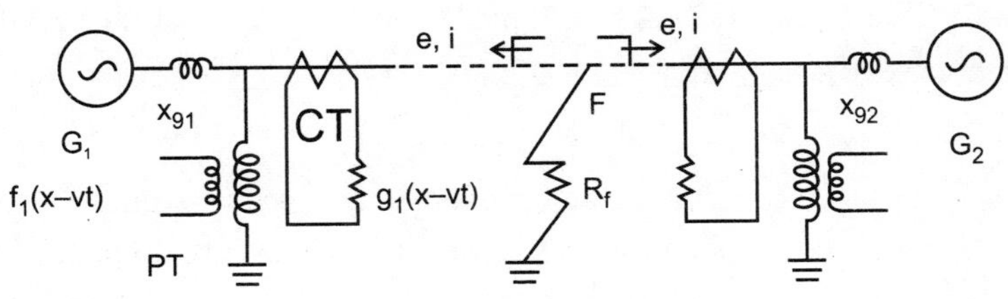

Fig. 15.8: Travelling-wave protection scheme for long transmission line.

A PT picks up the voltage $e(x, t)$ while a CT picks up the current $i(x, t)$ coming down the line to the ends. The current is multiplied or amplified by z times, and adding and subtractive circuits then filter out the functions $f_1(x - vt)$ and $f_2(x + vt)$ as shown above. A solid-state relay then can be fed with the signals and feed the trip coil of the breaker. All the functions can be carried out on a dedicated microprocessor.

Q.16. What is meant by Cathodic Protection? How is it carried out in practice?

Ans. Whenever a metallic component is buried in soil, such as a cable or pipe line, any current flowing in the ground will electrolyse the metal. Chemical corrosion takes place. This is very serious in urban distribution systems. Also, two dissimilar metals in contact in the presence of an electrolyte such as salts in moist earth cause galvanic or electro-chemical corrosion. Some if not all protective coatings will be eaten away. This occurs when one metal is at a higher potential than the other i.e., it becomes the anode of a cell. An anode gets eaten away by dissolving in the electrolyte, which may be rain-water soaked in the ground with salts in the air or even the metal of the cable protective sheath or other metallic components such as pipes.

In order to prevent such corrosion, sheaths made from extruded Lead are used for cables which are not attacked by chemicals, or pipes are painted suitably. However, a chemical process of preventing corrosion is very effective. This is based on the idea that when Zinc rod or plate is kept close to Iron or Steel in water, then the Zinc acquires a higher potential than the iron and becomes the anode in the electrolyte, while the Iron becomes the cathode. In this process the Zinc metal gets eaten away or "sacrifices itself". This is known as "Cathodic Protection".

In Cathodic Protection used in practice, a d.c. supply of only 0.5 volt is connected with its negative plate to the metal which is to be protected and the positive plate to a ground some distance away. In some cases, Mg or Zn rods are connected to the pipe and sunk in the ground. This is of course the "sacrificial" method of preventing corrosion and requires periodic inspection and replacement of the Mg or Zn rods. With solid-state rectifiers available, and domestic a.c. distribution also easily accessible, providing cathodic protection in urban areas is not a problem. The reader can observe these power supplies by the road side in cities and large towns where the rectifier units are housed in a small cabinet in just the same way as telephone terminal cabinets or mail-distribution cabinets.

❑❑

CHAPTER 16

TESTING SOURCES IN HIGH VOLTAGE LABORATORIES

Q.1. What are the types of sources required to perform tests on equipment in a high voltage laboratory?

Ans. These are:

(*a*) Power-frequency transformers with symmetrical voltages in the *two half-cycle* without wave-shape distortion;

(*b*) Impulse generators for lightning-surge waveshape of voltage;

(*c*) Impulse generators for switching-surge waveshape of voltage;

(*d*) Impulse generators for lightning waveshape of current; and

(*e*) D.C. generators for some million volts for insulation testing.

Q.2. What are the waveshapes of voltage and current used in high voltage testing?

Ans. These have already been given in Chapter 1. Some more details are given here:

(*a*) **Power-frequency Voltage:** These must be free of distortion. If the *two half-cycles* are not identical, the voltage will contain the fundamental component and all harmonics, both even and odd orders. Of these, the 2nd harmonic is predominant. Wave distortion is the result of the iron core of the transformer working in the non-linear portion of the magnetisation (B-H) characteristic. Therefore, testing transformers usually operate in the linear portion of their core's B-H curve. This results in the use of lower flux densities than in normal power transformers, which in turn requires large cross-section for the core and the transformer is more bulky than power transformers of comparable power rating.

(*b*) In impulse generators for **lightning-type voltage**, as will be discussed later on, the wave should be free from oscillations on the wavefront. They should also be free from "overshoot." These two phenomena are caused by the unavoidable presence of inductances in the generator circuit which is composed mainly of capacitors and resistors. But because of the large physical dimensions and long connecting leads which are necessary for the high voltage used (some millions of volts) inductances are present. We will discuss this aspect later on.

Q.3. What is the equation to lightning-type voltage and how is this generated?

Ans. The equation of lightning-type voltage is known as a "double-exponential". It is of the form shown in Fig. 16.1 and has the equation:

$$v(t) = V_0[\exp(-at) - \exp(-bt)]$$

Here, a and b are numerical constants which control the waveshape.

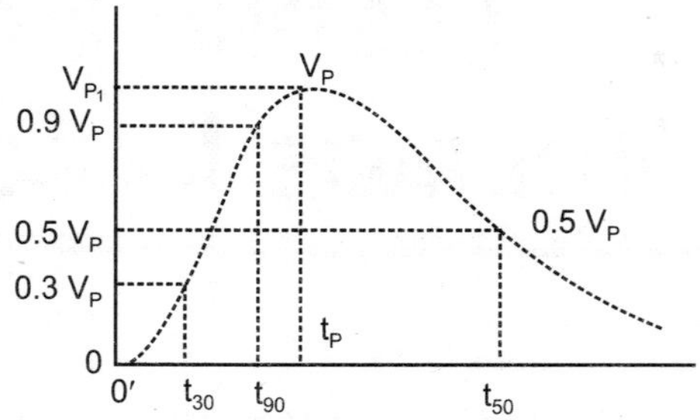

Fig. 16.1: Double-exponential voltage waveshape.

General properties: $v(t) = V_0(e^{-at} - e^{-bt})$

0: Real zero; 0′: Virtual zero;

V_p = peak value;

t_p = time from 0′ to peak value;

t_{30} = time from 0′ to 30% V_p;

t_{90} = time from 0′ to 90% V_p;

t_t = time to 50% V_p on tail from 0′. No overshoot.

The I.E.C. (International Electrotechnical Commission) and I.S. (Indian Standards) specify the lightning-impulse waveshape to have the following properties:

(*a*) **Wave-front time:** $t_f = 1.67(t_{90} - t_{30})$, where by definition,

t_{90} = time from "virtual zero" $0'$ to 90% of peak value = $0.9V_p$,

and, t_{30} = time from $0'$ to 30% peak value = $0.3V_p$.

(*b*) **Wave-tail time:** t_t = time from $0'$ to 50% of the peak on the falling portion of the voltage wave.

There is also a time to peak value t_p which is not specified by the Standards, but this will depend on t_f and t_t. When the I.E.C. and I.S. standard timings are used, they are t_f = 1.2 μs and t_t = 50 μs. For these timings, the constants turn out to be: $a = 14.5 \times 10^3$ and $b = 2.45 \times 10^6$. (Note that if time t is taken to be in micro-seconds $= 10^{-6}$ second, the values are: $a = 14.5 \times 10^{-3} = 0.0145$, and $b = 2.45$). The voltage

V_0 in the equation is not equal to the peak value V_p but a little higher. This is shown by an example in the next question.

Q.4. What are meant by "tolerances" in waveshape timings and magnitudes in lightning-impulse voltage waveshapes?

Ans. It is impossible to generate voltages from a generator with absolutely fixed timings and magnitudes for such fast-rising voltages as indicated in the previous question (1.2/50 μs) and double-exponential in waveshape because of the presence of inductances and because the waveshape depends to some extent on the capacitance of the test object which can vary from test to test. Therefore, standard specifications allow some margins or limits which are acceptable for testing purposes, namely, to assess the properties of a test object under stress. These are known as "tolerances". As per I.E.C. and I.S., the following tolerences are permissible:

Front Time: t_f = 1.2 μs ± 30% = 0.84 μs to 1.56 μs;

Tail Time: t_t = 50 μs ± 20% = 40 μs to 60 μs.

Voltage Overshoot: Because of the presence of inductance everywhere in the generator proper and testing arrangement, a series circuit consisting of *R-L-C* elements when excited suddenly by a source of d.c. voltage will exhibit oscillations in the voltage across the capacitor. Without the inductance being present, the voltage across the capacitor in only an *R-C* circuit is a double-exponential. The "overshoot" is defined as follows:

On the observed (photograph from an oscilloscope screen) voltage wave, draw a wave which would be obtained by calculation had the inductance been absent, as shown in Fig. 16.2. The peak value observed with oscillation in excess of the peak value of the ideal double-exponential wave is known as the "overshoot". As per Standard Specifications, the overshoot should not exceed 5% of the peak value of the pure double-exponential waveform.

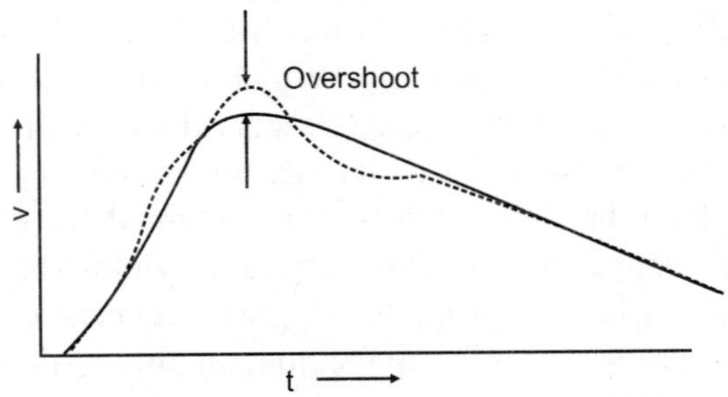

Fig. 16.2: Double-exponential voltage waveshape with oscillations and overshoot.

Since this is caused by the presence of inductance in the generator circuit and layout of other equipment, the overshoot or oscillation can be reduced or damped by adding more resistance in the generator and the measuring circuit. However, this of necessity lowers the voltage output and lowers the efficiency of the generator. We cannot discuss these mathematical details of transient electrical phenomena here.

Example: A double-exponential waveshape of voltage has the equation:

$$v(t) = 1.035[\exp(-14.5 \times 10^3\, t) - \exp(-2.45 \times 10^6\, t)]$$

Assume no overshoot. Examine this for t_f and t_t values.

Solution: For examining the wavefront, let us vary t from 0 to 1 μs in small steps and calculate v as follows:

t, (μs)	0.1	0.15	0.2	0.4	0.6	0.8	0.85	0.9	1.0
v	0.223	0.316	0.4	0.64	0.79	0.88	0.89	0.907	0.93

Note that a value of 0.3 falls between 0.1 and 0.15 μs. Further calculation will show that at t = 0.14 μs, v = 0.3.

Similarly, a value of 0.9 falls between 0.85 and 0.9 μs.

At t = 0.87 μs, v = 0.9

∴ t_{30} = 0.141 μs and t_{90} = 0.87 μs

The resulting front time is: t_f = 1.67(0.87 – 0.141) = 1.215 μs.

This value falls within the allowable "tolerance" of 1.2 ±30% μs.

At t = 50 μs, v = 0.501. This is allowable.

The peak value of $v = V_p = 1$ is obtained at t = 2.22 μs.

Therefore, the given equation satisfies the standards. [The choice of V_o = 1.035 is not explained here. But it has been obtained by tedious theoretical procedure].

Q.5. What are the timings and tolerances of a switching surge waveshape of voltage to be used for testing?

Ans. The importance of switching surges to high voltage insulation has been recognised only since 1960 when transmission voltages increased to 500 kV and higher. Prior to that date, all insulation coordination and airgap clearances were based on lightning voltages.

In 1963, John Kalb, Jr., of the Ohio Brass Co. of U.S.A., through very painstaking experiments discovered that positive-polarity switching surges created more havoc to large airgaps than d.c. or a.c. or lightning. He also discovered that the airgap clearance on a 500-kV tower had the lowest flashover strength when the wavefront was 250 μs, and the lowest strength for other gap lengths were obtained for other wavefront timings. Thus, there is no "one" shape of switching surge that yields the lowest insulation strength for all airgap lengths. It became a matter of grave concern on how to determine or specify a "standard switching surge" waveshape, in the same manner as was the case for specifying a "standard" lightning impulse waveshape when in actual practice lightning exhibited a very wide variation of waveshape in the field. However, some decision had to be taken in this regard even though lightning-impulse voltage and switching-surge voltage waveshapes show wide variation in timings in practice.

For the sake of testing purposes, standard switching surge voltage waveshapes are of two types:

1. For testing all h.v. equipment other than transformers:

Front time, t_f = 250 μs ± 30% = 175 μs to 325 μs;

Tail time, t_t = 2500 μs ± 20% = 2000 μs to 3000 μs

= 2 ms to 3 ms.

The equation is once again the double-exponential

$$v(t) = V_0(e^{-at} - e^{-bt})$$

The definitions of wavefront time and wave-tail time are the same as for the lightning impulse. Since the switching surge voltage is much longer than the lightning impulse, the values of the exponents, a and b, are much smaller than for lightning.

Example: A switching-surge impulse is described by the above equation with $V_0 = 1.1475$, $a = 326$, and $b = 11 \times 10^3$. Examine the wavefront and wavetail timings as well as the peak value V_p and the time to peak t_p.

Solution: $v(t) = 1.1475[\exp(-326t) - \exp(-11 \times 10^3 t)]$

The peak value is $V_p = 1.$

At the peak value $dv/dt = 0$

$\therefore$ $dv/dt = 0$

$= -ae^{-at} + be^{-bt}$

or, $(b - a)t_p = \ln(b/a)$

Thus, $t_p = \ln(b/a)/(b - a)$

$= \ln(11/0.326)/[(11 - 0.326) \times 10^3]$

$= 3.519 \times 10^{-3}/10.67$

$= 330 \times 10^{-6}$ sec $= 330$ μs

Check: $v = 1.1475(e^{-0.010758} - e^{-3.63})$

$= 1.1475(0.898 - 0.0265)$

$= 1$

Then, t_{30} is found from: $0.3 = 1.1475[\exp(-326t) - \exp(-11 \times 10^3 t)]$

A trial and error gives: $t_{30} = 28.7$ μs

In a similar manner,

t_{90} is found from: $0.9 = 1.1475[\exp(-326t) - \exp(-11 \times 10^3 t)]$

This gives, by trial and error, $t_{90} = 165$ μs

$\therefore$ The front time is: $t_f = 1.67(t_{90} - t_{30})$

$= (165 - 28.7)/0.6$

$= 227$ μs

This falls within the limits of 250 μs ± 30% = 175 μs to 325 μs as specified by standards.

Tail time: $0.5 = 1.1475[\exp(-326t) - \exp(-11 \times 10^3 t)]$

Once again, a trial and error gives, $t_t = t_{0.5} = 2550$ μs, which falls within the limits 2500 ± 20% = 2000 to 3000 μs.

Therefore, a switching surge voltage with standard timings of 250 μs ± 30% and 2500 μs ± 20% falls within the range described by the equation:

$v(t) = 1.1475[\exp(-326t) - \exp(-11 \times 10^3 t)]$ with t measured in seconds.

2. **Standard Switching-surge Waveshape for Testing Transformers:**

Transformers, unlike other h.v. equipment such as insulator strings, have a very high inductance. A transformer has two distinct types of inductance:

(*a*) **Magnetising Inductance:** When one winding is excited and the second is open i.e., it is unloaded, the inductance presented to the source is the magnetising inductance which is extremely high (of the order of 50 p.u. on transformer ratings as bases). In fact, for most purposes, the winding behaves as an open circuit, except under certain circumstances as will be described later on.

(*b*) **Leakage Inductance:** When both windings carry the load current, the transformer inductance falls to the low value of leakage inductance. Thus, when a large transformer is being tested for the switching-impulse withstand test of its insulation, the transformer presents a very high value of inductance, since it is unloaded. When such an inductance is connected to an impulse generator consisting of *R* and *C* only, it gives rise to oscillations on the tail of the impulse wave of voltage and not on the front of the wave. Considering this fact, the standard voltage waveshape to be used for switching surge testing of transformers has the following specifications according to I.E.C. and I.S. Refer to Fig. 16.3.

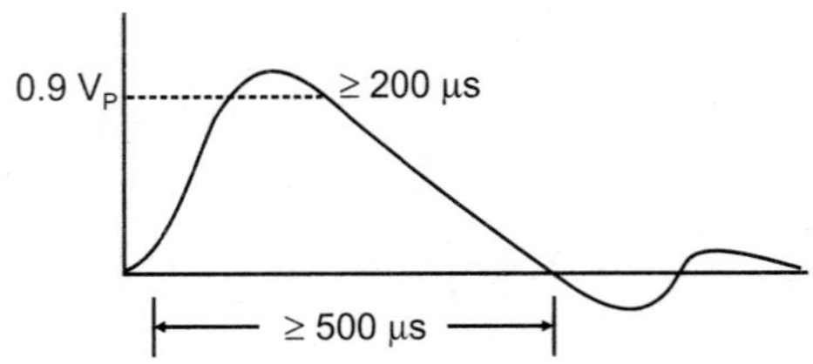

Fig. 16.3: Switching-surge voltage waveshape for transformer testing.

The wavefront should be at least 20 μs duration, the total length of the switching surge should be at least 500 μs to the first zero crossing, and the magnitude should be in excess of 90% of crest value for at least 200 μs. In the event saturation of the core occurs, the inductance of the winding collapses rapidly to zero so that the voltage across it is zero. (Inductance is defined as the slope of the *B-H* curve which is zero when *B* does not increase). Thus, the zero of the voltage is also the point where the magnetising current reaches the saturation value.

These restrictions are very stringent and it takes a lot of experience to generate such voltages. Thus, testing a transformer under longer-duration switching surges is more difficult than testing under shorter-duration lightning impulses, since for the very high rates of rise of voltage under lightning, the transformer winding undergoing test is always considered open. The flux does not normally penetrate into the core because of shielding offered by the metal at these high frequencies because of the eddy currents induced.

In special tests where the length of the surge cannot reach at least 500 μs, standards specify that a d.c. voltage of proper polarity must be used in order to prevent core saturation and hence collapse of voltage.

Q.6. How is a double-exponential waveshape generated? What are the components of an impulse generator and test circuit? Omit measuring equipment.

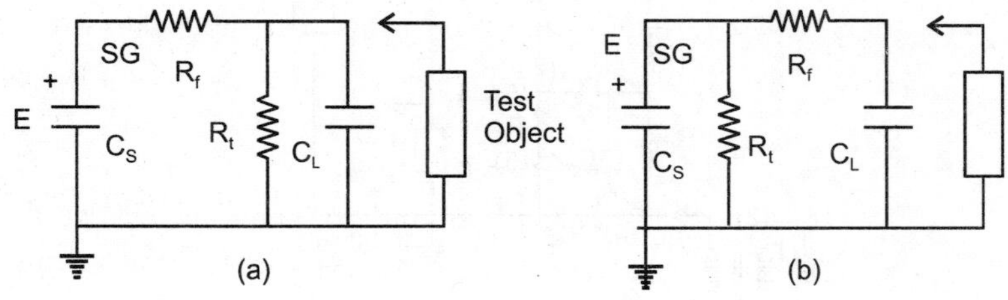

Fig. 16.4: Impulse generator equivalent circuits, alternatives.

Ans. Figure 16.4 shows the circuit. In principle, when the generator is discharging on to a test object, the equivalent circuit has the following components:

1. C_S is the source capacitance. It usually consists of several capacitors in series for obtaining the high voltage. The capacitors are all charged in parallel and discharged in series as will be described later. Thus, a d.c. voltage is suddenly impressed on the circuit.
2. R_f is called the "wavefront" resistor. It is in series with the generator capacitance C_S. By varying the value of R_f, the wavefront time t_f can be varied to suit the test object.
3. R_t is called the "wave tail" resistor. It is in parallel with the source capacitance C_S in some arrangements. In others it is placed in parallel with C_L as shown in Fig. 16.4(*b*). It is not possible to discuss the merits of these changes here, but where the physical dimensions of extremely-high voltage generators (above 1 MV) have high inductances in the leads inside the generator, it is usual to have R_f next to C_S as shown in Fig. 16.4(*a*).
4. C_L is called the "Loading Capacitor". It is required to obtain the double-exponential waveshape. It also serves, as will be explained later, as a voltage divider so that a low voltage can be tapped at the ground end in order to feed measuring instruments such as voltmeters, oscilloscopes, etc. This will be detailed in the next Chapter.
5. Test Object: This is in parallel with C_L. Usually it is a pure capacitance load on the generator. Therefore, the total capacitance load on the generator is the sum of the voltage-divider capacitance C_L and the capacitance of the test object itself.

Q.7. What is a Marx Multiplier Circuit and how does it function to generate very high impulse voltages?

Ans. Refer to Fig. 16.5. For generating impulse voltages above 100 kV or 200 kV, a single capacitor in the source is insufficient. Thus, for extremely high voltages a number of capacitors in the source, each rated for 100 kV or 200 kV in the laboratories, are charged

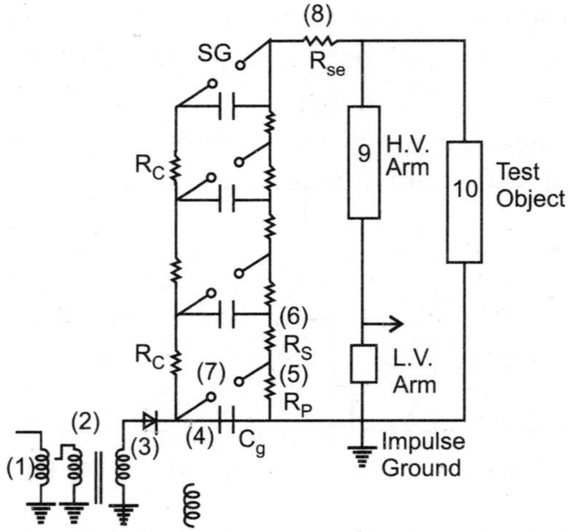

Fig. 16.5: Four-stage Marx Impulse Generator. (1) Excitation transformer; (2) Isolating and regulating transformer; (3) Diode; (4) Stage capacitors; (5) Parallel resistor; (6) Series resistor; (7) Spark gap; (8) External series resistance; (9) Voltage divider (loading capacitor); (10) Test object.

in parallel and they are connected in series by an arrangement of spark gaps. This multiplies the voltage, and the capacitors forming a total C_S then discharge into the test object. This is known as the Marx Impulse Generator circuit or multiplier. It functions as follows:

The 50-Hz (or 60 Hz) a.c. transformer (1) excites another transformer (2) called an isolating and regulating and charging transformer. The output voltage of the secondary can be varied to suit the testing-voltage level of the test object from 0 to 100% value. The charging voltage is normally 100 kV or 200 kV. This voltage charges the capacitor C_g, (4), through a diode (3). In the Figure, the polarity of the generator is positive. By reversing the diode, the polarity of voltage at the h.v. terminal can be reversed. In the Fig., four capacitors C_g are shown. The spark gap *SG* (7) is open and its gap-setting is such that at the voltage used for charging C_g, it does not flash over. C_g is called the "stage capacitance". Thus, in the Fig., the generator is shown as having 4 stages. When connected in series the total source capacitance will be $C_S = C_g/4$.

The other 3-higher stages are charged by the transformer (2) and diode (3) in parallel with the ground-stage capacitor C_g (4), but through resistors R_p (5) and R_S (6) denoting parallel and series resistors, respectively. At the top of the generator, a resistor R_{se} (8) is shown which is optional and is used for adding more series resistance to R_S inside the generator for changing the wavefront time, since the time t_f obtained from the generator circuit depends upon the capacitance of the test object which changes from test to test. Thus, it is imperative that the capacitance of the test object must be ascertained from other measurements.

The voltage divider (9) which is also known as the loading capacitor C_L has two parts: The h.v. arm and the l.v. arm. Thus, if we need to obtain a voltage ratio of say

1000 : 1, the l.v. arm must have 999 times the capacitance of the h.v. arm. With a usual value of 1 nF (10^{-9} F) for the h.v. arm, the l.v. arm has a capacitance of 999 nF (or nearly 1 μF) for obtaining a ratio of 1000 : 1. We will discuss the properties of voltage dividers in the next Chapter in more detail.

In the quiescent state, prior to the discharge of the generator stage capacitors into the test object, the spark gaps SG are open. All the stage-capacitors are charged in parallel to the output voltage of the charging transformer (2) through diode (3). This could be up to 100 kV or 200 kV, depending upon the voltage rating of the stage capacitors. When everything is ready and set to "go", and the test engineer needs to "fire" the generator, the lowest spark gap SG (7) is triggered by an external means by feeding a triggering pulse. As soon as the lowest gap fires, it places the capacitor in the next stage in series with the first-stage capacitor. The event is shown in Fig. 16.6.

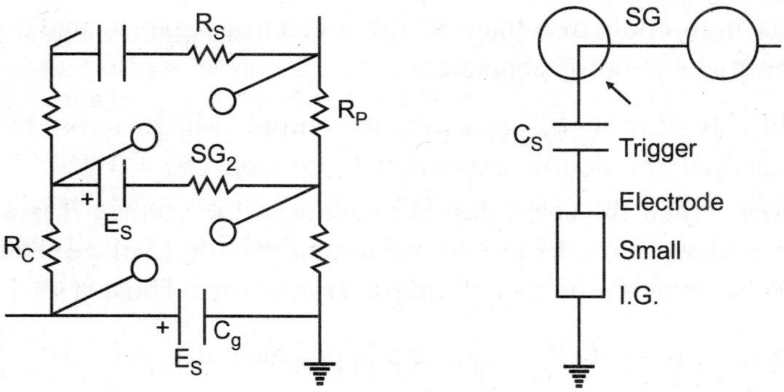

Fig. 16.6: Firing of series Spark Gap and 3-electrode trigger generator (or Trigatron).

If the stage voltage is V_S, to which the capacitors are charged then the voltage across the spark gap SG in the 2nd stage is $2V_S$. It therefore breaks down. This places $3V_S$ across the spark gap in the 3rd stage which flashes over. When this happens, the 3-lower stages are placed in series and the output voltage is $3V_S$ at this point. It is clear that the total output of the generator is $4V_S$. Thus, the initial firing of only the lowest spark gap (7) triggers all other spark gaps in turn and all capacitors are connected in series. If there are N stages with charging voltage of V_S = 100 kV or 200 kV per stage, say, when all the stages are connected in series the output voltage of the generator h.v. terminal becomes $N \cdot V_S$. For an output voltage of 1000 kV = 1 MV, there may be 10 stages when V_S = 100 kV per stage, or 5 stages when V_S = 200 kV. These are the usual standard charging voltages per stage adopted in the world. Note that these are the maximum voltages. By varying the tap on the ground transformer or charging transformer (1) or (2), the output voltage can be regulated to suit the test-voltage level.

When all stages are connected in series, $R_t = N \cdot R_p$ and $R_f = (N - 1)\ R_S + R_{se}$, which are shown in Fig. 16.4. One important operational requirement is that the gap spacings of SGs must also be varied to suit the charging voltage. Otherwise, the lower gap may not fire at all if it is set too large. This adjustment of gap spacing by the test engineer is done prior to the beginning of the test. This is usually carried out by having one sphere of every gap mounted firmly to the generator structure and the companions mounted on an insulated rod or column as tall as the impulse generator and rotating this column by a motor to vary the gap setting. (An impulse generator for 2 MV may be as tall as 5 metres).

There are several ways to initiate the triggering of the lowest spark gap SG: It can vary from the crude method of short-circuiting the gap by a hand-held insulation rod with a copper crosspiece attached to one end with which the operator can bridge the gap by holding the other end of the insulated rod, or, to the modern remote-controlled triggering arrangement where a 3rd electrode is introduced between the two spheres in the gap to which a voltage of suitable magnitude and polarity is applied. This is known as a 3-electrode gap or "trigatron" arrangement. It is schematically shown in Fig. 16.6(*b*). An auxiliary impulse generator of low voltage is fired by a push button from the control board which places a high enough voltage in the 3-electrode gap to make the gap flash over and trigger the main impulse generator.

Q.8. Referring to Fig. 16.4(*a*) or (*b*), show that the output voltage across the loading capacitor (voltage divider) has the double-exponential equation $v(t) = V_0(e^{-at} - e^{-bt})$.

Ans. See Fig. 16.4(*a*). When the spark gap SG conducts, the voltage E is suddenly impressed on the circuit and we have to use transient-analysis or Laplace Transforms. Thus, all quantities will be in terms of their Laplace Transforms. Thus, with Γ denoting Laplace Transforms, we have: $\Gamma(E) = E/s$, and impedance of C is $\Gamma[(1/C)\int_0^t dt\,] = 1/sC$. Also denote $\Gamma(i) = I(s)$, in general.

We can write the following two loop-current equations in Laplace-Transform notation for the transient performance:

$$[R_f + R_t + 1/sC_S] \cdot I_1(s) - R_t \cdot I_2(s) = E/s \qquad \text{... (1)}$$

since E is a d.c. voltage.

$$-R_t \cdot I_1(s) + (R_t + 1/sC_L) \cdot I_2(s) = 0 \qquad \text{... (2)}$$

The voltage $v(s)$ across C_L will be: $v(s) = (1/sC_L) \cdot I_2(s)$ ... (3)

∴ Eliminating $I_1(s)$, we obtain for $I_2(s)$:

$$I_2(s) = (E/R_f) \cdot s/(s^2 + A \cdot s + B) \qquad \text{... (4)}$$

where the reader can verify that the coefficients A and B are:

$$A = (1/R_tC_L) + (1/R_fC_L) + (1/R_fC_S)$$

and

$$B = 1/(R_fR_tC_SC_L) \qquad \text{... (5)}$$

∴ The operational form for $v(t)$ is, from Equation (3):

$$V(s) = (E/R_fC_L) \cdot 1/(s^2 + A \cdot s + B) \qquad \text{... (6)}$$

We can factor the quadratic in the denominator into two linear factors and write

$$(s^2 + A \cdot s + B) = (s + a)(s + b)$$

where $(-a)$ and $(-b)$ are the two roots of the quadratic. Then, Equation (6) can be written as:

$$V(s) = (E/R_f C_L) \cdot [1/(s + a)(s + b)]$$
$$= [E/\{R_f C_L(b - a)\}] \cdot [\{1/(s + a)\} - \{1/(s + b)\}]$$

The inverse transform or the time function is:

$$v(t) = [E/\{R_f C_L(b - a)\}] \cdot (e^{-at} - e^{-bt}) = V_0 (e^{-at} - e^{-bt})$$

This is the required double-exponential voltage output across the test piece or load.

Example: Normally, the stage capacitance of impulse generators is of the order of 0.5–1 μF for a 10-stage generator with charging voltage of 200 kV per stage. The output voltage is therefore of the order of 2 MV, maximum, when all 10 capacitors discharge in series. [We cannot go into a deeper discussion on why the stage capacitance is of this order. But we can observe that when all 10 stages are connected in series at the instant of discharge, the total capacitance C_g of the generator becomes 0.05 to 0.1 μF, which should normally be equal to 10 times the load capacitance for good efficiency].

Let, for this example, $C_S = 0.08$ μF = 80 nF, $R_f = 300$ ohms and the tail resistance $R_t = 600$ ohms. The loading capacitor (voltage divider + test piece) is $C_L = 2$ *n*F. (Normally, the ratio of C_S/C_L is in the neighbourhood of 40). Calculate the equation for voltage $v(t)$ for a charging voltage $E = 10 \times 200$ kV = 2 MV.

Solution: With
$$R_f = 300$$
$$R_t = 600$$
$$C_L = 2 \times 10^{-9}$$
$$C_S = 80 \times 10^{-9}$$

we obtain:
$$A = 1/(R_t C_L) + 1/(R_f C_L) + 1/(R_f C_S)$$
$$= 2.5417 \times 10^6$$

and
$$B = 1/(R_f R_t C_S C_L) = 3.4722 \times 10^{10}$$

∴
$$s^2 + As + B = s^2 + 2.5417 \times 10^6\, s + 3.4722 \times 10^{10}$$
$$= 0$$

This yields the two roots: $s = -1.2709 \times 10^6 \pm 1.2572 \times 10^6$

Thus,
$$a = 0.0137 \times 10^6$$

and,
$$b = 2.528 \times 10^6$$

giving,
$$(b - a) = 2.514 \times 10^6$$

Also,
$$V_0 = 1/[R_f C_L(b - a)]$$
$$= 1.3258 \text{ MV}$$

Consequently, the equation for the output voltage is:

$$v(t) = 1.3258[\exp(-13.7 \times 10^3 t) - \exp(-2.53 \times 10^6 t)]$$

with t in seconds. If t is measured in μs = 10^{-6} s, we have

$$v(t) = 1.3258\, (e^{-0.0137\, t} - e^{-2.53\, t})|_{t \text{ in } \mu s}$$

Referring back to Question 4 and the example worked out there, the reader can calculate:
(a) the front time $t_f = 1.67(t_{90} - t_{30})$,
(b) the tail time t_t,
(c) the peak value V_p of the impulse, and
(d) the time t_p when it attains V_p. The voltage efficiency of the generator is defined as $\varepsilon = V_p/E$ = peak value of output voltage/total charging voltage.

From the procedure given in Question 5, we observe that:

$$t_p = \{\ln(b/a)\}/(b - a) = 2.074 \times 10^{-6} \text{ sec} = 2.074\ \mu\text{s}$$

$$V_p = 1.3258(e^{-0.0284} - e^{-5.2472}) = 1.3258(0.972 - 0.00526) = 1.282 \text{ MV}$$

Total charging voltage E = 2 MV

$$\therefore \quad \varepsilon = 1.282/2 = 0.641 = 64.1\%$$

Q.9. Explain (non-mathematically) how the double-exponential waveshape is built up in the circuits of Fig. 16.4. What are the approximate equivalent circuits during the build-up of the wavefront and the wavetail. How can the front-and tail-timings be determined quickly from these approximate circuits for preliminary work?

Ans. When the spark gaps SG in Fig. 16.5 in the Marx Multiplier circuit are open, the stage capacitances are all charged in parallel by the regulating transformer and isolating transformer through the diode. When the gaps are triggered and they flash over, they are all connected in series. The total source or generator capacitance C_S will then commence to charge up the loading capacitor C_L. It does so through the series resistance R_f. The equivalent circuit during this stage is approximately C_S and C_L in series with R_f also in series. This builds up the wavefront. We omit the parallel resistance R_t. This is why R_f is called the "wavefront" resistor. Figure 16.7(a) shows the approximate equivalent circuit during the build-up of the wavefront of the voltage impulse.

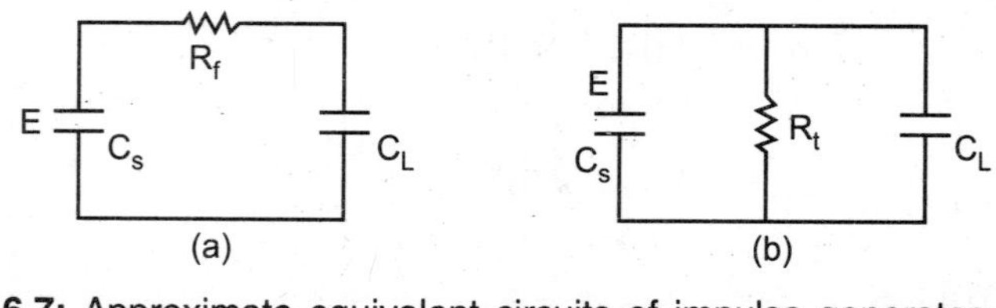

Fig. 16.7: Approximate equivalent circuits of impulse generator during: (a) Wavefront generation; and (b) Wavetail generation.

Once the charging of C_L is complete and the voltages of C_S and C_L become equal, they both discharge in parallel through R_t. This will make the voltage of C_L to decrease gradually and the wavetail is formed. The equivalent circuit during this stage is approximately as shown in Fig. 16.7(*b*). Now, C_S and C_L are in parallel with R_t.

1. Referring to Fig. 16.7(*a*), the final voltage across C_L is found as follows: Total charge in C_S at the instant of its discharge = $E \cdot C_S$ coulombs. If the voltage across C_S and C_L is V, their charge is $V(C_S + C_L)$, and this must equal the initial charge on C_S.

   $\therefore$ $$V = E \cdot C_S/(C_S + C_L)$$

   The voltage across C_L follows the usual charging equation:

   $$e_{LC} = V[1 - \exp(-t/\tau_C)]$$

   where t_C = time-constant of charging = $R_f \times$ (C_S and C_L in series)

   $$= R_f \times C_S \cdot C_L/(C_S + C_L)$$

   e_{LC} is an exponentially-rising voltage which corresponds to the wavefront of the impulse. Normally, we assume that the final voltage is reached in 5-time constants.

2. Refer to Fig. 16.7(*b*): When C_L, the loading capacitor, is completely charged, both C_L and C_S will discharge through R_t with initial voltage V and with the time constant

   $$\tau_d = R_t \times (C_S \text{ and } C_L \text{ in parallel}) = R_t(C_S + C_L)$$

   The equation for the voltage across C_L (or C_S or R_t) will be:

   $$e_{Ld} = V \cdot \exp(-t/\tau_d)$$

   This represents approximately the decaying portion of the voltage impulse. The initial time for this stage can be counted from the instant C_L attains its final voltage during the charging stage i.e., $5\tau_C$.

Example: In the previous example,

$$C_S = 80 \text{ nF}$$

$$C_L = 2 \text{ nF}$$

$$R_f = 300 \text{ ohm}$$

and, $$R_t = 600 \text{ ohm}$$

$$E = 2 \text{ MV}$$

1. $\therefore$ $$\tau_C = 300 \times (80 \times 2) \times 10^{-18}/\{(80 + 2) \times 10^{-9}\}$$

   $$= 0.585 \times 10^{-6} \text{ s}$$

   Also, $$V = E \cdot C_S/(C_S + C_L)$$

   $$= 0.9756\, E$$

   $$= 1.9512 \text{ MV}$$

   Thus, $$e_{LC} = V[1 - \exp(-1.708 \times 10^6 t)]$$

   since $$1/\tau_C = 1.708 \times 10^6$$

   This voltage reaches 30%V and 90%V in timings t_{30} and t_{90}, respectively.

   $\therefore$ $$0.3 = 1 - \exp(-1.708 \times 10^6 t_{30})$$

   giving, $$t_{30} = 0.209 \text{ µs}$$

   Also, $$0.9 = 1 - \exp(-1.708 \times 10^6 t_{90})$$

   giving, $$t_{90} = 1.349 \text{ µs}$$

The wavefront time will be approximately:

$$t_f = 1.67(t_{90} - t_{30})$$
$$= (1.348 - 0.209)/0.6$$
$$= 1.9\ \mu s$$

2. $$\tau_d = 600 \times 82 \times 10^{-9}$$
$$= 49.2 \times 10^{-6}\ \text{sec}$$
$$= 49.2\ \mu s$$

$\therefore$ $$e_{Ld} = V \cdot \exp(-t/49.2 \times 10^{-6})$$
$$= V \cdot \exp(-2.035 \times 10^4 t)$$

This will reach 0.5 in the time t_t

Thus, $$0.5 = \exp(-2.035 \times 10^4 t_t)$$

giving, $$t_t = 34.1 \times 10^{-6}\ \text{sec} = 34.1\ \mu s$$

If we add, $$5\tau_C = 5 \times 0.585 = 2.925\ \mu s,$$

the wavetail time will turn out to be

$$t_{0.5} = 34.1 + 2.925$$
$$= 37\ \mu s$$

According to these approximate methods, the generated wave has the timings 1.9/37 μs instead of the standard 1.2/50 μs. The timings obtained lie outside the tolerances, (0.84 to 1.56 μs)/(40 to 60 μs). The testing engineer will have to adjust the values of R_f and R_t by trial and error to obtain a standard wave within the allowable tolerances. Note that all this will depend upon the capacitance of the test object which is included in C_L which varies from test to test.

Q.10. How is a.c. power-frequency voltages of a few million volts generated in a cascade-connected testing transformer unit? What is the power factor of the transformer during testing of high voltage equipment?

Ans. The reader is referred back to Chapter 12 and Question 11 for details of cascade-connected testing transformer and Fig. 12.9.

The power factor of a h.v. testing transformer is always zero leading. It draws only a capacitive current from the source.

In order to minimise the current drawn from the source, inductive reactors have to be connected across the input to the first or ground unit. In many designs, these reactors are built into each unit of the cascade, since there is a magnetic core already present. It takes the form of a coil wound on the core but short-circuited on itself. Such a coil behaves as a pure inductance and compensates the capacitive current required by the transformer which is on capacitive load, such as insulator strings, open-circuited transformers, etc.

In order to reduce distortion of the output voltage waveshape, the core is designed to operate in the linear portion of the *B-H* characteristic and is not permitted to go into the non-linear part. Since the currents are low, the winding can be made from double-enamelled copper conductor wound very tightly in order to yield a high turn-to-turn capacitance. Since flashover of test object is very frequent during testing for flashover tests, which is a short-circuit on the unit, the leakage reactance must be high, about 0.1 per unit or 10%.

Because of the very high terminal voltages used, in the MV range, very large electrodes are necessary on top of the bushings. They are either spherical or hemi-spherical with sizes of approximately 3 metres in diameter for 400 kV units and 5 m for 2000 kV = 2 MV units.

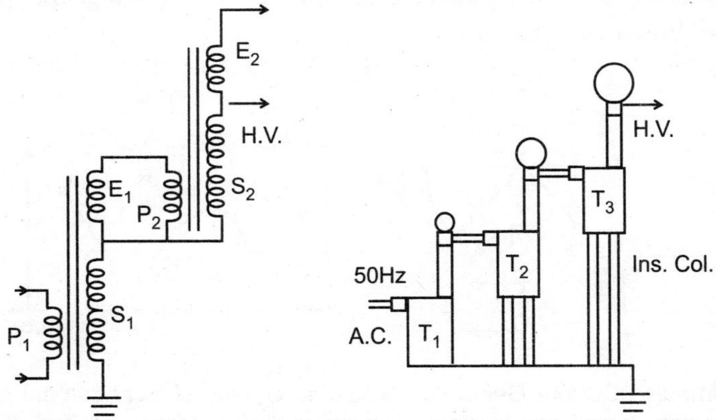

Fig. 16.8: Schematic of Primary (P), Secondary (S), and Excitation (E) windings of 2-unit cascade-connected testing transformer. (Right): Overall view of 3-unit tank cascade-connected transformer.

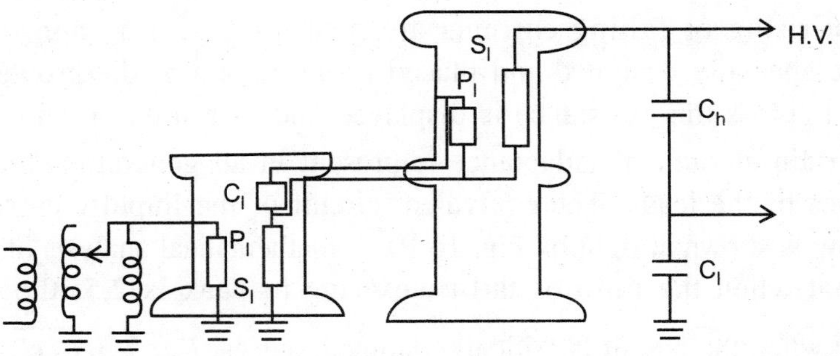

Fig. 16.9: Internal arrangement of 2-unit cascade-connected testing transformer inside insulated cylinders. Highvoltage electrodes removed.

Figure 16.8 shows overall sketch of a 3-unit cascade-connected testing transformer using sheet-steel tanks and Fig. 16.9, a 2-unit cascade-connected transformer using insulating cylinders. The internal arrangement of windings and the external measuring circuit are identical for both steel-tank and insulating-cylinder types.

The output voltage can be measured in several ways:

(*a*) The input voltage at the ground end can be calibrated to give the h.v. terminal voltage across the test object;

(*b*) A capacitive voltage divider can be used in a manner similar to impulse generators;

(*c*) A special voltmeter called the "peak voltmeter" is also used.

These will be discussed in the next Chapter under high voltage measurements.

Q.11. How are lightning-type impulse currents up to 200,000 amperes peak generated in a laboratory for testing lightning arresters?

Ans. Impulse currents are generated by charging several h.v. capacitors in parallel through a transformer and diode, and discharging them in parallel through the test object i.e., a lightning arrester or a h.r.c. fuse, etc. The schematic diagram of an impulse-current generator is shown in Fig. 16.10.

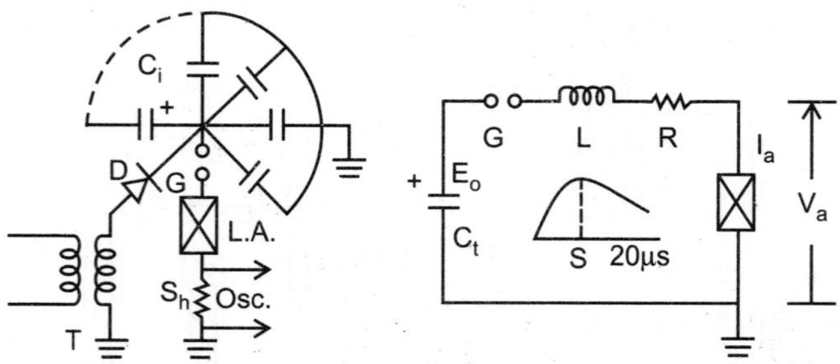

Fig. 16.10: Impulse Current Generator. Actual schematic of arrangement of components, equivalent circuit and current waveshape during testing of ZnO arrester.

Lightning arresters are tested at specified crest values of current having the waveshape of 8/20 μs, reaching crest value in 8 μs and falling to 50% of crest value in 20 μs. The large fast-rising or falling currents are measured with a "non-inductive shunt" or resistance specially designed and placed in the circuit at the ground end. The voltage developed across this resistance is displayed and measured on an oscilloscope.

A certain amount of inductance is present in all generators because of the liberal dimensions of the leads. The equivalent circuit of the impulse-current generator when discharging is shown at right on Fig. 16.10. A mathematical analysis of the electrical circuit shows that when the ratio of tail-times/time to peak is 2.5, the waveshape can be generated when the circuit is critically damped, that is, $R = 2\sqrt{(L/C)}$. For such a circuit, the current flowing through an arrester under test is given by:

$$i(t) = \{(E_0 - V_a)/L\} \cdot t \cdot \exp(-at)$$

where E_0 = charging voltage or voltage across each capacitor,

V_a = arrester voltage which we assume constant, that is, it is a ZnO gapless arrester,

$a = R/2L$,

R = circuit resistance = $2\sqrt{(L/C_t)}$

L = circuit inductance,

and C_t = total generator capacitance when all stages are in parallel.

This is the well-known transient response of a critically-damped *R-L-C* series circuit.

The peak value of current i_p occurs when $t = t_p$ at which $di/dt = 0$.

Now, $di/dt = \{(E_0 - V_a)/L\} \cdot e^{-at} \cdot (1 - at) = 0$ at $t = t_p$.

$\therefore \quad 1 - a \cdot t_p = 0$

giving $\quad a \cdot t_p = 1$

or, $\quad t_p = 1/a = 2L/R$

The resulting peak value of current is,

with
$$aL = R/2 = \sqrt{(L/C_t)}$$
$$i_p = \{(E_0 - V_a)/L\} \cdot (1/a) \cdot e^{-1} = \{(E_0 - V_a)/2.7183\}$$

In order to determine the time to 50% peak value on the tail, we must solve the equation:

$$\{(E_0 - V_a)/L\} \cdot t_t \cdot \exp(-at_t) = 0.5 i_p = 0.5\{(E_0 - V_a)/2.7183\} \cdot (1/La)$$

or, $\quad a \cdot t_t \cdot \exp(-at_t) = 0.5/2.7183$

or, $\quad 5.4366a \cdot t_t = \exp(a \cdot t_t)$

Taking logarithms, $\ln(5.4366 \cdot a \cdot t_t) = a \cdot t_t$

Using the property that $\quad (a \cdot t_p) = 1$

we can write: $\quad \ln(5.4366\, t_t/t_p) = t_t/t_p$

A trial and error solution yields, $t_t/t_p = 2.7$

When $\quad t_p = 8\ \mu s$

the time to 50% on tail, $\quad t_t = 2.7 \times 8 = 21.6\ \mu s$

$\therefore$ This is a standard waveshape of nearly 8/20 μs.

Therefore, the required properties in the impulse-current generator are as follows:

1. Several impulse capacitors are charged in parallel and discharged in parallel through the test object by firing a spark gap;
2. The inductance L of the generator circuit must be measured first, which stays nearly constant for all tests;
3. External resistance must be connected of such a value that the resulting R–L–C_t circuit is critically damped, which requires that $R = 2\sqrt{(L/C_t)}$;

4. The time to peak value will then be: $t_p = 2L/R$;
5. The peak value of impulse current is, with ZnO gapless arrester,

$$i_p = \{(E_0 - V_a)/2.7183\} \cdot \sqrt{(C_t / L)},$$

where E_0 = charging voltage of impulse capacitors,
V_a = constant voltage held by the ZnO gapless arrester,
C_t = total capacitance of generator with all stage-capacitors in parallel (which stays constant for all tests), and of course 2.7183 = base of natural logarithm, *e*.

6. The time to half value on the tail is: $t_t = 2.7t_p$

Example: An impulse current generator consists of 10 impulse capacitors each of 0.125 μF. The circuit inductance is 60 μH and the circuit is critically damped. The charging voltage is E_0 = 200 kV, and the arrester under test maintains a constant voltage of 25 kV peak. Calculate the following quantities:

(*a*) Value of series resistance required;
(*b*) Peak value of current through the arrester under test;
(*c*) Time to peak value, t_p; and
(*d*) Time to 50% peak value on tail, t_t.

Solution:

$$C_t = 10 \times 0.125 \times 10^{-6} = 1.25 \times 10^{-6} \text{ farad}$$

(*a*) ∴

$$R = 2\sqrt{(L/C_t)} = 2\sqrt{(60/1.25)} = 13.86 \text{ ohm}$$

(*b*)

$$i_p = \{(E_0 - V_a)/2.7183\} \cdot \sqrt{(C_t / L)} = \{(200 - 25)/(2.7183 \times 6.93)\} = 9.292 \text{ kA} \approx 10{,}000 \text{ amp}$$

(*c*)

$$t_p = 1/a = 2L/R = 2 \times 60 \times 10^{-6}/13.86 = 8.66 \text{ μs}$$

(*d*)

$$t_t = 2.7t_p = 23.4 \text{ μs}$$

The generated current wave is 8.66/23.4 μs or nearly 8/20 μs, having a peak value of nearly 10,000 amperes.

Q.12. What are the types of high voltage d.c. generators used in practice in laboratories for testing insulations and in other important fields?

Ans. There are two main types:

1. Voltage multiplier circuits using transformer, capacitors and diodes;
2. Belt-type electrostatic generators.

The voltage-multiplier circuit is severally known as (*a*) Cockroft-Walton generator, or, (*b*) Greinacher Chain. Belt-type electrostatic generator is known as van de Graaff generator. We will devote more attention to the former, since the latter is now out of date.

High voltage d.c. generators have become very widely used in the following fields:

(*a*) Testing of d.c. dielectric strength of insulations;

(*b*) Corona experiments in laboratory and in the field with full-scale d.c. transmission lines;

(*c*) X-ray apparatus for radiology in the medical profession;

(*d*) Particle accelerators in nuclear research application.

By far the most widely-used generator is the voltage-multiplier circuit since it has no moving parts and advances made in solid state diodes can be used very effectively and conveniently. But there are many installations using van de Graaff generators and their variations, especially with the use of high-pressure nitrogen as insulation for the generator even for some million volts. Voltage multiplier circuits are very large in size since they use dry air as insulation and the capacitors are large in dimensions. With the availability of high voltage diodes (for up to 15 kV per unit) and ease of change of polarity of the output polarity, they have acquired their own position (or niche') in the application of h.v. d.c. generators.

Q.13. How does a half-wave h.v. d.c. generator (power supply) function? What are the relations between d.c. output voltage, load current, ripple content and the applied a.c. voltage?

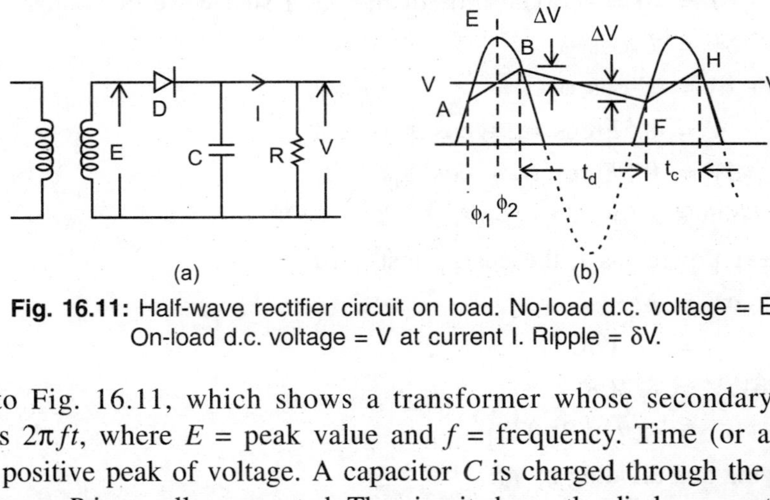

(a) (b)

Fig. 16.11: Half-wave rectifier circuit on load. No-load d.c. voltage = E. On-load d.c. voltage = V at current I. Ripple = δV.

Ans. Refer to Fig. 16.11, which shows a transformer whose secondary voltage is $e(t) = E \cos 2\pi ft$, where E = peak value and f = frequency. Time (or angle) is counted from a positive peak of voltage. A capacitor C is charged through the diode D. A load of resistance R is usually connected. The circuit shows the diode connection when ground is negative with respect to the source point S. By reversing the diode, ground can be made positive and terminal S negative.

First, considering no load on the capacitor i.e., with load current $I = 0$, it is clear that the capacitor will be charged to the peak value of voltage E and will remain constant, even though the source voltage follows the falling portion of the sine wave. The diode is reverse biased and cuts off. Thus, the d.c. voltage on no-load is $V = E$. A small resistance, not shown, is usually in the circuit behind the diode D so that when initially switched on, the current surge into C is not infinity.

Next consider the load-resistance R connected which draws a d.c. current I at voltage V.

$\therefore \quad I = V/R$

The charge acquired by C supplies this current when the diode is experiencing reverse bias, shown as point B when the capacitor voltage is more positive than the sine wave of source voltage. The charge of C will decrease over the discharge time shown as t_d. The voltage falls until the point shown as F when the sine wave becomes more positive. The capacitor gets charged from point F to H (or A to B).

Let the charging process from A to B (or F to H) take place linearly. The change in voltage during charging or discharging is shown as (2·δV). Then, the following relations hold:

E = peak value of sinusoidal a.c. voltage,

V = average value of d.c. voltage,

δV = ripple voltage,

t_C = charging time from angle ϕ_1 to ϕ_2,

and $T = 1/f$ = periodic time of the input sine wave of voltage.

$\therefore \quad V - \delta V = E \cos \phi_1$

and $V + \delta V = E \cos \phi_2$

Thus, $V = 0.5E(\cos \phi_1 + \cos \phi_2)$

and $\delta V = 0.5E(\cos \phi_2 - \cos \phi_1)$

Change in voltage across C is $2 \cdot \delta V$.

If the current flowing is I, the charge lost $= I \cdot t_d$.

$\therefore \quad 2C \cdot \delta V = I \cdot t_d$

$= t_d \cdot V/R,$

or, $\delta V/V = t_d/2CR$

$= (\pi f t_d)/(2\pi f CR)$

by multiplying and dividing by πf.

Here, $1/2\pi fC = X_C$ = reactance of C at frequency f, and as shown in Fig. 16.11(b),

$t_d/T = t_d . f$

$= 1 - (\phi_1 + \phi_2)/2\pi$

We also have the further properties:

1. $\delta V/V$ = fractional ripple $= (X_C/R) \cdot \pi \cdot \{1 - (\phi_1 + \phi_2)/2\pi\}$;
2. $I = V/R$;
3. $V - \delta V = E \cdot \cos \phi_1$,

   and $V + \delta V = E \cdot \cos \phi_2$;
4. $V/E = 0.5(\cos \phi_1 + \cos \phi_2)$.

We now have 3 dimensionless or normalised quantities: (*i*) Ripple, $\delta V/V$; (*ii*) V/E; and (*iii*) X_C/R. When two of these quantities are specified, we can determine the third if E and f are known. An example will make this clear.

Example: In the half-wave rectifier circuit, the average d.c. voltage is to be 70% of the peak value of a.c. input voltage, i.e., $V/E = 0.7$. The ripple is therefore $\delta V/V = 15\% = 0.15$. Calculate the ratio required for X_C/R.

Solution: There are many steps to follow. We have:

$$V - \delta V = E \cdot \cos \phi_1$$

and

$$V + \delta V = E \cdot \cos \phi_2$$

With the given values of $V/E = 0.7$

and $\delta V/V = 0.15$

so that $\delta V/E = 0.7 \times 0.15 = 0.105$

we can calculate the angles ϕ_1 and ϕ_2, as follows:

$$\cos \phi_1 = (V - \delta V)/E = 0.7 - 0.105 = 0.595$$

giving $\phi_1 = 0.9335$ rad

Similarly, $\cos \phi_2 = 0.7 + 0.105 = 0.805$

and $\phi_2 = 0.635$ rad

∴ The charging time is given by:

$$t_C/T = (\phi_1 + \phi_2)/2\pi = (0.9335 + 0.635)/2\pi = 0.25$$

or, $t_C = \frac{1}{4}$ cycle

The discharge time is, consequently,

$$t_d/T = 0.75$$

or, $t_d = \frac{3}{4}$ cycle

$$\delta V/V = (X_C/R)\pi \cdot \{1 - (\phi_1 + \phi_2)/2\pi\}$$

or, $0.15 = \pi \times 0.75 \cdot (X_C/R)$

This finally gives: $X_C/R = 0.15/(0.75\pi) = 0.0637$

Note that except in $X_C = 1/2\pi fC$

the frequency of the input a.c. voltage source does not enter into any of the other calculations. In order to keep the capacitance low, it is normally a good practice to use high frequency. But in a high voltage laboratory, it is usual to use power frequency (50 or 60 Hz).

Let for a system, $E = 1000$ volts peak

Then, the d.c. voltage will be $V = 700$ volt

ripple voltage $\delta V = 0.15 \times 700 = 105$ volt

If we chose $R = 1000$ ohm

the average d.c. current is $I = V/R = 0.7$ ampere

Then, $X_C = 0.0637 \times 1000$

$= 63.7\ \Omega = 1/2\pi fC$

At, $f = 50$ Hz

$C = 1/2\pi f X_C$

$= 1/2\pi \times 50 \times 63.7 = 50\ \mu\text{F}$

At, $f = 1$ kHz

$C = 2.5\ \mu\text{F}$

At, 700 volts and 0.7 amp through R, its power rating is nearly 500 watts!

For a load of 10 kilohms,

$I = 0.07$ amp $= 70$ mA,

and $C = 5\ \mu\text{F}$ at 50 Hz

and $0.25\ \mu\text{F}$ at $f = 1$ kHz

The power rating of the resistor is now about 50 watts.

Q.14. What is the principle of action of a voltage-doubler circuit?

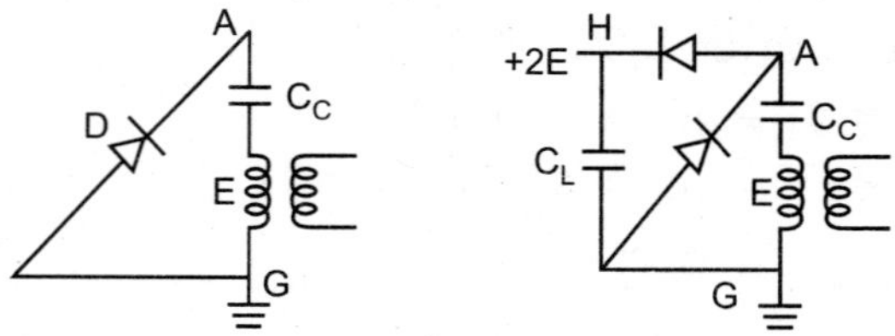

Fig. 16.12: Cockroft-Walton high voltage d.c. generator or Greinacher Chain. Connection of single stage with 2 capacitors, 2 diodes and excitation transformer.

Ans. The voltage-doubler circuit due to Greinacher and Cockroft-Walton as well designed by several others, is shown in Fig. 16.12. It functions as follows:

1. When the diode 1 is conducting, the capacitor C_C is charged to the voltage E of the transformer secondary with point A positive, in a manner similar to a half-wave rectifier circuit;
2. When C_C has acquired its full charge and voltage E, its charging stops;
3. During the next half-cycle, ground becomes negative and diode 1 is reverse biased. The voltage from A to G is now equal to a maximum of $2E$ with terminal A positive with respect to ground G;

4. Since the capacitor C_L has its bottom plate or electrode connected to G which is negative while point A is positive, the diode 2 is forward biased and conducts;
5. The voltage of C_C along with source in series will charge C_L to twice the peak value of the supply voltage;
6. During the next half-cycle, diode 1 conducts and makes potential of terminal A positive by E, whereas the point H remains at $2E$. The diode 2 is reverse biased and C_C charges again and regains the charge it lost to the capacitor C_L during the previous half cycle;
7. Thus, by using 2 capacitors with C_C = charging capacitor and C_L = loading capacitor, and 2 diodes, twice the value of supply voltage is obtained at the high voltage terminal H to ground G. When a load resistance R is connected, there is a drop in the d.c. output voltage.

   Let I = load current. This current is supplied by the load capacitor C_L.

   The charge lost in one cycle is $Q = I \cdot T = I/f$

   The voltage dip δV is given by: $C_L \cdot \delta V = I/f$

   or, $\delta V = I/fC_L$

   The voltage dip in C_C is also of the same amount. Therefore, the regulation is

$$\delta V/V = I/fC_LV = 1/fC_LR$$

   where $R = V/I$ = load resistance value.

   The polarity of voltage of H is reversed by reversing the diode.

Q.15. How does a voltage-multiplier generator of the Cockroft-Walton type or Greinacher Chain function? What is the ripple in the d.c. voltage output of such a generator with N stages and all capacitors have equal values? Explain first with a generator having 2 stages and then extend it to N stages.

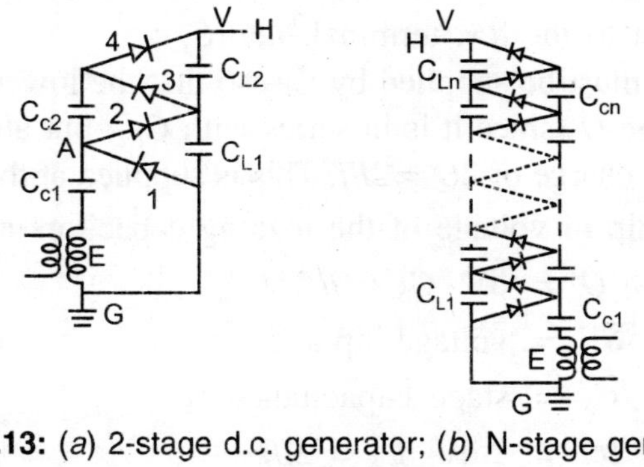

Fig. 16.13: (*a*) 2-stage d.c. generator; (*b*) N-stage generator.

Ans. Refer to Fig. 16.13(*a*). The action is as follows:

1. The capacitors C_{C1} and C_{C2} are known as charging capacitors, while C_{L1} and C_{L2} are loading capacitors. In a 2-stage Greinacher Chain or Cockroft-Walton generator, for 2 stages, there are 4 capacitors, 4 diodes and one source transformer. Thus, one stage consists of 2 capacitors and 2 diodes. Consequently, in an N-stage generator,

there will be $2N$ capacitors and $2N$ diodes with only one source transformer, as shown in Fig. 16.13(*b*);

2. In Fig. 16.13(*a*), during conduction of diode 1, C_{C1} is charged to a maximum voltage of E, the peak value of a.c. source voltage, with terminal A positive;
3. When the a.c. voltage decreases from its peak value on the sine wave, diode 1 is reverse biased and stops conducting. Point A is positive with respect to G and the potential of terminal C of capacitor C_{L1} is lower than that of point A, since it is not yet charged. Therefore, diode 2 is forward biased;
4. The capacitor C_{C1} and the secondary voltage E of the source transformer are in series and charge up C_{L1} to a voltage of $2E$. Then conduction of diode 2 stops;
5. When the source voltage decreases, point C is at a higher potential than A so that diode 2 is reverse biased. Diodes 1 and 3 conduct so that the charge and voltage of C_{L1} are transferred to capacitor C_{C2};
6. In the next half-cycle, diodes 2 and 4 are forward-biased. Thus, C_{L1} receives the charge it lost from C_{C1} and E. At the same time, C_{C2} transfers its charge and voltage to C_{L2};
7. Consequently, the voltage of H is positive with respect to G by the sum of the voltages of C_{L1} and C_{L2}. Since each of these is charged to $2E$ when no current is drawn from the generator, the generator output voltage on no load is $4E$.

   We observe that the lowest capacitor C_{C1} has a maximum voltage of E, while all others have $2E$. Here, all capacitors usually have the same value of capacitance, but in some designs the lowest capacitor C_{C1} has twice the capacitance value of all other capacitors;
8. When a load is connected between H and G and draws a current I, in one cycle time T the charge flowing out is $Q = I{\cdot}T$, coulomb $= I/f$. This is supplied by the capacitor closest to the h.v. terminal, i.e., C_{L2};
9. The charge lost must be supplied by C_{C2} so that the lower capacitor C_{L1} must supply not only a charge Q (since it is in series with C_{L2}) but also an equal amount to C_{C2}. Thus, it looses a charge of $2Q = 2I/f$. This is supplied at the expense of dip in voltage;
10. Therefore, the dip in voltage of the loading capacitors are as follows:

In $C_{L2},\ Q = \delta V \cdot C_S = I/f$

where δV = voltage dip and

C_S = stage capacitance

In $C_{L1},\ 2Q = 2{\cdot}\delta V \cdot C_S = 2I/f$

∴ Total dip in voltage on load is:

δV_2 = sum of voltage dips in C_{L1}

and $C_{L2} = (1 + 2){\cdot}I/fC_S$

If V = d.c. voltage output and

R = load resistance,

then $I = V/R$

We can therefore express the ripple in output voltage as:
For a 2-stage generator, $\delta V_2/V = (1 + 2)/(f \cdot C_S \cdot R)$

11. We can extend this argument for a generator with N stages. The total dip in output voltage will be:

$$\delta V_N = (1 + 2 + 3 + \cdots + N) \cdot I/fC_S = N(N + 1) \cdot I/2fC_S$$

Consequently, the total ripple as a fraction of the d.c. voltage is:

$$\delta V_N/V = N(N + 1)/2fC_SR$$

when all stage capacitors C_S are equal in value.
We can also write this in terms of $X_C = 1/2\pi fC_S$, the reactance of stage capacitor C_S at supply frequency f, as follows:

$$\delta V_N/V = \pi \cdot N(N + 1) \cdot X_C/R$$

when all C_Ss are identical.

12. We also observe that lower the capacitor is from the h.v. terminal, larger will be the dip in its voltage which it experiences. In order to reduce the dip and in order to supply the charge necessary, it is beneficial to have a higher value of capacitance as we approach the ground terminal. This is done in several designs, but we will not discuss the problem here.

Example: In a Greinacher Chain with 6 stages, the stage capacitance $C_S = 0.02$ μF each. The transformer secondary peak voltage is 120 kV. For each mA of load current I, find the ripple. The frequency used is $f = 750$ Hz, supplied by a motor-generator set.

Solution: $N = 6$, $I = 10^{-3}$ A, $f = 750$, $C_S = 0.02 \times 10^{-6}$, $E = 120$ kV

$$\therefore \quad \delta V = (1 + 2 + 3 + 4 + 5 + 6) \cdot I/fC_S$$
$$= 10^{-3} \times 21/(750 \times 0.02 \times 10^{-6})$$
$$= 1.4 \times 10^3 \text{ volt}$$
$$= 1.4 \text{ kV/mA}$$

$$\text{No-load voltage} = 2\,NE = 12 \times 120 \text{ kV}$$
$$= 1440 \text{ kV}$$
$$= 1.44 \text{ MV}$$

Q.16. What is the relation between the no-load voltage and the voltage V on load when the generator or chain is loaded to a current I?

Ans. On no-load, the voltage output is 2 NE. We will not give the detailed derivation of the voltage at the h.v. terminal on load, since it is considered beyond this tract. We will only give a formula here for the loss of voltage δV on load which the reader can use.

Bouwer's Formula: [Please see J.M. Craggs and J.M. Meek: High Voltage Laboratory Technology, Butterworth Scientific, London, 1954, Chapter 1]: When all stage capacitances C_S s are equal,

$$\delta V = (I/f\,C_S) \cdot [(2/3)N^3 + (1/2)N^2 - (1/6)N]$$

Consequently, the output d.c. voltage is: $V = 2NE - \delta V$

If the ripple rV is superimposed on this, the d.c. output voltage on load has the minimum value $(V - rV)$ and maximum value $(V + rV)$. Note that for large number of stages, N large, the leading term in δV is $(2/3).N^2$. Then the ratio

$$rV/\delta V = [N(N+1)/2]/(3/2N^3)$$
$$= (3/4) \cdot \{(N+1)/N^2\}$$
$$\approx (3/4)\cdot(N/N^2) = 3/(4N),\ N \text{ large.}$$

Example: In the previous example, $N = 6$, $E = 120$ kV

$$C_S = 0.02 \times 10^{-6}$$
$$= 2 \times 10^{-8}$$

and $f = 750.$

$\therefore$ The loss of voltage δV per mA of load current is:

$$\delta V = \{10^{-3}/(750 \times 2 \times 10^{-8})\} \cdot \{(2/3)\ 6^3 + (1/2)\ 6^2 - (1/6)\ 6\}$$
$$= 66.7(144 + 18 - 1)$$
$$= 10.74 \text{ kV}$$

$\therefore$
$$V = 2NE - \delta V$$
$$= 12 \times 120 - 10.74$$
$$= 1430 \text{ kV}$$
$$= 1.43 \text{ MV}$$

If the load current drawn is $I = 10$ mA,

$$V = 1440 - 107.4$$
$$= 1333 \text{ kV}$$
$$= 0.9257 \times 1440$$

[Regulation = 1 – 1333/1440 = 0.0743 = 7.43%].

Q.17. What is the principle of generation of very high d.c. voltages by van de Graaff belt-driven generator?

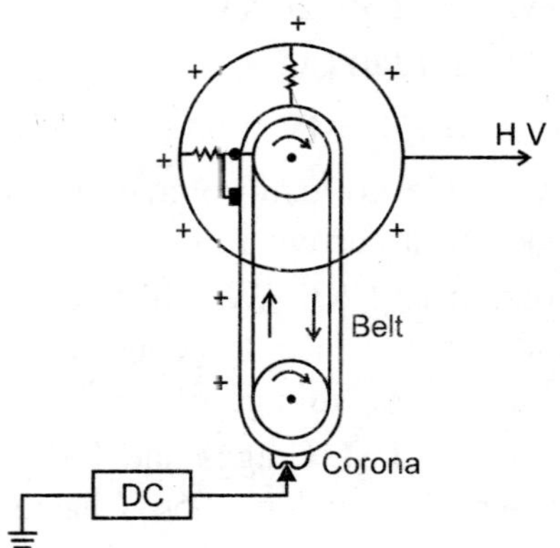

Fig. 16.14: van de Graaff h.v. belt-driven d.c. generator.

Ans. Figure 16.14 shows the schematic of a belt-driven van de Graaff electrostatic generator for some million volts d.c. These generators are used in nuclear physics laboratories and medical applications.

The principle of operation is very simple. At near ground end, a sharp pointed electrode is connected to a high voltage from 10 kV to 100 kV. This generates corona discharge which consists of charged particles or ions in the air or gas used. The point electrode is directed on to an insulated belt which is run by a pulley and driven by an electric motor. At the top end, where the belt runs over another pulley, electrodes in the form of brushes rubbing on the belt surface collect the charges and store them in a metallic electrode of very large size. This top metallic electrode is insulated from ground.

The voltage acquired is $V = Q/C$, where Q = stored charge collected from the belt and C = capacitance of electrode to ground.

The clearances required when air is used as the insulation i.e., when the generator is not enclosed in a pressurised medium such as nitrogen are very large. A generator for 1 MV may be 5 metres in height. The generator itself can be housed in a special room attached to X-ray clinics in hospitals or the entire generator with the belt and h.v. electrode can be enclosed in a metallic enclosure which could be grounded. But with the use of pressurised nitrogen in a tank the generator height can be brought down to 2 metres for 1 MV.

The size of the upper electrode is designed on the basis of yielding an electric stress not exceeding 3 MV/m = 30 kV/cm for radius of curvature for a spherical or hemi-spherical shape. These are manufactured by centrifugal spinning. For 1 MV, the radius of the sphere is 1/3 metre or a diameter of 2/3 m. Thus, a 1-metre diameter sphere or hemisphere can be used for 1.5 MV.

Q.18. What are the clearances to be used in a laboratory or in other high voltage installations from the generator h.v. terminal to ground or grounded objects?

Ans. These clearances determine the size of the laboratory dimensions. There are several types of voltage waveshapes to consider. These are: (*a*) d.c.; (*b*) Power-frequency a.c.; (*c*) Lightning impulse, (+ or –); (*d*) Switching impulse (+ or –).

Recommended clearances are as follows:

(*a*) **D.C.:** The flashover follows the expression:

$$V_{fo} = 500 \cdot d \text{ kV}$$

where d = gap length in metres. The withstand strength is nearly 80% of this flashover voltage. Thus, a gap of 1 metre can withstand 400 kV. With a further factor of safety, the recommended clearance for 1 MV is 4 metres. For other values of voltage, the clearance is linearly proportional. Thus, for every 250 kV, the clearance to be observed is 1 metre.

(*b*) **Power-frequency A.C.:** The flashover voltage of an air gap under 50 or 60 Hz a.c. is non-linear. The flashover voltage varies as $V_{fo} = 652 \cdot d^{0.576}$ kV, peak for rod-plane electrodes, and $V_{fo} = 500 \cdot d$ kV for rod-rod gaps, which is the same as the d.c. flashover value of a rod-plane gap.

The recommended clearances are as follows:
Up to 1 MV r.m.s., a linearly increasing clearance of 4 metres is required. Beyond 1 MV, the clearance is based on the withstand voltage varying as $d^{0.576}$.
Thus, for 1.5 MV r.m.s., the clearance required is

$$d = 1.5^{1/0.576} \times 4 = 4 \times 1.5^{1.736}$$
$$= 8.09 \text{ metres}$$

For 2 MV,
$$d = 4 \times 2^{1.736}$$
$$= 13.3 \text{ metres}$$

The general formula is $d = 4 \times (\text{MV})^{1.736}$, with MV in r.m.s. and $d \geq 1\text{m}$.

(*c*) **Lightning Impulse Voltage:** (1.2/50 μs). The required clearance increases linearly with voltage. A clearance of 2.5 metres will be necessary for each 1 MV. Thus, for 2 MV peak or crest value, the clearance required is 5 metres, etc.

(*d*) **Switching Impulse (250/2500):** A given electrode geometry has the lowest dielectric strength for positive-polarity switching surge impulse. The gap length or clearance required increases as the square of the voltage value (peak or crest).

The recommended clearance is: $d = 4 \times (\text{MV})^2$. This requires 4 metres for 1 MV, but 16 metres for 2 MV. Switching-surge testing of a transformer for 750-kV level can go up to 2.4 MV as per Standard Specifications. Thus, a laboratory required to test equipment at 750-kV transmission-voltage level will need a clearance of (4×2.4^2) = 23 metres! Such labs are huge and no intermediate pillars are permitted to support the roof, just as is the case for hangars for jumbo-jet planes or cinema halls.

As a matter of interest and example, the h.v. lab at the Central Power Research Institute (C. P. R. I.) at Bangalore is the largest hall of this type in Asia. There are no pillars inside the hall with only clear space between all four sides or walls. A lab meant only for lightning tests up to 3 MV will be smaller in size than a laboratory where switching-surge tests up to 2 MV have to be carried out.

It is, therefore, sometimes preferable to use outdoor generators to save on buildings as is done at the C. P. R. I for higher voltages. The High Voltage Laboratory at the Indian Institute of Science on the opposite side of the road in front of the C. P. R. I. has all its h.v. equipment indoors but was designed for only lightning-impulse tests, a.c. tests, and d.c. tests, as originally planned in 1948 when India attained political independence and engineering labs mushroomed at the I.I.Sc. The H.V. Lab was opened in 1951 by the first President of India, Dr. Rajendra Prasadji. (This author was a student there on that day! Later on, he was a scientist at the H.V. Division of the C. P. R. I. from 1978 to 1983!).

❑❑

CHAPTER 17

HIGH VOLTAGE MEASUREMENTS AND TECHNIQUES

Q.1. What are the major types of measurements to be carried out in a high voltage laboratory for testing equipment?

Ans. We will only list the measurements required in routine testing and not in research activities. Some of these will also apply for development of equipment. Following are the measurements required:

(*a*) Flashover voltages of an air-gap insulation and other resealable types of insulation. These are mainly: Insulator strings, bushings, lightning arresters, oil and other insulating media. Solid insulation in equipment such as transformers and cables are never tested for breakdown.

(*b*) Withstand strength of all types of insulation.

(*c*) Capacitance and resistance measurement of h.v. cable insulation (tan δ).

(*d*) Partial discharge measurement in h.v. transformers and cables.

(*e*) Radio noise measurement.

(*f*) Lightning arrester type tests. Type tests are carried out on a few random samples from a lot of arrester material. A routine test is a test performed on every disc manufactured. Such tests may be performed in manufacturing plants.

The above basic tests require the following quantities to be measured for which sources, instruments and techniques should be provided in laboratories:

1. Measurement of high and extremely-high voltages: These are of several types, namely: (*a*) D.C., (*b*) A.C., (*c*) Lightning impulse voltages, (*d*) Switching-surge voltages. Sources for these have already been described in Chapter 16 earlier.
2. Measurement of high lightning-type impulse currents:

3. Measurement of capacitances of h.v. cables and equipment, and loss angle tan δ at high voltages using the Schering Bridge;
4. Radio noise measurement with RN meter;
5. Partial discharge measurement using pulse discharges either visually on an oscilloscope or on a Radio Noise meter.

Q.2. How is power-frequency a.c. voltage measured by an electrostatic voltmeter?

Ans. The principle on which measurement of very high voltages is done depends upon the change in *e.s.* energy stored in a capacitor with distance of the electrodes. The *e.s.* energy stored in a capacitor is written as:

$$W_{es} = CV^2/2$$

$$= \tfrac{1}{2}\, \varepsilon_0 \varepsilon_r\, AV^2/x \text{ joule}$$

where V = applied voltage between the plates or electrodes,

C = capacitance,

A = area of the plates,

x = distance of separation between them,

ε_0 = permittivity of free space = 8.842 μF/m,

and ε_r = relative permittivity of the insulation between the plates.

ε_r = 1 for air and nitrogen.

Then, the force between the plates is given by

$$F_{es} = -\delta W_{eS}/\delta x$$

$$= \tfrac{1}{2}\, \varepsilon_0 \varepsilon_r\, AV^2/x^2 \text{ newton}$$

In an electrostatic voltmeter, one of the plates is fixed while the other is capable of movement, but is restrained by the force of an elongated spring. A pointer is attached to the movable plate (or a light spot) which moves on a scale calibrated in kV.

Since the force is proportional to V^2, the average force is the square of the root-mean-square (r.m.s.) value. The meter scale is thus non-uniform. Figure 17.1 shows the schematic of one type of meter.

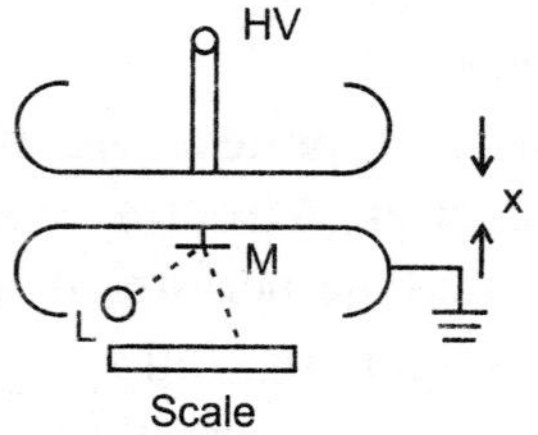

Fig. 17.1: Electrostatic voltmeter.

Single, directly-connected *e.s.* voltmeters can be used for voltages up to 100 kV, r.m.s., or d.c. For measuring higher voltages, voltage dividers are necessary which step the voltage down to suit the meter. Multi-range meters have a device for changing the gap length x between the two plates in proper steps. For a.c. work, potential transformers offer a convenient method of applying a lower voltage to the meter.

Q.3. How does a sphere gap measure a voltage? What is the technique to be followed in using this?

Ans. The gap between two perfect spheres (one of them at high voltage and the other usually grounded) makes a very convenient and accurate (to 3% tolerance) voltmeter for measuring extremely high d.c., a.c., and impulse voltages. But certain data must be available beforehand (calibration) and precautions have to be observed, chief among which are:

(*a*) The gap spacing should not exceed the diameter of the spheres. For spacing less than the radius, ±3% accuracy can be claimed. But for gap spacings exceeding the radius, ±5% accuracy is claimed. Thus, higher the voltage to be measured, larger will be the size of spheres.

(*b*) All grounded objects in the lab must be removed far away.

(*c*) Dust collection on the surface of spheres must be prevented.

(*d*) Atmospheric conditions in the laboratory must be carefully measured and the disruptive voltage corrected to N.T.P.

(*e*) Gap spacing must be measured accurately.

(*f*) A standard table of flashover voltage values against gap spacing must be available. These are published by all Standards Institutions, such as the I.E.C. and I.S. Some examples are provided below.

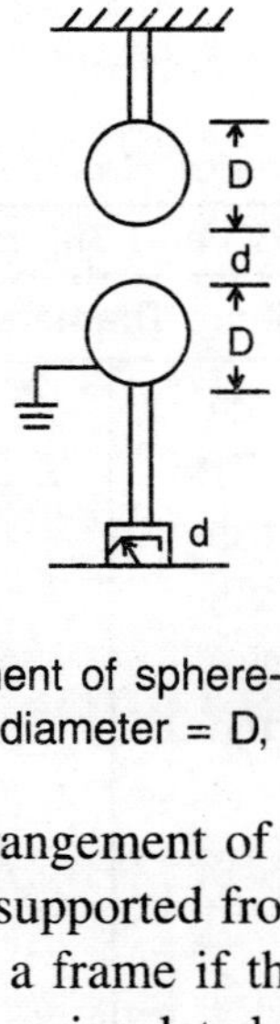

Fig. 17.2: Physical arrangement of sphere-gap for measurement of high voltages. Sphere diameter = D, Gap spacing = d.

Figure 17.2 shows the general arrangement of a sphere gap with sphere diameter D and gap spacing d. The h.v. sphere is supported from the top by an insulated rod fixed to the roof of the laboratory if fixed, or a frame if the sphere gap is to be made portable. The lower sphere is also mounted on an insulated rod and carries a threaded rod which can be rotated so as to vary the gap. This is carried out either manually or by a remote-controlled motor from the control panel. The gap spacing is indicated on a suitably-calibrated meter.

In order to determine the flashover voltage of a test object, such as an insulator string or bushing with spill gap (rod-to-plane gap or rod-to-rod gap) the test object is connected in parallel with the sphere gap. Let for example the flashover voltage of the test object be estimated to be nearly 500 kV. The gap setting of the spheres is chosen from the standard table of calibration for this voltage (500 kV and corrected to the atmospheric conditions in the laboratory). If the voltage is increased gradually from the source, and the test object (with resealable insulation) flashes over but not the sphere gap, then it is clear that the breakdown voltage of the test object is lower than 500 kV. The sphere-gap spacing is reduced a little and the experiment repeated. After a few trials, a gap setting is reached at which both the test object and the sphere gap flashover simultaneously. The corresponding voltage is read and corrected back to standard atmospheric conditions.

A similar procedure is followed in order to determine the withstand voltage of the insulation. In such cases, other standard specifications, such as IS 2026, give the withstand voltages at different system voltage levels. The sphere gap must be accurately set for this specified voltage or lower and made sure that the insulation does not flashover. The gap setting is changed gradually until the withstand voltage level is ascertained. Testing at very high voltages is a highly developed art and requires lot of experience in order that expensive equipments do not suffer damage during the testing procedure. This is a very big risk and cannot be learnt by reading books on this subject. But only some guidelines and directives can be given.

Some examples of standard sparkover voltage values of sphere gaps is given below in the following Table.

Table: Peak Values of Flashover Voltages in kV

AT 25°C and 760 mm Hg					
Gap Spacing	**Sphere Diameter, D, (cm)**				
d, (cm)	12.5	25	50	100	150
2.5	71.5	71.5			
5	127	135	136		
6	144	158			
10	192	240	259	262	262
12.5	210	280	315	320	
25			520	605	
50			720	1000	1110
75				1210	1420
100				1340	1630

All values are good within ±3%, except the ones underlined. They are good within ±5% and hold for $d > 0.5D$. In general, sphere gaps are accurate within these tolerances. Thus, as the voltage to be measured increases, larger spheres must be used. The largest sphere gaps used in the world have gone up to 200 cm = 2 metres in diameter.

Correction factors for atmospheric temperature and pressure have already been discussed in Chapter III. They depend upon the air-density factor δ. Recalling those properties here, we have:

If, V = actual measured sparkover voltage,

V_0 = corrected sparkover voltage for 25°C and 760 mm Hg,

p_0 = standard atmospheric pressure = 760 mm Hg,

p = actual measured pressure at test in mm Hg,

t = actual measured temperature at test in °C,

and t_0 = standard atmospheric temperature = 25°C,

then, $V_0 = V/\delta$

where, $\delta = (p/p_0) \cdot (T_0/T)$

$= \{p/(273 + t)\} \cdot \{(273 + 25)/760\}$

$= 0.392 \cdot p/(273 + t)$

Example: At the high voltage laboratories at Bangalore during a test, the temperature, pressure and flashover voltage of a test equipment were measured to be:

t = 18°C,

p = 745 mm Hg,

and, V = 765 kV, peak

If the test object is to be used in Chennai at sea level and in summer temperature of 45°C, what would be the flashover voltage of the same insulation?

Solution: At t = 18°C,

p = 745 mm Hg,

$\delta = 0.392 \times 745/(273 + 18)$

$= 1.0036$

$\therefore$ $V_0 = V/\delta$

$= 765/1.0036$

$= 762.3$ kV

at 25°C and 760 mm Hg. If the flashover voltage of the test object is V_m at Chennai,

then: $V_0 = V_m/\delta_m$

where, $\delta_m = 0.392 \times 760/(273 + 45)$

$= 0.9369$

$\therefore$ $V_m = V_0 \cdot \delta_m$

$= 762.3 \times 0.9369$

$= 714$ kV

[In general, if p_1, T_1 refer to the test location, then,

$$V_0 = V_1/\delta_1$$
$$= (1/0.392) \cdot (T_1/p_1) \cdot V_1$$

Similarly, if p_2, T_2 refer to the second location where the equipment will be used, then,

$$V_0 = (1/0.392) \cdot (T_2/p_2) \cdot V_2$$

Equating the two expressions for V_0, we obtain V_2 directly from V_1 , as follows:

$$V_2 = (T_1/T_2) \cdot (p_2/p_1) \cdot V_1$$
$$= (291/318) \cdot (760/745) \cdot 765$$
$$= 714 \text{ kV, peak}]$$

Note that the test result obtained at Bangalore (1000 m above sea level) and at its cool temperature of 18°C (745 mm Hg) must be corrected for the atmospheric condition at Chennai which is at sea level and very hot (760 mm Hg). [As a personal note, this author was born at Bangalore!]

Q.4. How can very high voltages be measured (steady-state only) with (*a*) a resistance, and (*b*) a capacitance?

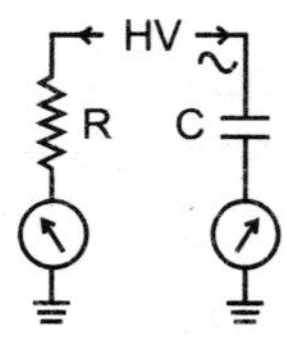

Fig. 17.3: Measurement of high voltages by High Resistance and Ammeter, and Capacitance and Ammeter (A.C. only).

Ans. (*a*) Figure 17.3 shows a high voltage resistance connected between the h.t terminal and ground through a milli-ammeter or micro-ammeter at the ground end. (The meter could also be located at the h.t. end and read either through a pair of binoculars or optical-fibre information-transmission methods can be used). Very simply, the reading on the ammeter can be calibrated to read kV, since $I = V/R$. This is possible for both d.c. and a.c. but not for impulse voltages. The entire resistance arrangement consists of a very large number of carbon resistors connected in series to give the required value and wound on a tall insulating cylinder or supported on flat cards.

Carbon resistors of the usual electronic kind are available up to 2 watts and 500 volts = 0.5 kV. The current loading permitted is therefore a maximum of 4 mA. Thus, for measuring 1 MV = 1000 kV, nearly 2000 such little resistors must be connected in series by soldering their pigtails! The total power dissipated will be 2 × 2000 = 4000 watts = 4 kW! If instead of 2 W resistors, ½ watt resistors are used, the power dissipated by 2000 of them is 1 kW while the maximum permitted current is 1 mA for a maximum voltage of 500 V across each. If sufficient cooling surface is provided or the entire assembly is enclosed in oil, this is a very easy and inexpensive way of measuring high voltages.

For measuring a.c. voltages, one of the serious limitations is the stray capacitance to ground of the resistor arrangement on the insulating cylinder or other supporting structure, especially as one nears the ground. This problem is not discussed here. Every such simple equipment must be calibrated when installing it by other means.

(*b*) For measuring steady-state a.c. voltages only, it is a simple matter to use a capacitor rated for the high voltage or connecting several capacitors in series along with a milli-ammeter or micro-ammeter at ground level (or h.t. end to eliminate stray capacitances) as shown in Fig. 17.3. The current flowing is $I = j\,2\pi fCV$. Thus, once again, the ammeter can be calibrated to read the applied high voltage. There is no necessity for using high-precision capacitors for such purposes, as is absolutely necessary for accurate Bridge Measurements in the high voltage laboratory. Normal capacitors are manufactured for 1 nF or that order for voltages up to 3 MV.

Q.5. How is a resistive voltage divider used for measuring high voltage?

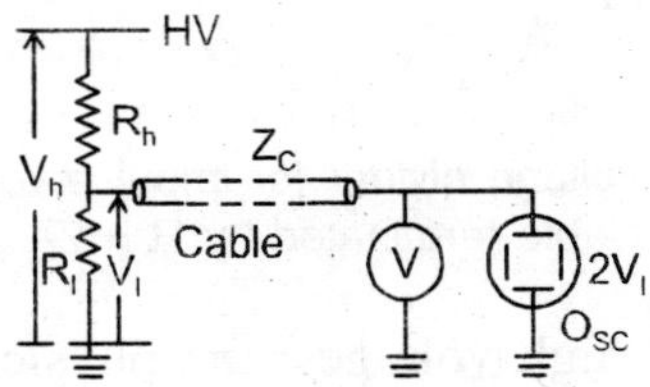

Fig. 17.4: Resistive voltage divider. General circuitry for all types of voltages.

Ans. Figure 17.4 shows two resistors R_h and R_l connected in series between the h.v. terminal and ground. A voltmeter or an oscilloscope is connected across the lower resistor R_l. The voltage applied to the meter is, by voltage division:

$$V_l = V_h \cdot R_l/(R_h + R_l),$$

or,
$$V_l/V_h = \text{voltage ratio}$$
$$= R_l/(R_h + R_l)$$

For example, let the voltage ratio required be 1/1000. Then for

$$V_h = 1 \text{ MV} = 1000 \text{ kV},$$

the low-voltage arm voltage will be $V_l = 1$ kV.

$$\therefore \quad R_h + R_l = 1000R_l,$$

or,
$$R_h = 999R_l \approx 1000R_l.$$

When the resistive voltage divider is used for measuring high impulse voltages which are not steady like d.c. or sinusoidal a.c., some special considerations must be given to the low-voltage circuit. Obviously, an indicating type of voltmeter cannot be used but only an oscilloscope, if the entire waveform is to be viewed or a special meter called the "Peak Voltmetre" has to be used. The output of the l.v. arm is led through a cable which has a characteristic impedance of Z_C ohms which behaves like a pure resistance. If the instrument end of the cable is connected to the vertical deflection plates of an oscilloscope, then the arrangement behaves as an open circuit for the surge. Therefore, the travelling wave of impulse voltage gets reflected totally and the indication on the oscilloscope screen is double the magnitude of voltage given by the voltage ratio. However, if the cable is terminated at the oscilloscope input terminals by a matching resistor whose value is equal to the cable impedance Z_C, the voltage indication on the oscilloscope screen will be identical with the voltage across the l.v. arm of the voltage divider. This is shown in Fig. 17.5. Usual coaxial cables have Z_C = 50 ohms in North America and 75 ohms elsewhere in the world, although this is not rigid. Therefore, the test engineer must know the surge impedance of the cable before designing the l.v. circuit.

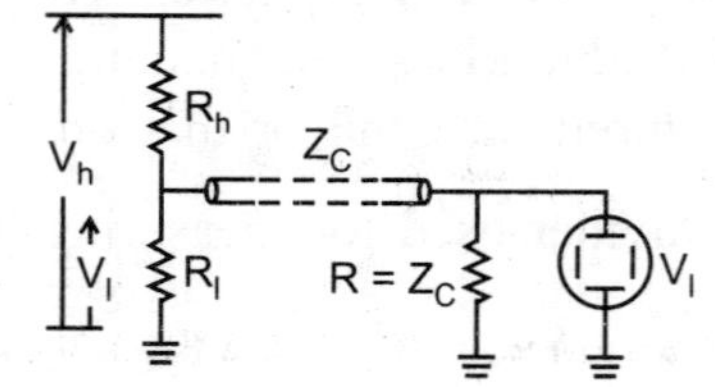

Fig. 17.5: Resistive voltage divider for impulse-voltage measurement. Cable terminated by R = Z_C.

Once again, for extremely high voltages, the physical size of the divider is very tall (6 to 10 metres in height) so that there will exist stray capacitance to ground, as well as inductance. However, the very high value of the divider resistance will damp out any oscillations that might be present in high-rate-of-rise voltages such as lightning impulses (1.2/50 μs). To eliminate inductance, it is usual to construct the high resistance by carbon film and not use wire-wound jobs.

Q.6. What is the principle of operation of a purely capacitive voltage divider and what are the precautions to be observed when using these, particularly in the low-voltage arm?

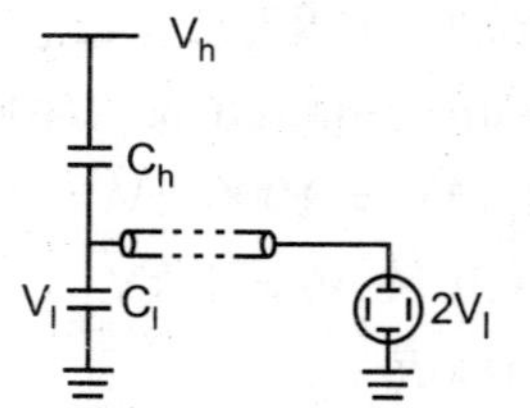

Fig. 17.6: Capacitive voltage divider.

Ans. In a manner similar to the resistive voltage divider, the capacitance voltage divider consists of two capacitors C_h and C_l in series as shown in Fig. 17.6. Since they are in series, their charges must be equal when voltage is applied.

$$\therefore \quad C_h(V_h - V_l) = C_l \cdot V_l$$

$$\Rightarrow \quad V_h/V_l = (C_h + C_l)/C_h$$

$$= 1 + C_l/C_h$$

If the voltage ratio is $V_h/V_l = 1000/1$,

then, $C_l = 999C_h \approx 1000C_h$.

Thus, the low-voltage arm has a high capacitance. For C_h = 1 nF, C_l = 1 μF. This is the inverse property as far as numbers are concerned from the resistive voltage divider. [Be careful! This author has been on many interview committees to select candidates for junior jobs in high voltage labs and has always been amused by the replies the candidates gave to this question!!]

Some of the precautions to be taken in the use of a capacitive voltage divider are as follows:

For impulse measurements, where the l.v. arm is located 6 to 10 metres away inside the large hall from the control desk where the measuring instruments are located, a length of cable is once again necessary as for any h.v. measurement considered earlier. If the end of cable with surge impedance Z_C is now terminated in a resistance of same value at the oscilloscope input terminals, then this resistance will short-circuit the entire l.v. arm. The voltage divider will now behave as if it consists of the high voltage arm capacitor C_h in series with the parallel combination of $R_C = Z_C$ and C_l. Therefore, the oscilloscope reading in practice will be practically zero. Consequently, where a capacitive voltage divider is used, the far end of the cable at the oscilloscope must be left open, i.e., connected directly to the input jack. Because of the resulting total reflection of the impulse voltage, the magnitude indicated on the scope screen is double the actual voltage picked up across the l.v. arm or C_l.

In those laboratories where the same voltage divider is capable of being used as both a pure resistive divider or a pure capacitive voltage divider, a changeover switch at the oscilloscope will ensure that in one case the matching resistor is connected to the input jack while for the latter case the input terminals are open. The calibration is also indicated on the front control panel of the oscilloscope. But these are very special voltage dividers. The capacitors are tubular and hollow, and the resistive portions of the divider are inserted inside this hollow cylinder, as is the case with the world-renowned Haefely voltage divider.

Once again, because of the physical size of purely capacitive voltage dividers, stray capacitances to ground are present as well as inductance. Unfortunately there is no resistance to damp out any oscillations (or ringing) that may occur under fast-rising impulse voltages. But there are several tests which are made in order to calibrate such dividers at the time of installation and before using, known as "Step Response", etc. These refined procedures are not discussed here.

These shortcomings are overcome in many cases by using what is known as a Resistive-Capacitive Voltage Divider, in which a small series resistance is inserted. The l.v. arm must also be matched by providing a resistance to preserve the voltage-division ratio, but in general its value is so low that it may be omitted. Even the lead resistance will provide this value.

One of the precautions to be observed in the l.v. arm is that the capacitance being of high value, it is advisable to generally construct the same with several capacitors in parallel. This will reduce the inductance since the current divides into several parts. This is also true for the resistive voltage divider in which the *l.v.*-arm resistor is constructed from several resistors in parallel and soldered to two rings.

Q.7. What is the equivalent circuit of a resistive-capacitive voltage divider? What is a Zaengel voltage-divider arrangement?

Ans. 1. When the size of capacitive voltage dividers becomes very large, usually in the form of a slender cylinder but very tall, the inductance of the electrodes affects the response of the divider to a fast-rising impulse wave in the micro-second range. The voltage waveshape at the l.v. arm output is not a faithful reproduction of the input waveshape. Consequently, errors will creep into the measurement of magnitude and waveshape.

In order to damp all oscillations in the *R-L-C* equivalent circuit of the "capacitive divider", a series resistance is purposely introduced in order to make the circuit critically damped. This requires the resistance to have the value $R = 2\sqrt{(L/C)}$. The proper value of resistance is selected through experiments called "calibration", where a low-voltage square wave is applied in order to determine the "step response". This is an advanced topic and is not discussed here.

As an example, for a 1 MV divider, the h.v. capacitance arm is 1 nF and its inductance may be 2.5 μH.

∴ The required value of resistor for critical damping is

$$R = 2\sqrt{(2.5 \times 10^{-6}/1 \times 10^{-9})} = 100 \text{ ohms}$$

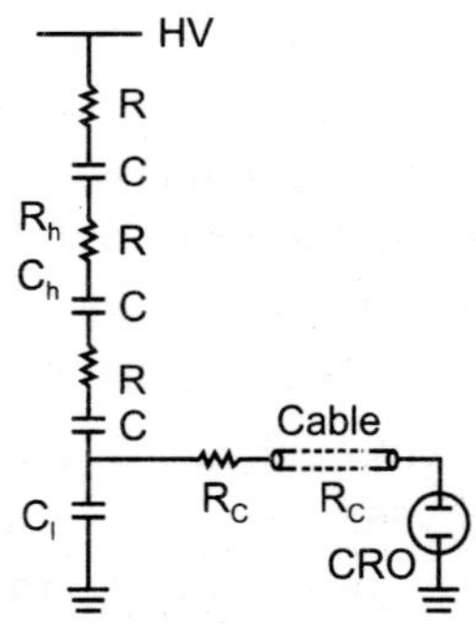

Fig. 17.7: R-C voltage divider.

In actual practice, the capacitance and required resistance are both fabricated in several parts in series as shown in Fig. 17.7. Formulae for inductance and surge impedance $Z = v \cdot l$ = velocity × inductance per unit length for a cylinder with dimensions r and H are as follows:

(*a*) Vertical cylinder: Height H and radius r: $Z = 60 \ln(2H/r)$ ohms;

(*b*) Horizontal cylinder: Height above ground plane = H, and radius r: $Z = 120 \ln(2H/r)$ ohm. This is the same as for an overhead conductor above ground in a single-phase transmission line.

The velocity of propagation is $v = c/\sqrt{\varepsilon_r}$, where c = light velocity in free space $= 3 \times 10^8$ m/s, and ε_r = relative permittivity of the dielectric. For air, $\varepsilon_r = 1$, and for plastic $\varepsilon_r = 2.5$. Capacitors used for measurement of high voltages have compressed nitrogen or plastic films or SF_6 gas as dielectrics.

2. **Zaengel Arrangement:** We have observed that inductance in h.v. measurements distorts fast-rising waveshapes through ringing or oscillations along with what are primarily resistive-capacitive circuits. There are three possible locations and sources for inductance to be present in an impulse-voltage system. We have already discussed the presence of inductance in the impulse generator and the voltage divider. The resulting oscillations can be damped by connecting sufficient series resistances inside the impulse generator and using a resistive-capacitive divider.

   The third source of inductance is present in the long horizontal lead connecting the impulse generator and the junction of voltage divider and test object. These two latter components, because of the high voltages used in the MV range are located at a large distance from the generator h.v. terminal. Those distances can range from 6 to 10 metres or more. The lead presents inductance and behaves as a surge impedance to the fast-rising impulse voltage originating from the generator. Therefore, the voltage travels on the lead and reflects at the junction of the lead and the load (voltage divider in parallel with the test object).

   In order to damp out these oscillations by preventing reflections, a resistance has to be inserted at the top of the voltage-divider column at the entry point of the h.v. lead. There are other inductances also present in the form of the mere physical dimensions of the laboratory and ground connections. All of them affect the measured waveform and are not generally amenable to precise calculations. Providing the series resistance in the h.v. lead in order to yield an ideal "step response" is known as the Zaengel arrangement or Zaengel Voltage Divider, named after the very famous Swiss engineer, Dr. Walter Zaengel at the Federal Institute of Technology, Zurich.

Q.8. What is a "Peak Voltmeter"? How does it measure the peak value of a high voltage a.c. and impulse?

Ans. One of the specifications during testing of h.v. equipment under power-frequency a.c. is that both half-cycles must be identical and that there should be no distortion in the waveshape. However, because of the magnetising current required in the h.v. transformer, some waveform distortion is always present and unavoidable. Thus, Standard Specifications take this into consideration and allow some distortion within limits. The engineer must consult the relevant Standards. It is said that the 2nd-harmonic distortion should not exceed 2%, but the details are not discussed here. Therefore, even if the r.m.s. value is measured on a moving-coil or e.s. voltmeter or by measuring the input voltage to the primary winding of the ground unit and multiplied by the overall turns ratio, the peak value of the output voltage at the h.t. terminal may not be equal to $V_P = \sqrt{2}\ V_{rms}$. On the other hand, the insulation strength (dielectric withstand or flashover) depends upon the peak value of voltage, although an oscilloscope is eminently suitable for all voltage measurements.

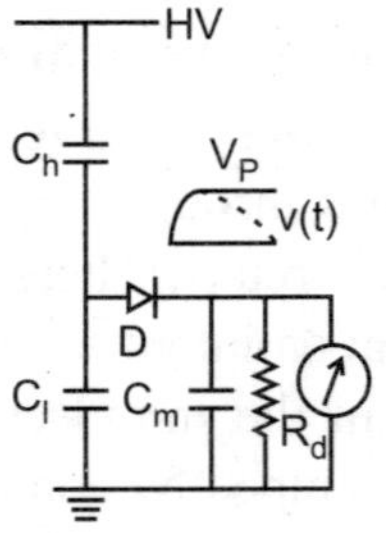

Fig. 17.8: Circuit arrangement of Peak Voltmeter.

Figure 17.8 shows the schematic diagram of a PVM of one type. There are a very large number of designs on the market. Basically, it consists of the usual capacitive voltage divider made up of capacitors C_h and C_l. Across the l.v. arm is connected a diode D which charges the capacitor C_m to the peak value of voltage output of the l.v. arm by voltage division. An electro-static voltmeter connected across C_m indicates the peak value of the voltage to be measured. A resistance R_d is required in parallel with C_m in order to discharge it for the next sequence of measurement. The same applies for impulse-type of voltages.

Q.9. What is the principle of measurement of high impulse and other types of voltages by means of a Kerr Cell?

Ans. It has been discovered that when a light beam passes through certain types of Glycerines in a container in which an electric field is applied, the plane of polarization of the light rotates, the amount of rotation being proportional to the intensity of the electric field. This principle can be utilised for measuring the high voltage in what is known as the Kerr Cell.

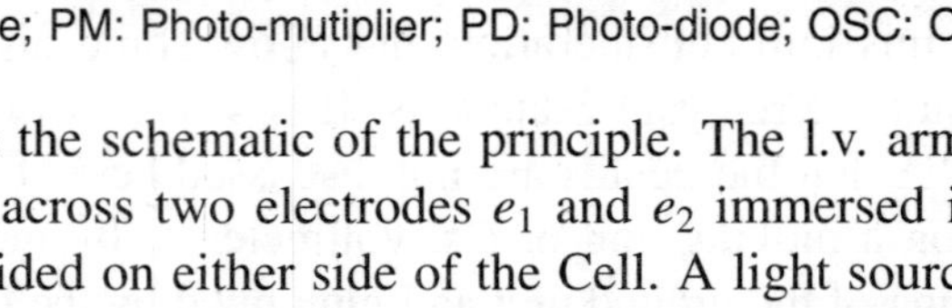

Fig. 17.9: Kerr Cell arrangement for measuring high voltages: LS: Light source; P: Polariser: KC: Kerr Cell with Glycerine; e_1, e_2: Electrodes with applied voltage V; OF: Optical Fibre; PM: Photo-mutiplier; PD: Photo-diode; OSC: Oscilloscope.

Figure 17.9 shows the schematic of the principle. The l.v. arm voltage from the voltage divider is applied across two electrodes e_1 and e_2 immersed in the Kerr Cell. Polarizer lenses P_L are provided on either side of the Cell. A light source directs light through the polariser. In modern Kerr Cells, the light source is a laser beam, but other types can also be used. The output of the polarizer is fed into an optical fibre with low attenuation.

Its other end feeds either a photo-multiplier or light amplifier. The resulting pattern is displayed on an oscilloscope screen. The rotation of the plane of polarization is an indication of the peak value of the applied voltage to be measured. Kerr Cell voltmeters are being used for measuring peak values of impulse voltages up to 500 kV.

Q.10. What are the characteristics to be incorporated in an Oscilloscope for use in high voltage laboratories?

Ans. Oscilloscopes, as is the case with all electronic equipment, are improving every year and the test engineer must become very familiar with progress in this most versatile of all measuring instruments. Oscilloscopes for h.v. work are mainly of two types: (*a*) Normal waveform display and measuring oscilloscopes; and (*b*) Storage oscilloscopes. These are more expensive than the normal oscilloscopes, but provide far more functions and facilities which are very versatile. We will not go into electronic details of these instruments in detail here.

We will confine the discussion to normal display-type scope with camera. There are special scopes manufactured particularly for high voltage impulse measurement which differ from normal general-purpose oscilloscopes only in the volt/cm deflection value. But the way they operate and used are the same. In fact, any laboratory-type general-purpose oscilloscope can be used for h.v. measurements.

HORIZONTAL AND VERTICAL DEFLECTION SIGNALS

Impulse voltages are transients which last for only a few micro-seconds and are termed "one-shot" voltages. They come and vanish. In the case of observing periodic or sustained signals, a horizontal-sweep arrangement (*X*-amplifier) is built into the instrument which uses a saw-tooth generator to sweep the beam in the horizontal direction and make it fly back to the beginning to repeat itself at a set frequency. However, this cannot be used when the instant of arrival of a signal on the screen is not known. Some arrangement must be provided whereby the voltage-transient itself should be used to trigger the electron beam in the horizontal direction (time axis) as also displaying the vertical pattern properly and no information should be lost. Thus, a time delay must be incorporated between the commencement of the horizontal sweep of the beam and its vertical deflection.

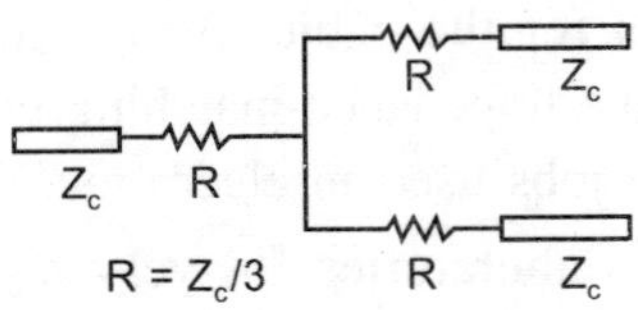

Fig. 17.10: T-Network for matching input cable from voltage divider to vertical and horizontal lead-in cables to oscilloscope.

Two methods are followed, the first one being the most usual and inexpensive:

1. At the l.v. arm of the voltage divider, the signal is led into a box by a cable from which two identical cables of surge impedance Z_c carry the signal out to the

oscilloscope by lengths of two cables. One cable is connected to the horizontal deflection plates (input jack) and the other, the longer, to the vertical plates, as shown in Fig. 17.10. Thus, a time delay is incorporated and the signal reaches the vertical plates after the same signal has reached the horizontal plates. By changing the lengths of the two cables the time delay can be varied very simply and easily. All cables are coaxial shielded ones having the same Z_c.

We have observed that the velocity of e.m. signal propagation in a cable is $v = c/\sqrt{\varepsilon_r}$. For normal low-attenuation coaxial plastic cables used in electronic communication (and in *h.v.* lab measurements),

$$\varepsilon_r = 2.25 \text{ to } 2.5$$

Thus,

$$v = 3 \times 10^8/\sqrt{2.5}$$

$$= 1.9 \times 10^8 \text{ m/sec}$$

So, 1 metre length of cable provides a signal delay of

$$1/(1.9 \times 10^8) = 5.27 \times 10^{-9} \text{ sec} = 5.27 \text{ ns}$$

20 metres of cable therefore delays the signal by about 100 ns = 0.1 μs. The test engineer must adjust the delay suitably by selecting the proper lengths of cables such that the horizontal triggering and the vertical triggering take place without loss of any information. Thus, this is a trial and error procedure to start with, but when once set initially there will be no need to change them for that particular oscilloscope.

We observe from the arrangement in Fig. 17.10 that if one cable with surge impedance Z_c enters the box and two identical cables leave, then there is an impedance discontinuity. The incoming signal is travelling on Z_c while the outgoing signals see $Z_c/2$. Thus, the three cables must be matched. This is carried out by a *T*-connection. The impedance viewed from any terminal must be the same. For this purpose, three equal resistors are connected, one in series with each cable. Now, the impedance looking from any terminal such as *A* is:

$$Z = R + (Z_c + R)/2 = Z_c$$

$$\therefore \quad R = Z_c/3$$

Thus, if Z_c = 75 ohms for the cable, then 3 resistors each of 25 ohms must be connected in *T* inside the impedance-matching box. These resistors can be ordinary general-purpose carbon jobs used in electronic circuitry.

2. In large and expensive laboratories for very high voltages, the second procedure used is to pick up the radiation from the spark gaps inside the impulse generator by a rod antenna or a loop antenna and feed the signal to the horizontal deflection plates. The vertical deflection plates will be fed in the normal manner from the l.v. arm of the voltage divider through a cable of sufficient length to provide a delay. This arrangement eliminates the impedance-matching box with the *T*-resistors.

Q.11. How are optical fibres used for the measurement of high voltages?

Ans. We observe that the l.v. arm of a voltage divider has always been spoken of as being connected to the h.v. arm at the ground end in order that the low voltage can be led by conventional shielded coaxial cables to the measuring equipment. However, this need not be the case always and the low-voltage arm can be located at the high voltage terminal of the voltage divider. Such a procedure will eliminate the effect of stray capacitances to ground of the tall voltage-divider h.v. arm. This can be achieved by the use of optical fibres which offer sufficient insulation for the measuring instruments which are near ground potential. The signal can be led from the high voltage end itself. Optical isolation can also be used to provide more electrical separation or insulation.

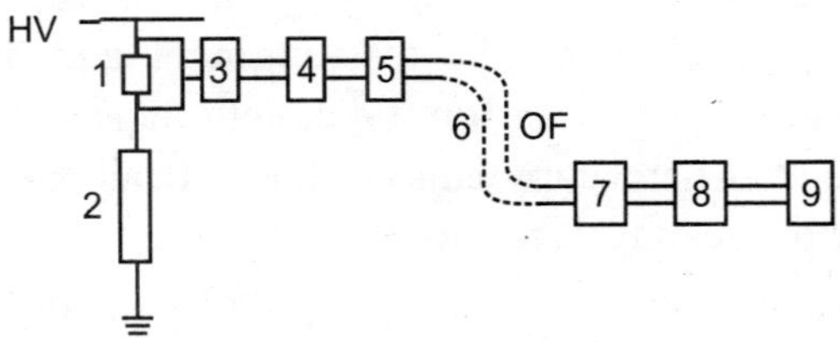

Fig. 17.11: General layout of voltage divider using optical fibre and low-voltage arm located at the high voltage terminal: (1) Low-voltage arm; (2) High-voltage arm; (3) A/D converter; (4) Driver; (5) LED; (6) Optical-fibre lead-in from l.v. arm to instruments at ground potential; (7) Photo-Diode; (8) Amplifier and D/A converter; (9) Oscilloscope.

The scheme is shown in Fig. 17.11. The low value of voltage developed across the l.v. arm (1) at the h.t. end of the voltage divider is fed to an A/D converter, (3) and an amplifier-driver (4) which excites the Gallium Arsenide LED, (5) The light intensity is modulated by the signal amplitude and is transmitted through the optical fibre (6) to instruments at ground level in the control cubicle (Faraday Cage). Here, the light signal is converted to electrical signal by means of a photo-diode (7) and taken by coaxial cable to the display device {Oscilloscope (9)} through a D/A converter (8) By further using a system of lenses and mirrors, the optical fibre can be made in two sections, isolated electrically for any desired voltage. At the h.v. end, all electronic equipment is battery-powered and they are all enclosed inside a metallic sphere to prevent discharges from sharp edges.

This type of optical-fibre method of measuring a current or voltage is not limited to h.v. laboratory testing, but can also be used with CTs for measuring currents flowing in overhead lines. Thus, the use of optical fibres is getting very widespread use in high voltage work.

Q.12. How is a Digital Recorder used for measurement and display of fast-rising impulse voltages?

Ans. Digital sampling of waveforms is a versatile means of storing very fast-rising voltages. This may replace high-speed oscilloscopes in the future.

A micro-processor can be programmed in order to calculate and display any of the properties of the waveform stored in the memory. Such properties of importance are:

(*a*) Peak value of the impulse;

(*b*) Time to peak value, t_P;

(*c*) Front time:

$$t_f = 1.67(t_{90} - t_{30});$$

(*d*) Tail time t_t to 50% peak value;

(*e*) Overshoot in %;

(*f*) Rate of rise of impulse in kV/μs;

(*g*) Time duration over which the voltage stays above 90% peak value; and any other desired property required by the test engineer and design engineer of the h.v. equipment. These can be stored on a magnetic tape unit or displayed on Computer Graphic screen, or a hard copy can be made, if all these auxiliaries or peripherals are provided in the testing laboratories.

With the use of 8-bit word length, it permits sampling $2^8 = 256$ discrete voltage levels of the waveform in a chosen interval of time. If the wave rises to peak value in 2 μs = 2000 ns, by sampling 256 points the front is displayed as a digital signal every 2000/256 = 8 ns. The resolution is thus equal to or superior to a conventional h.v. impulse oscilloscope. Front-of-the-wave oscillations are very clearly exhibited.

One precaution, and a very important and vital one, to be observed when using such instruments is that they are subject to pick-up of e.m. interference signals from external sources. The most serious one is the discharge originating from the spark gaps inside the impulse generator which act as very powerful radiators of e.m. energy, as well as from spark plugs of any vehicles passing near the laboratory. Therefore, the Digital Recorder must be placed inside a shielded enclosure made from thick metal, the usual ones being Aluminium, Soft Iron, Steel, with all joints at doors also sealed very well against intrusion of noise signals. A metallic shielded enclosure works on the principle that any high-frequency voltage impinging on the metal will induce heavy eddy currents. The thicker the metal higher will be the current flow because of lower resistance. These induced currents will generate a magnetic field that will repel the incoming radiation. Since there must be no obstructions to the current flow, holes or doors or gaps should not be present. The biggest offenders in this regard are the holes required to lead in the cables for both signal and power. But these drawbacks are not insurmountable and manufacturers specialising in shielded enclosures for all types of communication purposes have done an excellent job in this regard and are highly reliable and dependable.

Q.13. How are very high fast-rising impulse currents measured during testing of lightning arresters and other equipment?

Ans. Figure 16.10 shows the schematic of an impulse-current generator in which was shown a shunt for measuring and displaying the waveform of current. This shunt is a pure resistance of a very low value made from thick conductor for the current which can go up to 200 kA or more during testing of heavy-duty arresters and fuses. Because of the high rate of rise (di/dt), any little inductance in the resistance coil will distort the waveform. Thus, high-current shunts must be **totally free** of any trace of inductance, and thus are known as "Non-Inductive Shunts". Their design is very special and called Bifilar construction.

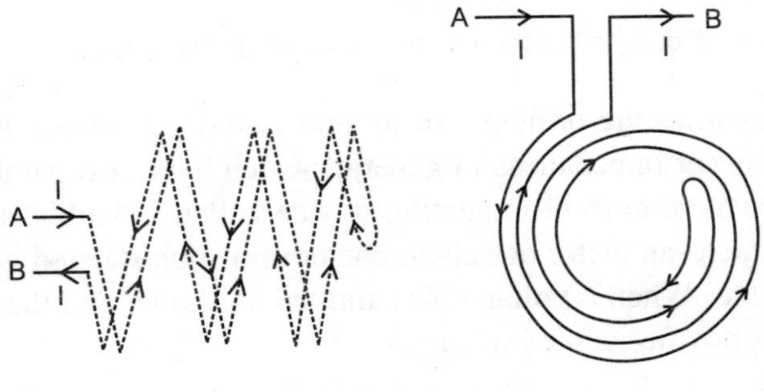

Fig. 17.12: Two examples of bifilar construction for non-inductive resistance shunt for measuring high impulse currents.

In principle and practice, a bifilar (bi = two, filar = filament or wire) resistance consists in taking two conductors (either round or any other shape) and winding them in close proximity on a former. Figure 17.12 shows two examples. The ends of the two wires at one end are kept open to carry the current in and out. But the two ends at the other end of the coil-former are connected together firmly. The reader can observe that when current enters terminal *A* and flows to the other end, it returns back to terminal *B*, but its direction is opposite to that of the other coil. Therefore, the magnetic fields are equal and opposite thereby cancelling each other. There is no net inductance and the entire bifilar coil behaves as a pure resistance.

Q.14. What is the principle of Schering Bridge? How does it measure capacitance and loss angle (tan δ) of high voltage equipment such as cable insulation?

Ans. The Schering Bridge is meant for measuring the capacitance and loss angle of h.v. cables and capacitors. It is a "null" bridge having 4 arms with a null detector, exactly the same as a d.c. Wheatstone Bridge for measuring resistances. However, certain extensions and precautions are found necessary, since the object under test is at very high voltages and it is an a.c. bridge arrangement.

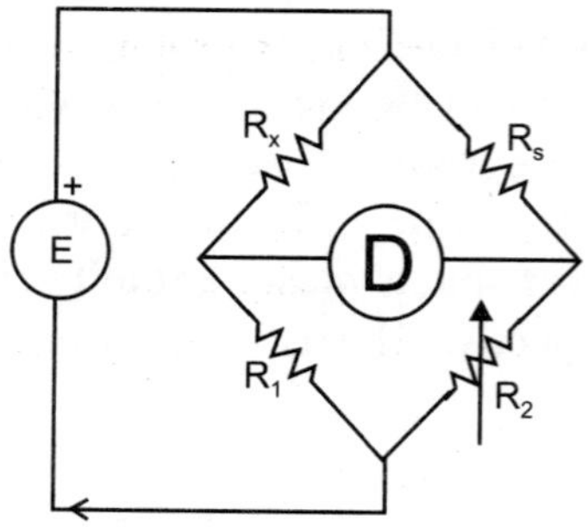

Fig. 17.13: General Wheatstone Bridge network.

Briefly recaputulating the principle of a "null bridge", consider Fig. 17.13 in which are shown 4 resistors (or impedances in general) which have two parallel arms bridged in the middle by a sensitive current-indicating device called a "null detector". This can range from a simple micro-ammeter or galvanometer or a sophisticated galvanometer movement with amplifier, etc. When a voltage E is applied as shown and the bridge current through D is zero, the following equations apply:

Current through R_x = current through $R_1 = I_1$;

Current through R_S = current through $R_2 = I_2$;

$$E = I_1(R_x + R_1)$$
$$= I_2(R_S + R_2);$$
$$I_1 \cdot R_x = I_2 \cdot R_S;$$
$$I_1 \cdot R_1 = I_2 \cdot R_2$$

$\therefore$ $$E \cdot R_x/(R_x + R_1) = E \cdot R_S/(R_S + R_2),$$

or, $$R_x(R_S + R_2) = R_S(R_x + R_1)$$

Consequently, $$R_x \cdot R_2 = R_S \cdot R_1$$

or, $$R_x = R_S \cdot R_1/R_2$$

Also, $$R_x/R_S = R_1/R_2$$

This is the "balance equation" of the bridge, when the current through the detector is zero.

In order to obtain balance or null condition, the unknown resistance R_x, whose value is to be determined, forms one arm of the bridge. A known standard resistance R_S forms a second arm. A known resistance R_1 is connected to the third arm. A variable resistance R_2 completes the fourth arm. A power source and null detector complete the bridge network. R_2 is varied until a null is detected and R_x calculated from known values of R_S, R_1 and R_2.

In high voltage a.c. bridges, since there are two components to an impedance, the unknown arm Z_x consists of capacitance C_x and resistance R_x in parallel (or a series-equivalent can be calculated at the frequency of the source). The standard arm consists of a loss-less precision h.v. capacitor of known value C_S. The arm Z_1 consists of a pure resistance R_1 which is at low voltage and at ground end which the operator can handle directly. The variable arm Z_2 also consists of a capacitor C_2 and resistor R_2 in parallel. Both the l.v. arms Z_1 and Z_2 must be protected by gas-gap which conducts in the event of a high voltage across them due to breakdown of any of the h.v. arms Z_x and/or C_S. The applied voltage must be such that neither the test object (a capacitor or a length of cable) nor the standard capacitor C_S breakdown.

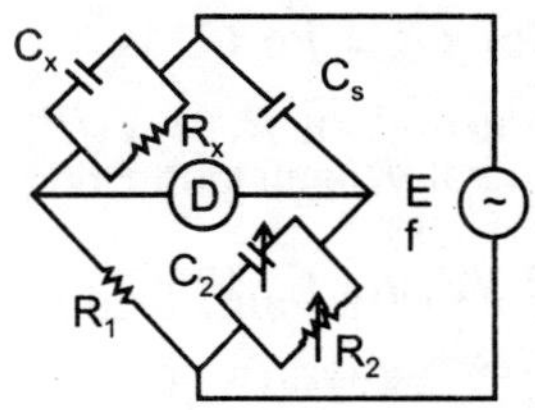

Fig. 17.14: Schering Bridge network for measuring capacitance C_x and its loss angle (tan δ) at high voltage.

The entire bridge arrangement invented by Schering is shown schematically in Fig. 17.14 with the null detector and high voltage power-frequency a.c. source.

The impedances of the four arms are as follows:

$$Z_x = C_x \,||\, R_x$$
$$= R_x(1/j\omega C_x)/(R_x + 1/j\omega C_x)$$
$$= R_x/(1 + j\omega C_x R_x),$$
$$\omega = 2\pi f$$
$$Z_S = 1/j\omega C_S$$
$$Z_1 = R_1$$

and,
$$Z_2 = C_2 \,||\, R_2$$
$$= R_2/(1 + j\omega C_2 R_2)$$

When the bridge is balanced and null is obtained on detector D the balance equation gives:

$$Z_x \cdot Z_2 = Z_S \cdot Z_1 \qquad \text{... (1)}$$

or, $\{R_x/(1 + j\omega C_x R_x)\} \cdot \{R_2/(1 + j\omega C_2 R_2)\} = R_1/j\omega C_S$ ... (2)

We thus obtain two equations for the unknown quantities C_x and R_x, by separating the real and j-parts of the above two equations:

$$j\omega\, C_S R_x R_2 = R_1[1 - \omega^2 C_2 R_2 C_x R_x + j\omega(C_2 R_2 + C_x R_x)] \qquad \text{... (3)}$$

$\therefore\ R_1(1 - \omega^2 C_2 R_2 C_x R_x) = 0,$

$\Rightarrow$
$$C_x R_x = 1/\omega^2 C_2 R_2 \qquad \text{... (4)}$$

and
$$C_S R_x R_2 = R_1(C_2 R_2 + C_x R_x)$$
$$= R_1(C_2 R_2 + 1/\omega^2 C_2 R_2)$$
$$= (R_1/\omega^2 C_2 R_2) \cdot (1 + \omega^2 C_2^2 R_2^2) \qquad \text{... (5)}$$

Consequently, $R_x = (R_1/\omega^2 C_2 C_S R_2^2) \cdot (1 + \omega^2 C_2^2 R_2^2)$ ... (6)

and $C_x = 1/\omega^2 C_2 R_2 R_x$

$= (C_S R_2/R_1)/(1 + \omega^2 C_2^2 R_2^2)$ ... (7)

In practice, we are not interested in knowing the value of resistance R_x of the unknown impedance but only tan δ, the loss-angle or power factor of the test object. This is determined as follows, and shown in Fig. 17.15.

Now, I_{CX} = charging current of $C_x = j\omega C_x V_x$,

and I_{rX} = loss-component of current in $R_x = V_x/R_x$.

$\therefore$ $\tan\delta = |I_{rx}/I_{cx}|$

$= (V_x/R_x)/\omega C_x V_x$

$= 1/\omega R_x C_x$ ... (8)

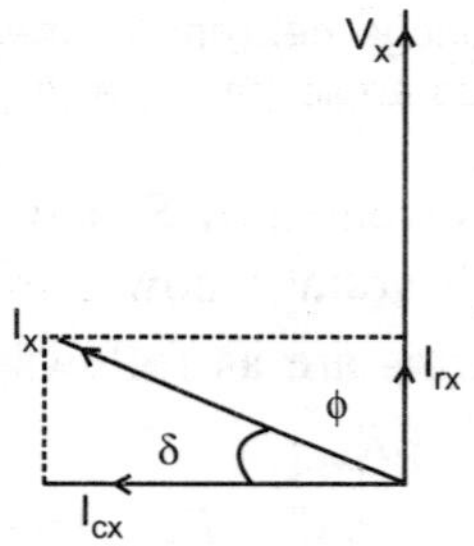

Fig. 17.15: Phasor diagram of currents through unknown capacitance C_x with loss.
I_{CX} = pure capacitive component of current,
I_{rX} = loss component of current. tan δ = I_{rX}/I_{CX}.

Inserting the values for R_x and C_x from equations (6) and (7):

$$\tan\delta = \{\omega^2 C_2 C_S R_2^2/R_1(1 + \omega^2 C_2^2 R_2^2)\} \times \{R_1(1 + \omega^2 C_2^2 R_2^2)/\omega C_S R_2\}$$

$$= \omega C_2 R_2 \quad \text{... (9)}$$

Using this value in equation (7), the value of the unknown capacitance becomes:

$$C_x = (C_S \cdot R_2/R_1)/(1 + \tan^2\delta) \approx C_S \cdot R_2/R_1,$$

when $\tan^2\delta \ll 1$ ...(10)

Usual values of tan δ ≈ 0.01.

$\therefore \tan^2\delta \ll 1$.

Note also that for this case,

$$C_x/C_S = R_2/R_1$$

in a manner similar to the Wheatstone resistance-bridge relation.

Equations (9) and (10) are the primary governing relations for determining the value of the unknown capacitance of the test object and the loss angle tan δ.

In the case of a d.c. resistance bridge, operating the dial to change R_2 in Fig. 17.13 will suffice to obtain null. However, in a.c. bridges, where the unknown impedance Z_x and balance arm Z_2 consist of resistance and capacitance elements, two balance procedures are necessary: (1) By changing C_2 the detector minimum reading is obtained first. (2) Then by changing R_2, the detector reading is again minimised. After a few such trials, the zero will result, since changing one element has an effect on the other. Balancing an a.c. bridge is an art and requires patience and experience. One of the mottos when dealing with high voltage equipment is "Never be in a hurry". Hurry is the cause of all accidents as in other day-to-day activities, involving certain degree of danger, such as driving an automobile!

Example: In a Schering Bridge experiment, null is obtained for the following values of C_S, R_1, C_2 and R_2, when testing a sample of h.v. cable at 100 kV, r.m.s., and 50 Hz, refer to Fig. 17.14:

$$C_S = 1 \text{ nF} = 10^{-9} \text{ F},$$
$$R_1 = 100 \text{ ohms}$$
$$R_2 = 1400 \text{ ohms}$$

and,
$$C_2 = 0.005\ \mu\text{F} = 5 \text{ nF}$$

Calculate the following:

(1) Capacitance of cable sample, C_x;
(2) tan δ of its insulation;
(3) Insulation resistance of cable, R_x; and
(4) Voltage across the low-voltage arms if the secondary voltage of the source transformer is 100 kV.

Solution:
$$C_S = 10^{-9} \text{ F}$$
$$R_1 = 100\ \Omega$$
$$R_2 = 1400\ \Omega$$
$$C_2 = 5 \times 10^{-9} \text{ F}$$
$$\omega = 314 \text{ rad/sec.}$$

From equation (9) we have:

$$\tan \delta = 314 \times 1400 \times 5 \times 10^{-9}$$
$$= 2.2 \times 10^{-3}$$
$$= 0.0022$$

1. From equation (10):

$$C_x = C_S \cdot R_2/R_1(1 + \tan^2 \delta) \approx C_S \cdot R_2/R_1$$
$$= 1 \times 10^{-9} \times 1400/100$$
$$= 14 \times 10^{-9} \text{ F} = 14 \text{ nF}$$

2. $$\tan\delta = 0.0022 = 0.22\%$$

3. From equation (8),

$$R_x = 1/(\omega C_x \tan\delta)$$
$$= 0.1034 \times 10^9\ \Omega$$
$$= 103.4 \text{ megohms}$$

4. By voltage division,

$$V_{lv}/V_{hv} = R_1/|Z_x|.$$

Now, $$Z_x = R_x/(1 + j\omega C_x R_x)$$
$$= R_x/(1 + j/\tan\delta),$$

since $$\tan\delta = 1/(\omega C_x R_x), \text{ from equation (8).}$$

$\therefore$ $$|Z_x| = R_x/\sqrt{(1 + 1/\tan^2\delta)}$$
$$= R_x \cdot \tan\delta/\sqrt{(1 + \tan^2\delta)} \approx R_x \cdot \tan\delta, \text{ for } \tan\delta \ll 1.$$

$\therefore$ $$V_{1v}/V_{hv} = 100/(103.4 \times 10^6 \times 0.0022)$$
$$= 440 \times 10^{-6}$$

For $$V_{hv} = 100 \text{ kV} = 10^5 \text{ volts},$$

the voltage across the low-voltage arms will be $V_{lv} = 44$ volts. This is a safe value for an operator.

Q.15. How are partial discharges in high voltage insulation detected and measured?

Ans. Partial discharges in high voltage solid and liquid dielectrics are indications of probable development of faults in the equipment, leading to what are known as "incipient or impending faults". These discharges are caused by the ionisation of trapped gas bubbles or voids. These are of corona-type discharges and have been discussed earlier in Chapter 10. Partial discharge measurement is not only a routine test during maintenance of equipment in service but also mandatory according to I.E.C. and I.S. Standard Specifications for all transformers above the 220 kV level which are subjected to test to ascertain insulation withstand strength. For voltages up to 220 kV, a power-frequency test at twice the normal frequency for 2 minutes is required with no necessity for detecting partial discharges. But for higher voltages, the test must be performed at 150% of rated voltage for 30 minutes with partial discharge measurement by suitable tests, at normal frequency. Note that when a transformer is subjected to a voltage at a value higher than the rated value, the flux in the core is correspondingly higher at the same frequency. In order to keep the flux at rated value at the higher voltage, the frequency of applied test voltage must be increased in the same ratio to keep V/f = constant. The design engineer is very nervous when a higher-than-normal frequency is used since it stresses the turn-to-turn insulation more severely than in normal operation at rated frequency. However, because of increased thickness of solid insulations, formation of voids is more prevalent and the p.d. test is mandatory.